S0-CVI-256

Quinolone Antimicrobial Agents

Quinolone Antimicrobial Agents

Edited by
John S. Wolfson and **David C. Hooper**
Infectious Disease Unit, Medical Services,
Massachusetts General Hospital, and Harvard Medical School,
Boston, Massachusetts

American Society for Microbiology
Washington, D.C.

1913 I Street, N.W.
Washington, DC 20006

Library of Congress Cataloging-in-Publication Data

Quinolone antimicrobial agents.

Includes bibliographies and index.
1. Quinolone antibacterial agents—Testing. 2. Bacterial diseases—Chemotherapy—Evaluation. I. Wolfson, John S. II. Hooper, David C.
RM666.Q55Q56 1989 616.9'2061 88-7866
ISBN 1-55581-007-1

Printed in the United States of America

Cover illustration by Edith Tagrin

To our spouses, Elizabeth and Sally, and to our children, Lauren, Eric, Brook, and Allison, for their understanding and continued support

Contents

Contributors

Arnold S. Bayer
UCLA School of Medicine, Los Angeles, CA 90024, and Division of Infectious Diseases, LAC Harbor-UCLA Medical Center, Torrance, CA 90509

Gina A. Dallabetta
Johns Hopkins University School of Medicine and Division of Infectious Diseases, Department of Medicine, Johns Hopkins Hospital, Baltimore, MD 21205

George L. Drusano
Division of Infectious Diseases, University of Maryland School of Medicine, Baltimore, MD 21201

Herbert L. DuPont
Program in Infectious Diseases and Clinical Microbiology, Department of Internal Medicine, Medical School, University of Texas Health Science Center, Houston, TX 77225

C. T. Eliopoulos
Infectious Disease Section, Department of Medicine, New England Deaconess Hospital, Boston, MA 02215

G. M. Eliopoulos
Harvard Medical School and Department of Medicine, New England Deaconess Hospital, Boston, MA 02215

Edward W. Hook III
Johns Hopkins University School of Medicine and Division of Infectious Diseases, Department of Medicine, Johns Hopkins Hospital, Baltimore, MD 21205

David C. Hooper
Harvard Medical School and Infectious Disease Unit, Medical Services, Massachusetts General Hospital, Boston, MA 02114

Robert C. Moellering, Jr.
Harvard Medical School and New England Deaconess Hospital, Boston, MA 02115

S. Ragnar Norrby
Department of Infectious Diseases, University of Lund, and University Hospital of Lund, Lasarettet S-22185 Lund, Sweden

Brian E. Scully
Division of Infectious Diseases, Department of Medicine, College of Physicians and Surgeons, Columbia University, New York, NY 10032

Morton N. Swartz
Harvard Medical School and Infectious Disease Unit, Medical Services, Massachusetts General Hospital, Boston, MA 02114
Francis A. Waldvogel
Clinique Médicale Thérapeutique, University Hospital, 1211 Geneva 4, Switzerland
Drew J. Winston
Department of Medicine, UCLA Medical Center, Los Angeles, CA 90024
John S. Wolfson
Harvard Medical School and Infectious Disease Unit, Medical Services, Massachusetts General Hospital, Boston, MA 02114

Chapter 1

Introduction

John S. Wolfson and David C. Hooper

Infectious Disease Unit, Medical Services,
Massachusetts General Hospital, Boston, Massachusetts 02114

In recent years there has been a logarithmic increase in new information about the quinolone antimicrobial agents. These drugs, also called fluoroquinolones, 4-quinolones, and quinolone carboxylic acids, include norfloxacin, ciprofloxacin, ofloxacin, pefloxacin, enoxacin, amifloxacin, difloxacin, fleroxacin, temafloxacin, lomefloxacin, and compounds which are as yet only numbers, such as CI-934, S-25930, DR-3355, and others (structures given in Fig. 1 of chapter 2).

The quinolones are analogs of the earlier developed agent nalidixic acid. Nalidixic acid was originally isolated by Lesher and associates (1) from a distillate during chloroquine synthesis and thus was a by-product of antimalarial research (2). Additional older analogs include oxolinic acid, pipemidic acid, and cinoxacin. These older analogs will not be considered further in this book except for purposes of comparison with the new agents.

The new quinolones are substantially more potent in vitro and broader in antibacterial spectrum than nalidixic acid but maintain the favorable property of being well absorbed after oral administration. The new agents have additional advantageous pharmacokinetic properties, including relatively long half-lives in serum, allowing generally for twice daily administration, excellent penetration into many tissues, and permeation into human cells, resulting in antimicrobial activity against intracellular pathogens.

Nalidixic acid and all quinolone agents are completely synthetic, and structurally related compounds have not been identified as products of living organisms. More than a thousand quinolones and analogs have been synthesized and evaluated for antimicrobial activity, including compounds with substitutions and additions to many parts of the molecule. Of particular importance to increased potency are moieties attached to N-1, a fluorine attached to C-6, and moieties attached to C-7, often a piperazinyl or methylpiperazinyl group.

Nalidixic acid and enoxacin contain a nitrogen rather than a carbon in position 8 of the double ring and thus are 1,8-naphthyridines. They are, however, often referred to as quinolones, as will be the convention in this book.

The favorable potency in vitro and the pharmacologic properties of the quinolones predicted potential for treatment of a variety of bacterial infections, and clinical studies, most of which have been published in the past 3 years, have borne out this promise.

The purpose of this book is to bring together this rapidly expanding body of knowledge into a concise volume for use by clinicians, clinical microbiologists, pharmacologists, basic scientists, and others needing information about these drugs.

The format of this book is summarized as follows.

Chapter 2 details the mechanisms of action of and bacterial resistance to the quinolones. Included are discussions of the bacterial enzyme DNA gyrase (topoisomerase II), which is the intracellular target of the quinolones, and the effects of quinolones on this enzyme and bacterial cells. Also discussed are the two documented mechanisms of resistance: mutation in DNA gyrase and mutation in genes which decrease drug accumulation. Finally, the clinical relevance of these resistance mechanisms is addressed.

Chapter 3 presents a detailed description of the potency in vitro of quinolone agents against virtually all bacterial species. These data are grouped to suggest types of infections particularly suited to treatment with quinolones. The chapter also includes discussion of technical variables that alter MICs, bacterial killing, bacterial resistance, and effects of combining quinolones with other antibacterial agents.

Chapter 4 presents a detailed account of the pharmacokinetic properties of the quinolone agents discussed in groupings of drugs eliminated by primarily renal, nonrenal, and both renal and nonrenal routes. Also considered are effects of renal failure and hepatic failure on pharmacokinetics. Finally, dosing recommendations are presented.

Chapters 5 through 13 present data on therapy of specific groups of infections. Each of these chapters begins with an overview of the potency in vitro and pharmacokinetic properties of the quinolones relevant to the pathogens and diseases to be discussed, making this important information readily accessible to readers in a directed fashion.

Chapter 5 considers treatment of infections of the urinary tract, including cystitis, pyelonephritis, and prostatitis. The quinolones have proved highly efficacious in the therapy of many of these infections. The drugs notably offer the option for oral therapy of urinary tract infections with difficult-to-treat organisms such as *Pseudomonas aeruginosa*, which in many cases previously required parenteral administration of antibacterial agents.

Chapter 6 considers sexually transmitted diseases. The quinolones show excellent efficacy against uncomplicated gonorrhea caused by penicillin-susceptible or penicillin-resistant strains. The drugs also are effective against infections with *Haemophilus ducreyi*, the agent of chancroid. Certain quinolones, such as ofloxacin, also show promise for the therapy of chlamydial urethritis.

Chapter 7 considers therapy of upper and lower respiratory tract infections. The agents have been effective against a variety of pathogens, including *Haemophilus influenzae*, members of the family *Enterobacteriaceae*, and *P. aeruginosa*, and also certain gram-positive bacteria, including *Staphylococcus aureus*. Quinolones have poor activity against anaerobic bacteria and therefore are not appropriate for therapy of anaerobic aspiration pneumonia. Efficacy in treatment of pneumonia due to *Streptococcus pneumoniae* is under evaluation, with some favorable preliminary findings, even though this pathogen is only moderately susceptible to quinolones in vitro. Of particular importance is the usefulness of the quinolones for therapy of exacerbations of respiratory infections caused by *P. aeruginosa* in patients with cystic fibrosis.

Chapter 8 considers therapy of bacterial gastroenteritis. The quinolones are potent in vitro against almost all species known to cause bacterial gastroenteritis and achieve very high drug concentrations in stool. The drugs in preliminary studies have proved efficacious in bacterial gastroenteritis caused by many species of bacterial pathogens and also in prophylaxis against traveler's diarrhea. The agents also have shown promise in the treatment of salmonella infections, including typhoid fever and the clearing of chronic carriers.

Chapter 9 considers therapy of osteomyelitis and arthritis. Although data are still preliminary, the quinolones show much promise in therapy of chronic osteomyelitis caused by gram-negative bacilli, a particularly difficult infection to treat.

Chapter 10 considers the role of quinolone agents in treatment of immunocompromised patients. The use of the agents for prophylaxis against bacterial infection in neutropenic patients is examined. As well, the use of quinolones to treat febrile neutropenic patients and immunocompromised patients with defined bacterial infection is discussed.

Chapter 11 considers therapy of infectious endocarditis and meningitis. Little experience with therapy of these infections in humans exists. This information is reviewed in the context of a critical analysis of the more extensive data on therapy of animal models of these infections.

Chapter 12 considers use of the quinolone agents in treatment of skin and soft tissue infections. The drugs show promise for therapy of infections caused by gram-negative and certain gram-positive pathogens, although the preferred agents for the latter bacteria remain beta-lactam drugs.

Chapter 13 considers quinolones and ophthalmologic infections. Data are limited but suggest promise for topical treatment of bacterial conjunctivitis. The drugs also appear to merit further evaluation for therapy of keratitis and endophthalmitis.

Chapter 14 presents discussion of the adverse effects of the quinolone agents. Established and possible toxicities are reviewed, including adverse interactions with other drugs. While still in their adolescence, the new quinolones have been administered to many thousands of patients, and a profile of drug toxicity is emerging. The drugs appear for the most part to be well tolerated (6 to 13% toxicity), with adverse effects being no more common and in several instances less common than comparison conventional agents used in nine double-blind, comparative trials.

With chapter 15, the book concludes with an overview of the current and future roles of the quinolone agents in the therapy of bacterial infections.

Literature Cited

1. **Lesher, G. Y., E. D. Froelich, M. D. Gruet, J. H. Bailey, and R. P. Brundage.** 1962. 1,8-Naphthyridine derivatives. A new class of chemotherapeutic agents. *J. Med. Pharm. Chem.* **5**:1063–1068.
2. **Neu, H. C.** 1987. Ciprofloxacin: an overview and prospective appraisal. *Am. J. Med.* **82**(Suppl. 4A):395–404.

Chapter 2

Mechanisms of Action of and Resistance to Quinolone Antimicrobial Agents

John S. Wolfson, David C. Hooper, and Morton N. Swartz

Infectious Disease Unit, Medical Services, Massachusetts General Hospital, Boston, Massachusetts 02114

Introduction

In recent years, there has been considerable excitement concerning the development and clinical use of the newer quinolone agents (38, 71–73, 105, 112, 163, 163a). These agents include norfloxacin, ciprofloxacin, ofloxacin, enoxacin, amifloxacin, difloxacin, fleroxacin, temafloxacin, lomefloxacin, and others, many of which are designated only by compound numbers (Fig. 1, Table 1). Nalidixic acid and oxolinic acid were the first marketed quinolone agents. In contrast to these earlier drugs, the newer agents are more potent, broader in spectrum of activity in vitro, and less prone to selection of resistant strains (163).

In this chapter we review information on the mechanisms of action of and bacterial resistance to quinolone agents. DNA gyrase (bacterial topoisomerase II) (49) is a primary target of the quinolone agents and is central to mechanisms of action. DNA gyrase and also the bacterial outer membrane are important to mechanisms of resistance. Topics to be discussed include problems with the structure of DNA, topoisomerases and how they solve the problems, the structure and functions of DNA gyrase, effects of quinolone agents on DNA gyrase and on viable bacteria, killing of bacteria by quinolones, and mechanisms of bacterial resistance and their clinical relevance.

Six reviews containing detailed discussions of topoisomerases and mechanisms of action of and resistance to quinolone agents have recently been published (34, 47, 74, 137, 157, 158).

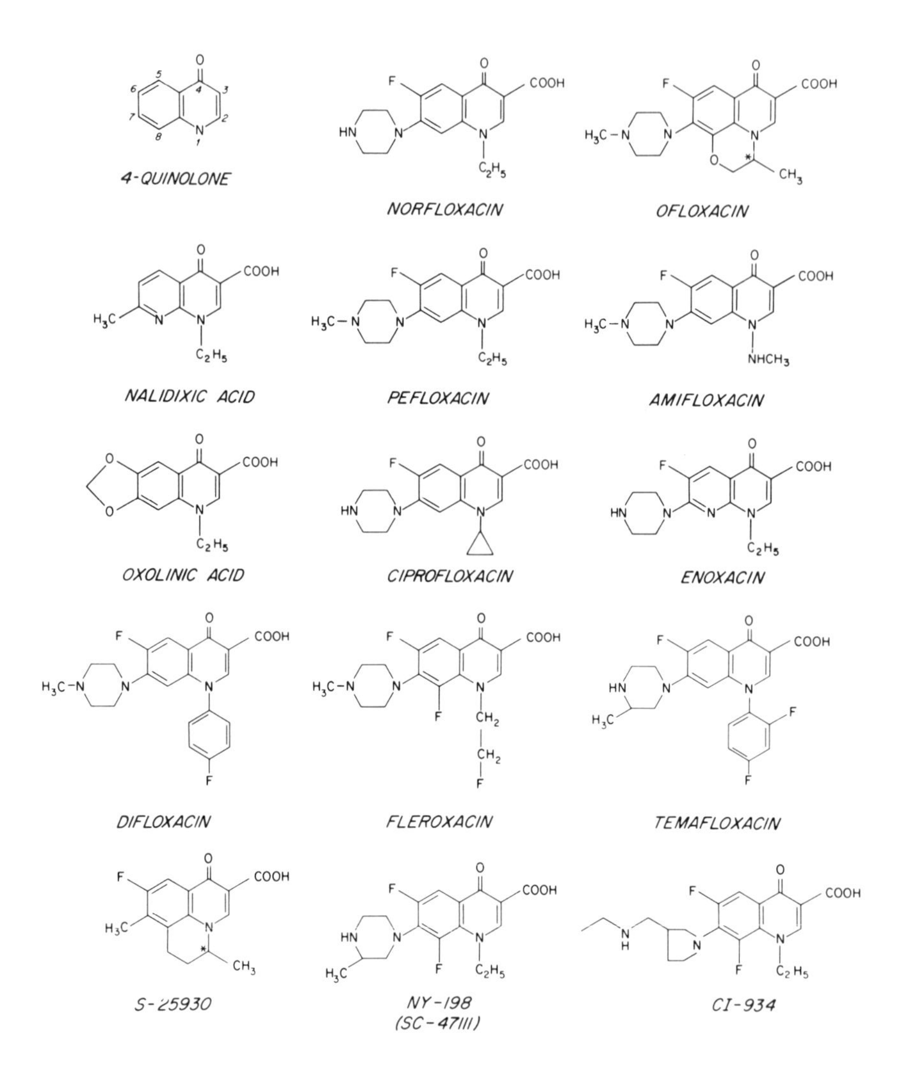

Figure 1. Structures of quinolone agents.

Structural analogs of quinolones, such as the 1,8-naphthyridines nalidixic acid and enoxacin, will for simplicity be referred to as quinolones in this chapter.

Table 1. Partial list of quinolones and structurally related agents

Older agents	Newer agents		
Nalidixic acid	Norfloxacin	Fleroxacin (AM-833, Ro23-6240)	A-62824
Oxolinic acid	Pefloxacin	Temafloxacin (A-62254)	CI-934
Pipemidic acid	Ciprofloxacin	Irloxacin (E-3432)	DR-3355 (S-ofloxacin)
Cinoxacin	Ofloxacin	Lomefloxacin (NY-198, SC-47111)	EN-272
Flumequine	Enoxacin		S-25930
Rosoxacin	Amifloxacin		T-3262, A-60969
	Difloxacin (A-56619)		

Comments on the Structure of DNA

The linear, double helical structure of DNA (159) in a beautiful way encodes genetic information, allows mutation and recombination, and serves as a template for semiconservative replication, DNA repair, and transcription (160).

The configuration of the DNA molecule, however, leads to certain potential difficulties. One problem arises from the condensed state of DNA within the cell. For the bacterium *Escherichia coli*, the chromosome is a circular DNA molecule 1,100 μm in length (16) present in a cell of only 1 to 2 μm in length. This DNA molecule, despite its 1,000-fold condensed state, must, however, be able to replicate, segregate into daughter chromosomes, and allow transcription of individual genes, without becoming lethally entangled.

A second problem occurs because of the helical nature of the DNA duplex. With each turn of the helix, which occurs on the average every 10.4 base pairs, two single strands are wrapped around each other. In the *E. coli* chromosome, which contains 4×10^6 base pairs, strands are intertwined about 400,000 times, to generate a linking number of 400,000. For *E. coli* the two strands must then unwind 400,000 times to allow semiconservative replication. In 1963, Cairns (16) recognized the magnitude of the unwinding problem, when he first visualized the replicating chromosome of *E. coli* and deduced that a swivel was required to permit untwisting of the DNA double helix during strand separation.

A third special situation exists for procaryotes, because negative supertwists are present in DNA isolated from bacteria (7, 47). These negative supertwists indicate that bacterial DNA contains slightly less than one helical turn for each 10.4 base pairs (47) and therefore has a linking number lower than that of eucaryotic DNA. This slightly underwound state of intracellular bacterial DNA is thought to facilitate strand separation required for DNA replication and transcription. Negative supertwists are energetically unfavorable, and therefore within the bacterial cell an energy-requiring process is needed for their generation.

Topoisomerases

For procaryotic DNA, the problems of entanglement, strand unwinding, and negative supertwisting are solved, at least in part, by topoisomerases (34, 47, 157, 158). A topoisomerase is an enzyme that alters the number of times one single strand of a DNA duplex winds around its complementary strand; that is, a topoisomerase alters the linking number of a double-stranded DNA molecule. The DNA molecule is otherwise unchanged, except for increases or decreases in linking number. Such DNA molecules which differ only in linking number are called topological isomers (topoisomers).

Topoisomerases may be categorized into three groups: type II topoisomerases (called DNA gyrase in procaryotes), type I topoisomerases, and special topoisomerases (such as enzymes catalyzing transposition or integration into and excision of bacteriophage DNA from the bacterial chromosome).

Topoisomerase II and topoisomerase I differ in enzymatic activity. Type II enzymes transiently cleave both strands of the double helix and pass another double segment through this break, while type I enzymes transiently break one strand of a double helix and pass through another single strand. As was originally demonstrated by Brown and Cozzarelli (12), DNA topoisomers are altered in linking number in steps of two when acted on by topoisomerase II and in steps of one when acted on by topoisomerase I.

Type II and type I topoisomerases have been isolated from many species of bacteria and eucaryotes, including human cells. Presumably every living organism encodes at least one of each type of topoisomerase.

In 1971 Wang (156) isolated the first topoisomerase, topoisomerase I from *E. coli*. The enzyme is a 110-kilodalton protein and is encoded by the *topA* gene, which is located at 28 min on the *E. coli* genetic map. Topoisomerase I from both procaryotic and eucaryotic sources catalyzes removal of negative supertwists from DNA. Eucaryotic topoisomerase I has the additional ability to remove positive supertwists from DNA. Within bacteria, topoisomerase I, along with DNA gyrase, regulates the level of negative supertwisting of intracellular DNA and is required for transcription of certain operons.

DNA Gyrase (Bacterial Topoisomerase II)

In 1976 Gellert and associates (49) isolated the first topoisomerase II, DNA gyrase, from *E. coli* (34, 47, 157, 158). DNA gyrase (Table 2) contains two A subunits, 100 kilodaltons in mass and encoded by the *gyrA* gene located at 48 min on the *E. coli* map, and two B subunits, 90 kilodaltons in mass and encoded by the *gyrB* gene located at 83 min on the *E. coli* map. Both genes have been cloned and sequenced. The B subunit is a target of the antibiotics novobiocin, coumermycin A1, and clorobiocin, which are structurally related to each other but not to quinolones.

Reactions catalyzed by purified DNA gyrase (Table 3) include introduction of negative supertwists into DNA, formation and resolution (catenation and decatenation) of covalently closed circular DNA molecules interlocked like links in a chain, and formation and removal of knots within a duplex DNA molecule. These reactions each require ATP, which the enzyme hydrolyzes to ADP and P_i. DNA gyrase also requires divalent cation for activity, optimally magnesium, although manganese will substitute. A or B subunits in the absence of the other subunit have no detectable enzymatic activities.

Table 2. *E. coli* DNA gyrase

Subunit	Characteristics	Genetics
A	Two subunits per enzyme molecule (100-kilodalton proteins); a target of quinolone agents	Encoded by the *gyrA* (formerly *nalA*) gene located at 48 min on the *E. coli* map
B	Two subunits per enzyme molecule (90-kilodalton proteins); a target of novobiocin, coumermycin A1, and clorobiocin	Encoded by the *gyrB* (formerly *cou*) gene located at 83 min on the *E. coli* map

Table 3. Reactions catalyzed by *E. coli* DNA gyrase

DNA gyrase	Reactions
Purified enzyme	Introduction of negative supertwists into DNA; catenation and decatenation of interlocked DNA circles; knotting and unknotting of DNA; hydrolysis of ATP; cutting of DNA in the presence of quinolone, sodium dodecyl sulfate, and protease
Enzyme within viable bacteria	Introduction of negative supertwists into DNA; decatenation of interlocked DNA circles; separation of interlocked DNA during chromosome segregation; involvement in DNA replication and aspects of transcription, DNA repair, recombination, and transposition

Reactions catalyzed by DNA gyrase within living bacteria (Table 3) include introduction of negative supertwists and decatenation of interlocked circles. Presumably other intracellular strand passage events are also catalyzed by DNA gyrase. The enzyme is involved in the initiation, elongation (growing point propagation), and termination phases of DNA replication; transcription of certain operons; and aspects of DNA repair, recombination, and transposition. The involvement in initiation and elongation likely represents enhancement of DNA strand unwinding and also the requirement for negative supertwists so that proteins involved can bind properly to the double helix. The requirement for termination appears to represent the need for decatenation, because dividing bacteria in which DNA gyrase has been inactivated accumulate partially segregated nucleoids which can fully separate upon the addition of purified DNA gyrase (139).

DNA gyrase and topoisomerase I within living bacteria set the net level of negative supertwisting of DNA, with the introduction of negative supertwists by DNA gyrase and their removal by topoisomerase I. The amount of intracellular DNA gyrase and topoisomerase I is regulated at the transcriptional level in response to the superhelicity of DNA: decreasing the negative superhelical state of intracellular DNA stimulates transcription of the *gyrA* and *gyrB* genes and suppresses transcription of the *topA* gene. For the *gyrA* and the *gyrB* genes, a DNA sequence 20 base pairs long that includes the −10-base-pair consensus regions within the promoters at which transcription begins and the first few transcribed bases is responsible for induction of transcription of these genes by DNA relaxation (99).

Quinolone Agents: Effects on DNA Gyrase and Viable Bacteria

Early studies

Quinolone agents have marked effects on DNA gyrase and the bacterial cell. Many important consequences of exposure of bacteria to quinolones were determined for nalidixic acid and oxolinic acid prior to the discovery of DNA gyrase. Goss and co-workers (22, 29, 57, 58) reported in the mid-1960s that nalidixic acid selectively antagonized DNA synthesis, caused DNA degradation, and induced filamentation of bacteria. The drug was rapidly bactericidal, but this killing effect was blocked in the presence of chloramphenicol, dinitrophenol, or amino acid starvation, three conditions that share inhibition of protein synthesis. Crumplin and Smith (25) additionally evaluated DNA synthesis in the presence of nalidixic acid and found accumulation of intermediate-sized DNA fragments.

DNA gyrase is a primary target of the quinolones

In 1977 Gellert and associates (48) and Sugino et al. (141) reported experiments that defined the A subunit of DNA gyrase as a primary target of nalidixic acid and oxolinic acid, using a combined biochemical and genetic approach. Purified DNA gyrase containing an A subunit isolated from a bacterial strain carrying a nalidixic acid resistance mutation in the *gyrA* (formerly *nalA*) gene was compared with purified DNA gyrase containing the same B subunit and an A subunit isolated from an isogenic wild-type strain. The enzyme containing the A subunit from the resistant strain was found to be active in the presence of drug concentrations that inhibited the wild-type enzyme. Similar findings in other bacterial species, including *Bacillus subtilis* (140), *Pseudomonas aeruginosa* (81, 102, 124), *Haemophilus influenzae* (133), and *Citrobacter freundii* (1), and for newer quinolone agents, including norfloxacin (23, 69, 70, 76), ofloxacin (129, 130, 164), and ciprofloxacin (75, 124), proved that the A subunit of DNA gyrase is also a primary target in different bacterial species and for the newer drugs.

Quinolone resistance mutations affecting DNA gyrase most frequently occur for *E. coli* in the *gyrA* gene (23, 48, 69, 75, 76, 129, 130, 141, 164), but two mutations encoding nalidixic acid resistance have also been identified in the *gyrB* gene (80, 165, 166). One of these mutations, *nalD*, causes increased resistance to newer quinolone agents, while interestingly the other, *nalC*, produces increased susceptibility to these newer drugs. Thus, the B subunit, as well as the A subunit, is a target of quinolones, perhaps by direct interaction of drug with the B subunit or alternatively by indirect interaction via the A subunit.

Studies in which genetically well-defined sets of *E. coli* strains containing mutations in the *gyrA* and *gyrB* genes have been used also indicate that both the A and the B subunits are involved in action of the newer quinolone agents (118, 137).

Quinolones: effects on purified DNA gyrase

Nalidixic acid, oxolinic acid, and other quinolone agents antagonize almost all of the activities of purified DNA gyrase (34, 47, 157, 158), including introduction of negative supertwists, catenation-decatenation, and unknotting (Table 4). These drugs also have an important additional effect on purified DNA gyrase. They stabilize double-strand breaks in DNA initiated by the enzyme at specific sites which can be revealed by exposure to detergent (sodium dodecyl sulfate) and proteinase K (48, 141). Staggered single-strand breaks 4 base pairs apart are generated, and the A-subunit protein is found covalently attached to the protruding 5′ DNA ends at the tyrosine (77) within the active site of the enzyme. Breakage occurs at preferred DNA sites

Table 4. Effects of quinolones on DNA gyrase and bacteria

Site	Effects
Purified enzyme	Inhibition of almost all enzymatic activities, including introduction of negative supertwists, catenation-decatenation, and unknotting; induction of double-strand cleavage of DNA by enzyme in the presence of detergent (sodium dodecyl sulfate) and protease (proteinase K), with A subunit left covalently attached to 5′ ends of cut DNA
Bacteria	Decrease in introduction of negative supertwists into DNA; impairment of decatenation of interlocked DNA circles; damage to DNA; inhibition of DNA synthesis; antagonism of RNA and protein syntheses at high drug concentrations; filamentation of cells; induction of the SOS DNA repair system and certain heat shock proteins; rapid cell death

that fit into a broadly defined consensus sequence and are found within the 120- to 150-base-pair DNA segment that binds to enzyme. This cleavage event likely relates to trapping of a reaction intermediate before completion of the DNA rejoining event crucial to the catalytic activity of DNA gyrase. This reaction also seems likely to represent the mechanism of DNA damage induced by quinolones within viable bacteria.

The structure of DNA gyrase-DNA complexes was recently studied by electric dichroism (122). Observations suggest that a single turn of DNA is wrapped around the enzyme, with DNA entry and exit points in proximity. Addition of ATP or a nonhydrolyzable ATP analog induces a structural change in the complex, consistent with increased wrapping of DNA around the enzyme. Remarkably, this ATP-induced structural change does not occur in the presence of norfloxacin. It was postulated that norfloxacin might stabilize the enzyme-DNA complex in the less wrapped configuration and block supertwisting by preventing structural transition. This possibility is consistent with previous studies showing that DNA gyrase-DNA complexes are more stable in the presence of quinolone agents (67). An alternative interpretation is that norfloxacin might uncouple binding of unwrapped DNA segments to the enzyme from double-strand cleavage.

Shen and Pernet (135) have studied mechanisms of quinolone antagonism at the molecular level, specifically investigating binding of norfloxacin to DNA and to DNA gyrase. They reported binding of norfloxacin to DNA, with a plateau at a drug concentration similar to the inhibitory constant (K_i) for inhibition of the supertwisting activity of purified enzyme. Denaturated DNA and negatively supertwisted DNA both bound more drug than did relaxed double-stranded DNA circles, suggesting higher drug affinity for single-stranded than double-stranded DNA. Under the conditions used, little or no norfloxacin bound to DNA gyrase. In contrast, in preliminary studies, Le Goffic (92) reported alterations in ^{19}F nuclear resonance of pefloxacin in the presence of DNA gyrase but not DNA alone. More studies are required for clarification of the interaction of quinolones, DNA gyrase, and DNA.

Quinolones and DNA gyrase: structure-activity relationships

Many studies indicate that modification of the quinolone structure affects its antibacterial activity (26, 32, 33, 74, 103, 169). Information on the role of different portions of the molecule in antagonism of DNA gyrase, however, has only recently emerged. Activity can be markedly altered by substitutions attached to N-1 (Fig. 1). Particularly potent is a cyclopropyl ring substitution (ciprofloxacin). N-1 with C-8 can also serve as an attachment site for a third saturated ring. For ofloxacin and S-25930, this third ring has an asymmetric carbon, the exact structure of which affects potency, as illustrated by the fact that the *S* isomer of each drug is 10-fold more potent in inhibiting the DNA gyrase than is the *R* isomer (79, 104, 164). For the carboxylic acid group attached to C-3, only recently have alterations that enhance enzyme inhibition been identified (D. T. W. Chu, P. B. Fernandes, A. K. Claiborne, L. L. Shen, and A. G. Pernet, *27th Intersci. Conf. Antimicrob. Agents Chemother.*, abstr. no. 250, 1987). The oxygen double bonded to C-4 is present in all active quinolone agents. Substitution of a fluorine atom to C-6 has markedly augmented potency. Substitutions on C-7 have also enhanced activity and have been extensively evaluated. Antibacterial activity is enhanced by a piperazinyl moiety, at least in part because of improved enzyme antagonism. In several series of compounds, C-8 is replaced with N-8 (nalidixic acid and enoxacin), resulting in 1,8-naphthyridines, an alteration that appears in some instances to decrease antagonism of DNA gyrase. Recently it has been found that inhibition of DNA gyrase does not require an intact quinolone ring, as the enzyme is also inhibited by monosubstituted rings structurally related to quinolones (51).

Information on effects of quinolone structure on activity against DNA gyrase and living bacteria was recently incorporated into a computer program (84). When challenged, this program predicted with considerable accuracy the effects of structural variations on drug activity. This program seems likely to be useful in the design of new quinolone agents.

Effects of quinolones on viable bacteria and bacteriophage

Antagonism of bacteria by quinolones results in decreased introduction of negative supertwists into DNA, decreased decatenation of interlocked DNA circles, and damage to DNA (Table 4) (10, 22, 23, 29, 34, 47, 57, 58, 80, 93, 114, 122, 137, 138, 157, 158, 165). Induction of these changes requires higher drug concentrations in quinolone-resistant *gyrA* mutants, indicating that DNA gyrase is involved in these effects. The result of drug exposure is interference with DNA synthesis indicated by an immediate halt to incorporation of precursors into DNA, suggesting interference with DNA growing point propagation. At high drug concentrations, RNA and protein syntheses are also impaired. Filamentation occurs, at least in part because of antagonism of DNA synthesis and induction of the SOS DNA repair system (117, 119). In addition, expression of *groEL* and *dnaK*, two genes which are part of the generalized heat shock protein system that protects the cell against adverse conditions, is induced (88, 149).

Kreuzer and Cozzarelli reported an important experiment that suggests that antagonism of bacteriophage T7 by quinolones involves more than inhibition of DNA gyrase catalytic activity (87). DNA gyrase does not appear to be required for growth of phage T7 in *E. coli*, because a normal phage burst size occurs at an elevated temperature in host bacteria containing a thermosensitive DNA gyrase. Strikingly, in this same bacterial host in the presence of nalidixic acid, phage T7 multiplies to give a normal burst size at an elevated but not at a permissive temperature. Thus, inhibition of phage growth by quinolone would not be explained simply by antagonism of DNA gyrase enzymatic activity. As an explanation of these results, Kreuzer and Cozzarelli suggest formation of an irreversible complex of drug and enzyme that functions as "poison."

Marked variations in antibacterial potency of different quinolones have been well documented. These variations represent, in general, differences in inhibitory activity of individual quinolones against DNA gyrase, as has been demonstrated, for instance, for *Micrococcus luteus* (169) and for *E. coli* (32, 75). Antibacterial activity does not always correlate exactly with antagonism of DNA gyrase, however, indicating the presence of other factors, such as permeation, which may affect drug potency (32, 169).

A puzzling observation is that quinolone concentrations needed to inhibit purified DNA gyrase, to decrease introduction of negative supertwists within viable cells, or to decatenate interlinked DNA circles within viable cells are 10- to 100-fold higher than concentrations that inhibit bacterial growth. These discrepancies suggest that something other than DNA gyrase may be the intracellular target of quinolones. This interpretation is open to question, however, for three reasons. First, conditions used to evaluate inhibition of purified enzyme may not reflect the intracellular environment. Second, the inhibitory event within a cell might be a subtle perturbation of enzymatic activity, such as one or a few DNA cleavage events or a slight reduction of negative supertwisting. Related to this latter possibility, a series of quinolones has been shown to inhibit enzyme supertwisting activity by 10% at concentrations of drug similar to those that inhibit cell growth (74). Third, genetic and biochemical studies described above indicate that DNA gyrase is an intracellular target of quinolone agents. Thus, it does not seem necessary to invoke a quinolone target in addition to DNA gyrase, although the existence of additional targets has not been excluded.

Recently it has been reported that quinolones inhibit purified *E. coli* topoisomerase I at concentrations 10-fold higher than those inhibiting purified DNA gyrase (143). The significance of these findings is not as yet clear.

Killing of Bacteria by Quinolones

Exposure of most susceptible bacterial species to quinolone agents results in rapid cell death (23, 24, 29, 34, 57, 82, 137, 168). For example, for *E. coli* cells in logarithmic growth, CFU fall by a factor of 10^4 in 60 to 120 min. While the precise mechanism of bacterial killing is poorly understood, some facts about the phenomenon, in particular for *E. coli*, have emerged and are enumerated below.

(i) Killing involves DNA gyrase, because resistance mutations mapping in *gyrA* result in proportional increases in concentrations of drug needed to inhibit bacterial growth and to kill bacteria.

(ii) Killing is rapid but approaches a plateau at a viable cell count about 10^4-fold below the initial inoculum.

(iii) Killing is markedly diminished by chloramphenicol, rifampin, dinitrophenol, or amino acid starvation (23, 29, 168). These treatments have in common the capacity to inhibit protein synthesis, suggesting that killing requires new synthesis of a protein(s).

(iv) Killing tends to be maximal at approximately 30- to 60-fold above the MIC of the drug (137). Killing may be reduced at still higher drug concentrations (24, 137), and thus the quinolone agents can exhibit a paradoxical or Eagle effect (36), as has been documented for some bacterial strains exposed to penicillin. For quinolones, the paradoxical effect may be the consequence of the partial inhibition of protein synthesis known to occur in the presence of high concentrations of drug (23) and may thus have similarity to antagonism of killing by chloramphenicol, rifampin, dinitrophenol, and amino acid starvation.

(v) Quinolone agents differ in ability to kill the same bacterial strain (137). Ofloxacin and ciprofloxacin, for instance, are more effective in killing *E. coli* KL16 than is norfloxacin. Killing by ofloxacin and ciprofloxacin is also less effectively blunted by chloramphenicol and rifampin in comparison with other quinolone agents.

(vi) Killing appears to be inhibited in bacteria growing under strictly anaerobic conditions (C. S. Lewin and J. T. Smith, *27th ICAAC*, abstr. no. 473, 1987).

(vii) Killing of certain bacterial species, such as *Staphylococcus saprophyticus*, may be slower than killing of *E. coli* (46).

The mechanism of bacterial killing by quinolones is unclear. One possible mechanism (22, 34, 117, 119, 137) is that quinolones might cause DNA gyrase to damage DNA by induction of a lesion which is either nonrepairable or which is rendered nonrepairable when altered by repair enzymes. Observations that quinolones induce purified DNA gyrase to break DNA in certain settings, damage DNA in living cells, induce the SOS DNA repair system, and have increased potency against bacteria that have mutations in SOS repair genes are consistent with this hypothesis.

An approach to the study of killing of bacteria by quinolones is identification of mutant loci that reduce killing but not inhibition of bacterial growth and thus cause bacterial tolerance. One such *gyrA* allele has been reported (23); this interesting finding awaits confirmation.

Quinolones: Bacterial Resistance

Bacterial resistance to quinolone agents will be considered in subsections on acquired resistance, intrinsic resistance, and clinical relevance.

Acquired resistance

Topics on acquired resistance will include selection of resistance, mechanisms of resistance, and scarcity of plasmid-mediated resistance.

Selection of resistant bacteria. The newer quinolone agents are 100- to 1,000-fold less prone than nalidixic acid to selection of single-step spontaneous highly resistant organisms for many bacterial species (164), with the exception of *P. aeruginosa*, for which the difference is only 10-fold (27, 39). Additionally, selection of such mutants for many bacterial species can be detected only at concentrations two- to eightfold above the MICs of the drugs. At higher concentrations, it has not been possible to select resistant bacteria within the sensitivity of methods used, indicating a mutation frequency of $<10^{-10}$ (27, 75).

Selection of strains highly resistant to newer quinolone agents is possible, nevertheless, by serially passing bacteria in the presence of increasing drug concentrations (6, 74, 76, 127, 144). Strains so selected may have either stable or unstable resistance upon passage in the absence of drug. These strains typically are cross resistant to other quinolone agents. These strains also, in some instances, exhibit pleiotropic resistance to low concentrations of nonquinolone antimicrobial agents, such as chloramphenicol, tetracycline, β-lactam drugs, and trimethoprim (61, 75, 76, 108, 127, 147).

Mechanisms of quinolone resistance. A number of genes that can mutate to cause quinolone resistance have been identified. These genes in *E. coli* include *gyrA* (*nalA* [64], *nfxA* [76], *norA* [69], *cfxA* [75], and *ofxA* [164]) and *gyrB* (80, 137, 165, 166) encoding DNA gyrase (Table 5) (discussed above). Other mutant loci (Table 6), in addition to those encoding DNA gyrase, include *nfxB* (76), *norB* (69), *norC* (69), and *cfxB* (75), which appear to alter drug permeation (discussed below). Mutations in additional genes (Table 6) produce low-level resistance to nalidixic acid by as yet undefined

Table 5. Quinolone resistance mutations in genes encoding DNA gyrase in *E. coli*

DNA gyrase gene	Resistance mutations
gyrA	*nalA, nfxA, ofxA, norA, cfxA;* 48 min on the *E. coli* map; cause resistance to quinolone agents only
gyrB	*nalC* (resistant to nalidixic acid and hypersusceptible to certain fluoroquinolones); *nalD* (resistant to nalidixic acid and fluoroquinolones); 83 min on the *E. coli* map

Table 6. Quinolone resistance mutations in *E. coli* caused by mutations in genes other than *gyrA* and *gyrB*

Selected with:	Resistance mutations[a]	Characteristics
Nalidixic acid	*nalB* (58), *nalD* (89), *crp* (74), *cya* (85), *icd* (25), *purB* (25), *ctr* (52)	
Newer quinolone agents		
Norfloxacin	*nfxB* (19), *norB* (34), *norC* (8)	Likely permeation mutants; resistant to quinolones and structurally unrelated drugs; decreased OmpF porin protein in outer membrane; decreased norfloxacin accumulation; altered lipopolysaccharide and hypersusceptible to detergent *(norC)*
Ciprofloxacin	*cfxB* (34)	
Tetracycline or chloramphenicol	*marA* (34)	Decreased accumulation of tetracycline; nalidixic acid resistant

[a]Numbers within parentheses indicate map locations in minutes on the *E. coli* chromosome.

mechanisms and include *crp* (89), *cya* (90), *icd* (90), *purB* (66, 90), *ctr* (66), and *marA* (50). Genetic or other data have thus far not demonstrated drug destruction as a mechanism of bacterial resistance to quinolones.

Recently, nucleotide sequences have been determined for *gyrA* genes isolated from two strains of *E. coli* selected for spontaneous resistance to nalidixic acid and two strains selected for resistance to the structurally related agent pipemidic acid (167). In each mutant gene, a single base pair was changed. The four mutations were located close to each other, in a sequence near the amino terminus of the *gyrA* polypeptide. The four mutation sites were located in hydrophilic sections of the peptide, suggesting that the sites reside near the surface of the *gyrA* protein. The mutations were also relatively near the codon for tyrosine 122, which has been shown to be the site of covalent attachment of the A subunit to DNA when enzyme cleaves DNA in the presence of quinolone (77).

Nucleotide sequences have also been determined for *nalC* and *nalD*, the two *gyrB* mutant genes selected for resistance to nalidixic acid (80, 165, 166). The *nalD* locus causes increased resistance to newer quinolone agents, particularly those with piperazinyl substitutions, while interestingly the *nalC* locus produces increased susceptibility to these newer drugs. Each mutation resulted in a distinct single-nucleotide change in the midportion of the coding sequence. The two substitutions encoded single amino acid changes that altered protein charge in opposite directions, suggesting that charge is in some way related to drug-enzyme interaction.

E. coli strains highly resistant to norfloxacin, ofloxacin, and ciprofloxacin have been selected through in vitro exposure to increasing concentrations of one or another of these drugs; transfer of single mutant genes back to wild-type strains by phage P1 transduction has identified two classes of quinolone-resistant mutants (75, 76, 164). One class of mutations is characterized by low-level resistance to quinolones only and maps in the *gyrA* gene. *nfxA* (76) and *ofxA* (164), selected by exposure to norfloxacin and ofloxacin, respectively, are examples of this class of mutations. Purified DNA gyrase containing an A subunit isolated from an *nfxA* or *ofxA* strain confers increased quinolone resistance to the enzyme, indicating that the resistance mutation resides in the structural gene for the DNA gyrase A subunit.

The second class of mutations is characterized by low-level resistance to both quinolones and structurally unrelated drugs. *nfxA* (76) and *cfxB* (75), which were selected with norfloxacin and ciprofloxacin and map at 19 and 34 min on the *E. coli* chromosome, respectively, are examples of such mutations. Decreased drug permeation is suggested by the pleiotropy of these mutations. Impaired permeation seems very likely, because these two loci are associated with decreased binding of norfloxacin and decreased expression of OmpF, an outer membrane porin protein. OmpF forms pores in the outer membrane through which small hydrophilic compounds, including certain antimicrobial agents, diffuse into the bacterial periplasmic space. These observations confirm observations that bacteria carrying mutations in the OmpF gene (*ompF*, 21 min on the *E. coli* map) exhibit low-level resistance to quinolones and structurally unrelated drugs (68, 76).

norB and *norC*, two additional mutants selected with norfloxacin in *E. coli*, also appear to decrease drug permeation (69). *norB* has the same characteristics as *cfxB* and also maps at 34 min, but it is unknown whether the mutations are allelic. *norC* differs from *nfxB*, *cfxB*, and *norB*. This mutation causes hypersusceptibility to hydrophobic quinolones such as nalidixic acid, detergents, and dyes, exhibits both decreased OmpF protein and also altered lipopolysaccharides in the outer membrane, and maps at 8 min on the *E. coli* chromosome.

Similar to norfloxacin, an enoxacin-resistant strain of *E. coli* selected by serial passage in the presence of increasing drug concentrations also exhibited decreased OmpF protein in outer membrane preparations and decreased binding of enoxacin (9).

The inner membrane, as well as the outer membrane, of *E. coli* appears to play an important role in accumulation of norfloxacin (21). Studies of everted inner membrane vesicles (in which the inner surface faces outward) revealed energy-dependent uptake of norfloxacin into the vesicles. The process was carrier mediated and driven by proton motive force. OmpF-deficient intact cells were twofold less susceptible to norfloxacin and showed twice as much energy-dependent reduction in drug uptake. Active uptake into everted vesicles from the OmpF-deficient strain, however, was similar to that with the parent wild-type strain. These findings suggest that in the OmpF mutant, decreased outer membrane permeability, combined with the active efflux at the inner membrane, by some mechanism results in decreased accumulation of norfloxacin and lower drug susceptibility.

Studies with bacterial species other than *E. coli* also have shown that mutation in the A subunit of DNA gyrase and decreased drug permeation are associated with the development of quinolone resistance. Mutations in the nalidixic acid resistance locus *nalA* of *P. aeruginosa* increase bacterial resistance to newer quinolone agents. Purified DNA gyrase containing A subunits isolated from a *nalA* strain is more resistant to nalidixic acid (102) and newer quinolones (81) than is enzyme containing A subunits isolated from the isogenic wild-type strain. This result proved that the *nalA* mutation is in the A subunit of DNA gyrase and that mutation in the A subunit can cause resistance. In a similar fashion, *P. aeruginosa* strains selected for resistance to norfloxacin (70) and ciprofloxacin (124) have been found to contain an *nfxA* mutation and a *cfxA* mutation, respectively, both of which map in the *nalA* locus and encode quinolone-resistant A subunits.

In contrast, the *nalB* gene of *P. aeruginosa* causes resistance to quinolones and structurally unrelated antimicrobial agents, suggesting decreases in drug permeation (123). Mutations in the *nalB* gene have also been selected with ciprofloxacin (124). In addition, mutation in another gene, *nfxB* (70), which was selected with norfloxacin, causes resistance to newer quinolone agents, alters susceptibility to structurally unrelated drugs, and decreases binding of norfloxacin, all properties implicating decreased drug permeation. Outer membrane studies of *nfxB* cells have demonstrated the presence of a new 54-kilodalton protein, rather than a decrease in amount of a previously existing protein, indicating a mechanism different from that in *E. coli*.

Additional examples of pleiotropic drug resistance that include quinolone agents have been reported for *Enterobacter* sp., *E. coli, Klebsiella pneumoniae, P. aeruginosa, Salmonella paratyphi, Serratia marcescens,* and other bacterial species (8, 50, 60, 61, 108, 127, 128, 136, 145, 147), and for some strains changes in outer membranes suggesting a decrease in porinlike proteins have been documented. Significantly, some of these resistant isolates were selected by exposure to quinolones, while others were selected by exposure to structurally unrelated agents, including β-lactams, tetracycline, and chloramphenicol.

Scarcity of plasmid-mediated resistance to quinolones. With a single, as yet unconfirmed, apparent exception (110), mutations causing quinolone resistance are located on the bacterial chromosome and not on naturally occurring plasmids, despite an extensive search (13). Four attributes of quinolones and quinolone resistance might contribute to this scarcity of plasmid-mediated resistance.

(i) Wild-type susceptible genes are dominant to many *gyrA* mutations encoding quinolone resistance (64, 91, 142, 167). This dominance pattern might reflect subunit mixing in which the hybrid enzyme remains sensitive to drug action. This dominance pattern suggests that transfer of a mutant *gyrA* gene into a susceptible cell would likely not generate phenotypic resistance in the absence of gene conversion by recombination of mutant and wild-type alleles.

(ii) Expression of mutations which decrease outer membrane porins is delayed when introduced into wild-type cells (45). Thus, delayed expression of quinolone resistance mediated by genes altering outer membrane proteins may also occur.

(iii) Plasmid conjugation is inhibited by quinolone agents (14, 111).

(iv) Some plasmids are eliminated from bacteria at quinolone concentrations below those completely inhibiting bacterial growth (120, 161).

For these reasons, the spread of quinolone resistance by a mechanism involving plasmid-mediated transfer of resistance genes is likely to be suppressed in the presence of a quinolone agent.

Intrinsic resistance

In addition to acquired resistance, intrinsic resistance to clinically achievable concentrations of quinolone agents exists for certain microbial organisms. Intrinsically resistant microbes may include exceptional isolates of bacterial species usually susceptible to quinolones or all the isolates of a species, such as *Pseudomonas* spp. other than *P. aeruginosa*, anaerobic bacteria, and fungi (72, 163). The mechanisms of quinolone resistance have not been defined in these organisms. It is anticipated, however, that resistant DNA gyrase or altered drug permeation or both will be important in some of these organisms, based on our current understanding of mechanisms of acquired resistance to quinolone agents. For yeasts, it seems likely that topoisomerase II will be drug resistant, because topoisomerase II purified from *Saccharomyces cerevisiae*, a nonpathogenic yeast species, is antagonized by only high concentrations of nalidixic acid or oxolinic acid (59).

Clinical importance

Certain aspects of mechanisms of quinolone resistance have known or potential clinical significance. Problems related to acquired resistance to quinolones might be anticipated in patients infected with bacteria for which maximally achievable drug concentrations in tissues do not substantially exceed drug concentrations inhibiting bacterial growth (132). In this setting, selection of single-step spontaneous low-level resistance might result in resistant bacteria that are associated with failure of drug therapy. *P. aeruginosa* is of particular concern for its potential for acquiring resistance, because of lower susceptibility to quinolones in vitro and because of a higher frequency of mutation to quinolone resistance relative to many other bacterial species (27, 39, 163).

Examples of proved or likely emergence of quinolone resistance during therapy of organisms which tend to require for inhibition MICs close to achievable drug concentrations have been documented (Table 7), particularly with *P. aeruginosa*, but also with other bacterial species, including *Staphylococcus aureus*, *Staphylococcus epidermidis*, *Streptococcus pneumoniae*, *Acinetobacter* sp., *Enterobacter* sp., *Serratia* sp., *Klebsiella* sp., *Campylobacter* sp., and *Mycobacterium tuberculosis*. For *P. aeruginosa*, it is noteworthy that in patients with exacerbation of respiratory infection in the setting of cystic fibrosis, clinical improvement occurred despite emergence of resistant bacteria (121, 132).

Table 7. Apparent emergence of quinolone resistance during therapy

Study (reference)	Drug	Infection	Species (% resistant)
Maesen et al. (94)	Pefloxacin	Bronchitis	*P. aeruginosa* (43)
Tsukamura et al. (148)	Ofloxacin	Pulmonary tuberculosis	*M. tuberculosis* (63)
Scully et al. (132)	Ciprofloxacin	Cystic fibrosis, respiratory, skin	*P. aeruginosa* (26)
Follath et al. (41)	Ciprofloxacin	Osteomyelitis, respiratory, skin	*P. aeruginosa* (30)
Giamarellou et al. (52)	Ciprofloxacin	Varied infections	*P. aeruginosa* (13)
Finch et al. (40)	Ciprofloxacin	Cystic fibrosis, skin, abscess	*P. aeruginosa* (27)
Boerema and Van Saene (11)	Norfloxacin	Complicated urinary tract infection	*P. aeruginosa* (13)
Desplaces et al. (31)	Pefloxacin, ciprofloxacin	Varied types	Varied (6)
Goldfarb et al. (55)	Ciprofloxacin	Cystic fibrosis	*P. aeruginosa* (83)
Shalit et al. (134)	Ciprofloxacin	Cystic fibrosis	*P. aeruginosa* (45)
Gilbert et al. (54)	Ciprofloxacin	Osteomyelitis	*P. aeruginosa* (31)
Mulligan et al. (109)	Ciprofloxacin	MRSA[a] carriers	*S. aureus* (32)

[a]MRSA, Methicillin-resistant *S. aureus*.

For treatment of bacterial infections caused by other species of gram-negative bacilli, particularly members of the family *Enterobacteriaceae*, data suggest that single-step mutation to increased resistance would result in organisms still susceptible to clinically achievable drug concentrations, particularly for infections of the lower urinary tract and the gastrointestinal tract, two body sites at which quinolone concentrations are high (73, 163). Consistent with this possibility is the observation that emergence of resistance occurs in less than 2% of patients with uncomplicated urinary tract infections treated with fluoroquinolone agents (163a).

In two instances, mechanisms of resistance of clinical strains isolated from patients treated with a quinolone agent have been studied. A strain of *E. coli* isolated from a patient treated with norfloxacin was found to have properties suggesting both a drug-resistant DNA gyrase and decreased drug permeation (1). Another strain of *E. coli* isolated from a patient treated with ofloxacin was found to encode a DNA gyrase A subunit that conferred quinolone resistance to purified enzyme (129). These investigations suggest that resistance in the clinical setting may be mediated by both altered enzyme and decreased drug permeation, as was earlier established in studies with laboratory strains.

A potential clinical problem might occur with superinfection of colonizing microbial species, such as *Pseudomonas* spp. other than *P. aeruginosa*, certain gram-positive bacteria, or yeasts, because these organisms exhibit intrinsic resistance to quinolone agents. In fact, superinfection is to date uncommon (3) but has been reported with *Pseudomonas* spp., staphylococci, enterococci, streptococci, *Klebsiella* spp., *Serratia* spp., *Citrobacter* spp., and yeasts with different quinolone agents.

Relevant to the problem of emergence of quinolone resistance are studies evaluating effects in vitro of combinations of quinolones with other antimicrobial agents. The results of many studies are consistent: different quinolones in combination with many β-lactams or aminoglycosides are usually additive or indifferent, occasionally synergistic, and only rarely antagonistic in inhibition of growth of most bacterial species (4, 5, 15, 18–20, 28, 35, 53, 62, 83, 95, 97, 106, 107, 150, 153, 162). Exceptions are some combinations of a quinolone and a β-lactam that are synergistic against *P. aeruginosa* in up to one-half of strains tested (4, 15, 19, 96, 106). Other exceptions include antagonism by some, but not all, combinations of a quinolone with rifampin (23, 86), chloramphenicol (23, 97), tetracycline (97), or vancomycin (115, 151, 152). In animal models, some data suggest that drug combinations can suppress emergence of quinolone resistance (5, 100). Use of drug combinations may well prove useful clinically (31) but is unlikely to suppress emergence of resistance completely. Even now there have been several reports of emergence of quinolone resistance, despite treatment with two drugs, not only in animal models (100) but also in human infections (31, 131).

Coumermycin A1, an inhibitor of the B subunit of DNA gyrase, has also been evaluated in combination with a quinolone against gram-positive bacteria. Coumermycin was synergistic in vitro in combination with ciprofloxacin (113) or CI-934 (154) but antagonized ciprofloxacin in an animal model system (116).

It is important to realize that our clinical experience with the quinolone agents is relatively limited and that the extent to which emergence of quinolone resistance or superinfection with quinolone-resistant organisms will be a problem is not yet known. Therefore, it will be important in the future to monitor carefully for the development of bacterial resistance during clinical use of the new fluoroquinolones.

Effects of Quinolone Agents on Human Cells

The quinolones are relatively nontoxic to human cells (98, 105, 163a), an observation that might at first seem puzzling, because both bacterial and human cells contain type II topoisomerases (47, 157, 158). This difference nevertheless appears to reflect differences in enzyme sensitivity to drug, as quinolones are 100- to 1,000-fold more potent against purified DNA gyrase than against mammalian topoisomerase II (78). The eucaryotic enzyme also is known to differ from DNA gyrase both structurally and functionally. Eucaryotic topoisomerase II is encoded by one gene rather than two and removes, rather than introduces, supertwists into DNA.

The lower toxicity of quinolones for human cells is noteworthy in the face of excellent penetrations into these cells. Quinolones achieve intracellular concentrations equal to or greater than extracellular drug concentrations (17, 37, 85). This property is likely to be important for treatment of infections with bacterial pathogens such as *Legionella* spp. (65, 126, 155), which grow within human cells, or *S. aureus* (37, 101) or gram-negative bacilli (146), which are ingested by polymorphonuclear leukocytes.

Quinolones at high concentrations have additional effects on human lymphoid cells (30, 42–44, 56, 125) (see also chapter 14). Ciprofloxacin at 100 μg/ml inhibits growth of lymphoblastoid cells in tissue culture (78) and at 200 μg/ml inhibits proliferation of mitogen-stimulated human lymphocytes (42). Difloxacin (25 μg/ml) (56) or pefloxacin, ciprofloxacin, or ofloxacin (50 μg/ml) (125) antagonizes the proliferative response of mitogen-stimulated human mononuclear cells. In contrast, quinolones at lower concentrations (1.6 to 6.25 μg/ml) appear to increase DNA synthesis in mitogen-stimulated human lymphocytes (43), although this effect may be an artifact of alterations in cellular pyrimidine nucleotide pools. Both ofloxacin and norfloxacin (30 μg/ml) antagonize lymphocyte blastogenesis, and norfloxacin decreases gamma interferon production (44). The clinical importance of these and other effects (44) of quinolones on human cells is as yet unknown.

Conclusions

Studies indicate that the mechanism of quinolone action involves antagonism of DNA gyrase and that mechanisms of resistance involve mutations in genes of DNA gyrase and in genes that decrease drug accumulation. Despite the amount of information on these mechanisms reviewed here and elsewhere (34, 47, 74, 137, 157, 158), there remain, however, many important gaps in our knowledge. To understand more fully mechanisms of drug action, we need substantial additional information on (i) how quinolones, DNA gyrase, and DNA interact at the molecular level; (ii) the nature of DNA damage induced in bacteria by quinolones; and (iii) the events required for quinolone-mediated bacterial killing. For understanding mechanisms of quinolone resistance, additional information is needed on (i) the molecular details of how mutant *gyrA* and *gyrB* proteins confer drug resistance and (ii) the routes of drug penetration into the bacterial cell and the nature of changes that alter these pathways. Hopefully, new findings on these and other aspects of quinolone action and resistance will be forthcoming and will prove useful for the development of more potent agents and approaches to cope with bacterial resistance.

Literature Cited

1. **Aoyama, H., K. Sato, T. Fujii, K. Fujimaki, M. Inoue, and S. Mitsuhashi.** 1988. Purification of *Citrobacter freundii* DNA gyrase and inhibition by quinolones. *Antimicrob. Agents Chemother.* **32:**104–109.
2. **Aoyama, H., K. Sato, T. Kato, K. Hirai, and S. Mitsuhashi.** 1987. Norfloxacin resistance in a clinical isolate of *Escherichia coli. Antimicrob. Agents Chemother.* **31:**1640–1641.
3. **Arcieri, G., E. Griffith, G. Gruenwaldt, A. Heyd, B. O'Brien, N. Becker, and R. August.** 1987. Ciprofloxacin: an update on clinical experience. *Am. J. Med.* **82**(Suppl. 4A):381–386.
4. **Baltch, A. L., C. Bassey, G. Fanciullo, and R. P. Smith.** 1987. In-vitro antimicrobial activity of enoxacin in combination with eight other antibiotics against *Pseudomonas aeruginosa,* Enterobacteriaceae and *Staphylococcus aureus. J. Antimicrob. Chemother.* **19:**45–48.
5. **Bamberger, D. M., L. R. Peterson, D. N. Gerding, J. A. Moody, and C. E. Fasching.** 1986. Ciprofloxacin, azlocillin, ceftizoxime, and amikacin alone and in combination against gram-negative bacilli in infected chamber model. *J. Antimicrob. Chemother.* **18:**51–63.
6. **Barry, A. L., and R. N. Jones.** 1984. Cross-resistance among cinoxacin, ciprofloxacin, DJ-6783, enoxacin, nalidixic acid, norfloxacin, and oxolinic acid after in vitro selection of resistant populations. *Antimicrob. Agents Chemother.* **25:**775–777.
7. **Bauer, W. R.** 1987. Structure and reactions of closed duplex DNA. *Annu. Rev. Biophys. Bioeng.* **7:**287–313.
8. **Bayer, A. S., L. Hirano, and J. Yih.** 1988. Development of β-lactam resistance and increased quinolone MICs during therapy of experimental *Pseudomonas aeruginosa* endocarditis. *Antimicrob. Agents Chemother.* **32:**231–235.
9. **Bedard, J., S. Wong, and L. E. Bryan.** 1987. Accumulation of enoxacin by *Escherichia coli* and *Bacillus subtilis. Antimicrob. Agents Chemother.* **31:**1348–1354.
10. **Bliska, J. B., and N. R. Cozzarelli.** 1987. Use of site-specific recombination as probe of DNA structure and metabolism in vivo. *J. Mol. Biol.* **194:**205–218.
11. **Boerema, J. B. J., and H. K. F. Van Saene.** 1986. Norfloxacin treatment in complicated urinary tract infections. *Scand. J. Infect. Dis. Suppl.* **48:**20–26.
12. **Brown, P. O., and N. R. Cozzarelli.** 1979. A sign inversion mechanism for enzymatic supercoiling of DNA. *Science* **206:**1081–1083.
13. **Burman, L. G.** 1977. Apparent absence of transferable resistance to nalidixic acid in pathogenic gram-negative bacteria. *J. Antimicrob. Chemother.* **3:**509–516.
14. **Burman, L. G.** 1977. R-plasmid transfer and its response to nalidixic acid. *J. Bacteriol.* **131:**76–81.
15. **Bustamante, C. I., G. L. Drusano, R. C. Wharton, and J. C. Wade.** 1987. Synergism of the combinations of imipenem plus ciprofloxacin and imipenem plus amikacin against *Pseudomonas aeruginosa* and other bacterial pathogens. *Antimicrob. Agents Chemother.* **31:**632–634.
16. **Cairns, J.** 1963. The chromosome of *Escherichia coli. Cold Spring Harbor Symp. Quant. Biol.* **28:**43–46.
17. **Chadwick, P. R., and A. R. Mellersh.** 1987. The use of a tissue culture model to assess the penetration of antibiotics into epithelial cells. *J. Antimicrob. Chemother.* **19:**211–220.

18. **Chalkley, L. J., and H. J. Koornhof.** 1985. Antimicrobial activity of ciprofloxacin against *Pseudomonas aeruginosa, Escherichia coli,* and *Staphylococcus aureus* determined by the killing curve method: antibiotic comparisons and synergistic interactions. *Antimicrob. Agents Chemother.* **28:**331–342.
19. **Chin, N. X., K. Jules, and H. C. Neu.** 1986. Synergy of ciprofloxacin and azlocillin in vitro and in a neutropenic mouse model of infection. *Eur. J. Clin. Microbiol.* **5:**23–28.
20. **Chin, N.-X., and H. C. Neu.** 1987. Synergy of imipenem—a novel carbapenem, and rifampin and ciprofloxacin against *Pseudomonas aeruginosa, Serratia marcescens,* and *Enterobacter* species. *Chemotherapy* (Basel) **33:**183–188.
21. **Cohen, S. P., D. C. Hooper, J. S. Wolfson, K. S. Souza, L. M. McMurry, and S. B. Levy.** 1988. Endogenous active efflux of norfloxacin in susceptible *Escherichia coli. Antimicrob. Agents Chemother.* **32:**1187–1191.
22. **Cook, T. M., W. H. Deitz, and W. A. Goss.** 1966. Mechanism of action of nalidixic acid on *Escherichia coli.* IV. Effects on the stability of cellular constituents. *J. Bacteriol.* **91:**774–779.
23. **Crumplin, G. C., M. Kenwright, and T. Hirst.** 1984. Investigations into the mechanisms of action of the antibacterial agent norfloxacin. *J. Antimicrob. Chemother.* **13**(Suppl. B): 9–23.
24. **Crumplin, G. C., and J. T. Smith.** 1975. Nalidixic acid: an antibacterial paradox. *Antimicrob. Agents Chemother.* **8:**251–261.
25. **Crumplin, G. C., and J. T. Smith.** 1976. Nalidixic acid and bacterial chromosome replication. *Nature* (London) **260:**643–645.
26. **Culbertson, T. P., J. M. Domagala, J. B. Nichols, S. Priebe, and R. W. Skeean.** 1987. Enantiomers of 1-ethyl-7-[3-[(ethylamino)methyl]-1-pyrrolidinyl]-6,8-difluoro-1,4-dihydro-4-oxo-quinoline carboxylic acid: preparation and biological activity. *J. Med. Chem.* **30:**1711–1715.
27. **Cullman, W., M. Stieglitz, B. Baars, and W. Opferkuch.** 1985. Comparative evaluation of recently developed quinolone compounds—with a note on the frequency of resistant mutants. *Chemotherapy* (Basel) **31:**19–28.
28. **Davies, G. S. R., and J. Cohen.** 1985. In-vitro study of the activity of ciprofloxacin alone and in combination against strains of *Pseudomonas aeruginosa* with multiple antibiotic resistance. *J. Antimicrob. Chemother.* **16:**713–717.
29. **Deitz, W. H., T. M. Cook, and W. A. Goss.** 1966. Mechanism of action of nalidixic acid on *Escherichia coli.* III. Conditions required for lethality. *J. Bacteriol.* **91:**768–773.
30. **DeSimone, C., L. Baldinelli, M. Ferrazzi, S. DeSantis, L. Pugnaloni, and F. Sorice.** 1986. Influence of ofloxacin, norfloxacin, nalidixic acid, piromidic acid and pipemidic acid on human gamma-interferon production and blastogenesis. *J. Antimicrob. Chemother.* **17:**811–814.
31. **Desplaces, N., L. Gutmann, J. Carlet, J. Guibert, and J. F. Acar.** 1986. The new quinolones and their combinations with other agents for therapy of severe infections. *J. Antimicrob. Chemother.* **17**(Suppl. A):25–39.
32. **Domagala, J. M., L. D. Hanna, C. L. Heifetz, M. P. Hutt, T. F. Mich, J. P. Sanchez, and M. Solomon.** 1986. New structure-activity relationships of the quinolone antibacterials using the target enzyme. The development and application of a DNA gyrase assay. *J. Med. Chem.* **29:**394–404.
33. **Domagala, J. M., C. L. Heifetz, T. F. Mich, and J. B. Nichols.** 1986. 1-Ethyl-7-[3-[(ethylamino)methyl]-1-pyrrolidinyl]-6,8-difluoro-1,4-dihydro-4-oxo-quinoline carboxylic acid. New quinolone antibacterial with potent gram-positive activity. *J. Med. Chem.* **29:**445–448.
34. **Drlica, K.** 1984. Biology of bacterial deoxyribonucleic acid topoisomerases. *Microbiol. Rev.* **48:**273–289.

35. **Drugeon, H. B., J. Caillon, M. E. Juvin, and J. L. Pirault.** 1987. Dynamics of ceftazidime-pefloxacin interaction shown by a new killing curve-chequerboard method. *J. Antimicrob. Chemother.* **19:**197–203.
36. **Eagle, H., and A. D. Musselman.** 1948. The rate of bactericidal action of penicillin in vitro as a function of its concentration, and its paradoxically reduced activity at high concentrations against certain organisms. *J. Exp. Med.* **88:**99–131.
37. **Easmon, C. S. F., and J. P. Crane.** 1985. Uptake of ciprofloxacin by human neutrophils. *J. Antimicrob. Chemother.* **16:**67–73.
38. **Fass, R. J.** 1985. Quinolones. *Ann. Intern. Med.* **102:**400–401.
39. **Fernandes, P. B., C. W. Hanson, J. M. Stamm, C. Vojtko, N. L. Shipkowitz, and E. St. Martin.** 1987. The frequency of in-vitro resistance development to fluoroquinolones and the use of murine pyelonephritis model to demonstrate selection of resistance in vivo. *J. Antimicrob. Chemother.* **19:**449–465.
40. **Finch, R., M. Whitby, C. Craddock, A. Holliday, J. Martin, and R. Pilkington.** 1986. Clinical evaluation of treatment with ciprofloxacin. *Eur. J. Clin. Microbiol.* **5:**257–259.
41. **Follath, F., M. Bindschedler, M. Wenk, R. Frei, H. Stalder, and H. Reber.** 1986. Use of ciprofloxacin in the treatment of *Pseudomonas aeruginosa* infections. *Eur. J. Clin. Microbiol.* **5:**236–240.
42. **Forsgren, A., A.-K. Bergh, M. Brandt, and G. Hansson.** 1986. Quinolones affect thymidine incorporation into the DNA of human lymphocytes. *Antimicrob. Agents Chemother.* **29:**506–508.
43. **Forsgren, A., A. Bredberg, A. B. Pardee, S. F. Schlossman, and T. F. Tedder.** 1987. Effects of ciprofloxacin on eucaryotic pyrimidine nucleotide biosynthesis and cell growth. *Antimicrob. Agents Chemother.* **31:**774–779.
44. **Forsgren, A., S. F. Schlossman, and T. F. Tedder.** 1987. 4-Quinolone drugs affect cell cycle progression and function of human lymphocytes in vitro. *Antimicrob. Agents Chemother.* **31:**768–773.
45. **Foulds, J.** 1976. *tolF* locus in *Escherichia coli*: chromosomal location and relationship of loci *cmlB* and *tolD*. *J. Bacteriol.* **128:**604–608.
46. **Garlando, F., S. Rietiker, M. G. Täuber, M. Flepp, B. Meier, and R. Lüthy.** 1987. Single-dose ciprofloxacin at 100 versus 250 mg for treatment of uncomplicated urinary tract infections in women. *Antimicrob. Agents Chemother.* **31:**354–356.
47. **Gellert, M.** 1981. DNA topoisomerases. *Annu. Rev. Biochem.* **50:**879–910.
48. **Gellert, M., K. Mizuuchi, M. H. O'Dea, T. Itoh, and J. Tomizawa.** 1977. Nalidixic acid resistance: a second genetic character involved in DNA gyrase activity. *Proc. Natl. Acad. Sci. USA* **74:**4772–4776.
49. **Gellert, M., K. Mizuuchi, M. H. O'Dea, and H. A. Nash.** 1976. DNA gyrase: an enzyme that introduces superhelical turns into DNA. *Proc. Natl. Acad. Sci. USA* **73:**3872–3876.
50. **George, A. M., and S. B. Levy.** 1983. Gene in the major cotransduction gap of the *Escherichia coli* K-12 linkage map required for the expression of chromosomal resistance to tetracycline and other antibiotics. *J. Bacteriol.* **155:**541–548.
51. **Georgopapadakou, N. H., B. A. Dix, P. Angehrn, A. Wick, and G. L. Olson.** 1987. Monocyclic and tricyclic analogs of quinolones: mechanism of action. *Antimicrob. Agents Chemother.* **31:**614–616.
52. **Giamarellou, H., N. Galanakis, C. Dendrinos, J. Stefanou, E. Daphnis, and G. K. Daikos.** 1986. Evaluation of ciprofloxacin in the treatment of *Pseudomonas aeruginosa* infections. *Eur. J. Clin. Microbiol.* **5:**232–235.
53. **Giamarellou, H., and G. Petrikkos.** 1987. Ciprofloxacin interactions with imipenem and amikacin against multiresistant *Pseudomonas aeruginosa*. *Antimicrob. Agents Chemother.* **31:**959–961.

54. **Gilbert, D. N., A. D. Tice, P. K. Marsh, P. C. Craven, and L. C. Preheim.** 1987. Oral ciprofloxacin therapy for chronic contiguous osteomyelitis caused by aerobic gram-negative bacilli. *Am. J. Med.* **82**(Suppl. 4A):254–258.
55. **Goldfarb, J., R. C. Stern, M. D. Reed, T. S. Ymashita, C. M. Myers, and J. L. Blumer.** 1987. Ciprofloxacin monotherapy for acute pulmonary exacerbations of cystic fibrosis. *Am. J. Med.* **82**(Suppl. 4A):174–179.
56. **Gollapudi, S. V. S., B. Vayuvegula, S. Gupta, M. Fok, and H. Thadepalli.** 1986. Aryl-fluoroquinolone derivatives A-56619 (difloxacin) and A-56620 inhibit mitogen-induced human mononuclear cell proliferation. *Antimicrob. Agents Chemother.* **30:**390–394.
57. **Goss, W. A., W. H. Deitz, and T. M. Cook.** 1964. Mechanism of action of nalidixic acid on *Escherichia coli. J. Bacteriol.* **88:**1112–1118.
58. **Goss, W. A., W. H. Deitz, and T. M. Cook.** 1965. Mechanism of action of nalidixic acid on *Escherichia coli.* II. Inhibition of deoxyribonucleic acid synthesis. *J. Bacteriol.* **89:**1068–1074.
59. **Goto, T., P. Laipis, and J. C. Wang.** 1984. The purification and characterization of DNA topoisomerases I and II of the yeast *Saccaromyces cerevisiae. J. Biol. Chem.* **259:**10422–10429.
60. **Gutmann, L., D. Billot-Klein, R. Williamson, F. W. Goldstein, J. Mounier, J. F. Acar, and E. Collatz.** 1988. Mutation of *Salmonella paratyphi* A conferring cross-resistance to several groups of antibiotics by decreased permeability and loss of invasiveness. *Antimicrob. Agents Chemother.* **32:**195–201.
61. **Gutmann, L., R. Williamson, N. Moreau, M.-D. Kitzis, E. Collatz, J. F. Acar, and F. W. Goldstein.** 1985. Cross-resistance to nalidixic acid, trimethoprim, and chloramphenicol associated with alterations in outer membrane proteins of Klebsiella, Enterobacter, and Serratia. *J. Infect. Dis.* **151:**501–507.
62. **Haller, I.** 1985. Comprehensive evaluation of ciprofloxacin-aminoglycoside combinations against *Enterobacteriaceae* and *Pseudomonas aeruginosa* strains. *Antimicrob. Agents Chemother.* **28:**663–666.
63. **Handwerger, S., and A. Tomasz.** 1985. Antibiotic tolerance among clinical isolates of bacteria. *Rev. Infect. Dis.* **7:**368–386.
64. **Hane, M. W., and T. H. Wood.** 1969. *Escherichia coli* K-12 mutants resistant to nalidixic acid: genetic mapping and dominance studies. *J. Bacteriol.* **99:**238–241.
65. **Havlichek, D., L. Saravolatz, and D. Pohlod.** 1987. Effect of quinolones and other antimicrobial agents on cell-associated *Legionella pneumophila. Antimicrob. Agents Chemother.* **31:**1529–1534.
66. **Helling, R. B., and B. S. Adams.** 1970. Nalidixic acid-resistant auxotrophs of *Escherichia coli. J. Bacteriol.* **104:**1027–1029.
67. **Higgins, N. P., and N. R. Cozzarelli.** 1982. The binding of gyrase to DNA: analysis by retention by nitrocellulose filters. *Nucleic Acids Res.* **10:**6833–6847.
68. **Hirai, K., H. Aoyama, T. Irikura, S. Iyobe, and S. Mitsuhashi.** 1986. Differences in susceptibility to quinolones of outer membrane mutants of *Salmonella typhimurium* and *Escherichia coli. Antimicrob. Agents Chemother.* **29:**535–538.
69. **Hirai, K., H. Aoyama, S. Suzue, T. Irikura, S. Iyobe, and S. Mitsuhashi.** 1986. Isolation and characterization of norfloxacin-resistant mutants of *Escherichia coli* K-12. *Antimicrob. Agents Chemother.* **30:**248–253.
70. **Hirai, K., S. Suzue, T. Irikura, S. Iyobe, and S. Mitsuhashi.** 1987. Mutations producing resistance to norfloxacin in *Pseudomonas aeruginosa. Antimicrob. Agents Chemother.* **31:**582–586.
71. **Hoiby, N.** 1986. Clinical uses of nalidixic acid analogues: the fluoroquinolones. *Eur. J. Clin. Microbiol.* **5:**138–140.
72. **Holmes, B., R. N. Brogden, and D. M. Richards.** 1985. Norfloxacin. A review of its antibacterial activity, pharmacokinetic properties, and therapeutic use. *Drugs* **30:**482–513.

73. **Hooper, D. C., and J. S. Wolfson.** 1985. The fluoroquinolones: pharmacology, clinical uses, and toxicities in humans. *Antimicrob. Agents Chemother.* **28:**716–721.
74. **Hooper, D. C., and J. S. Wolfson.** 1988. Mode of action of quinolone antimicrobial agents. *Rev. Infect. Dis.* **10**(Suppl. 1):S14–S21.
75. **Hooper, D. C., J. S. Wolfson, E. Y. Ng, and M. N. Swartz.** 1987. Mechanisms of action of and resistance to ciprofloxacin. *Am. J. Med.* **82**(Suppl. 4A):12–20.
76. **Hooper, D. C., J. S. Wolfson, K. S. Souza, C. Tung, G. L. McHugh, and M. N. Swartz.** 1986. Genetic and biochemical characterization of norfloxacin resistance in *Escherichia coli*. *Antimicrob. Agents Chemother.* **29:**639–644.
77. **Horowitz, D. S., and J. C. Wang.** 1987. Mapping the active site tyrosine of *Escherichia coli* DNA gyrase. *J. Biol. Chem.* **262:**5339–5344.
78. **Hussy, P., G. Maass, B. Tümmler, F. Grosse, and U. Schomburg.** 1986. Effect of 4-quinolones and novobiocin on calf thymus DNA polymerase α primase complex, topoisomerase I and II, and growth of mammalian lymphoblasts. *Antimicrob. Agents Chemother.* **29:**1073–1078.
79. **Imamura, M., S. Shibamura, I. Hayakawa, and Y. Osada.** 1987. Inhibition of DNA gyrase by optically active ofloxacin. *Antimicrob. Agents Chemother.* **31:**325–327.
80. **Inoue, S., T. Ohue, J. Yamagishi, S. Nakamura, and M. Shimizu.** 1978. Mode of incomplete cross-resistance among pipemidic, piromidic, and nalidixic acids. *Antimicrob. Agents Chemother.* **14:**240–245.
81. **Inoue, Y., K. Sato, T. Fujii, K. Hirai, M. Inoue, S. Iyobe, and S. Mitsuhashi.** 1987. Some properties of subunits of DNA gyrase from *Pseudomonas aeruginosa* PAO1 and its nalidixic acid-resistant mutant. *J. Bacteriol.* **169:**2322–2325.
82. **Ito, A., K. Hirai, M. Inoue, H. Koga, S. Suzue, T. Irikura, and S. Mitsuhashi.** 1980. In vitro antibacterial activity of AM-715, a new nalidixic acid analog. *Antimicrob. Agents Chemother.* **17:**103–108.
83. **Johnson, M., P. Miniter, and V. T. Andriole.** 1987. Comparative efficacy of ciprofloxacin, azlocillin, and tobramycin alone and in combination in experimental Pseudomonas sepsis. *J. Infect. Dis.* **155:**783–788.
84. **Klopman, G., O. T. Macina, M. E. Levinson, and H. S. Rosenkranz.** 1987. Computer automated structure evaluation of quinolone antibacterial agents. *Antimicrob. Agents Chemother.* **31:**1831–1840.
85. **Koga, H.** 1987. High-performance liquid chromatography measurement of antimicrobial concentrations in polymorphonuclear leukocytes. *Antimicrob. Agents Chemother.* **31:**1904–1908.
86. **Korvick, J., V. L. Yu, and J. A. Sharp.** 1987. Interaction of rifampin or rifapentine with other agents against *Pseudomonas aeruginosa*. *J. Antimicrob. Chemother.* **19:**847–848.
87. **Kreuzer, K. N., and N. R. Cozzarelli.** 1979. *Escherichia coli* mutants thermosensitive for deoxyribonucleic acid gyrase subunit A: effects on deoxyribonucleic acid replication, transcription, and bacteriophage growth. *J. Bacteriol.* **140:**424–435.
88. **Krueger, J. H., and G. C. Walker. 1984.** *groEL* and *dnaK* genes of *Escherichia coli* are induced by UV irradiation and nalidixic acid in an *htpR*+-dependent fashion. *Proc. Natl. Acad. Sci. USA* **81:**1499–1503.
89. **Kumar, S.** 1976. Properties of adenyl cyclase and cyclic adenosine 3′,5′-monophosphate receptor protein-deficient mutants of *Escherichia coli*. *J. Bacteriol.* **125:**545–555.
90. **Kumar, S.** 1980. Types of spontaneous nalidixic acid resistant mutants of *Escherichia coli*. *Indian J. Exp. Biol.* **18:**341–343.
91. **Lampe, M. F., and K. F. Bott.** 1984. Cloning the *gyrA* gene of *Bacillus subtilis*. *Nucleic Acids Res.* **12:**6307–6323.

92. **Le Goffic, F.** 1985. Les quinolones, mecanisme d'action, p. 15–23. *In* J. J. Pocidalo, F. Vachon, and B. Regnier (ed.), *Les Nouvelles Quinolones.* Editions Arnette, Paris.
93. **Lockshon, D., and D. R. Morris.** 1985. Sites of reaction of *Escherichia coli* DNA gyrase on pBR322 in vivo as revealed by oxolinic acid-induced plasmid linearization. *J. Mol. Biol.* **181:**63–74.
94. **Maesen, F. P. V., B. I. Davies, and J. P. Teengs.** 1985. Pefloxacin in acute exacerbations of chronic bronchitis. *J. Antimicrob. Chemother.* **16:**379–388.
95. **Manek, N., R. Wise, and J. Andrews.** 1986. In-vitro activity of pefloxacin in combination with other antimicrobial agents. *J. Antimicrob. Chemother.* **17:**827–828.
96. **Marino, P., M. Venditti, B. Valente, C. Brandimarte, and P. Serra.** 1985. N-formimidoyl-thienamycin and norfloxacin against multiple-resistant *Pseudomonas aeruginosa* strains. Combined in vitro activity and comparison with 14 other antibiotics. *Drugs Exp. Clin. Res.* **11:**247–251.
97. **Mascellino, M. T., A. Lorenzi, M. Bonanni, and F. Iegri.** 1986. Antimicrobial activity of norfloxacin in enteric and urinary tract infections: combined effect of norfloxacin with aminoglycosides, tetracycline and chloramphenicol. *Drugs Exp. Clin. Res.* **12:**319–323.
98. **McQueen, C. A., and G. M. Williams.** 1987. Effects of quinolone antibiotics in tests for genotoxicity. *Am. J. Med.* **82**(Suppl. 4A)**:**94–96.
99. **Menzel, R., and M. Gellert.** 1987. Modulation of transcription by DNA supercoiling: a deletion analysis of the *Escherichia coli gyrA* and *gyrB* promoters. *Proc. Natl. Acad. Sci. USA* **84:**4185–4189.
100. **Michéa-Hamzehpour, M., J.-C. Pechère, B. Marchou, and R. Auckenthaler.** 1986. Combination therapy: a way to limit emergence of resistance? *Am. J. Med.* **80**(Suppl. 6B)**:**138–142.
101. **Milatovic, D.** 1986. Intraphagocytic activity of ciprofloxacin and CI 934. *Eur. J. Clin. Microbiol.* **5:**659–660.
102. **Miller, R. V., and T. R. Scurlock.** 1983. DNA gyrase (topoisomerase II) from *Pseudomonas aeruginosa. Biochem. Biophys. Res. Commun.* **110:**694–700.
103. **Mitscher, L. A., P. N. Sharma, D. T. W. Chu, L. L. Shen, and A. G. Pernet.** 1986. Chiral DNA gyrase inhibitors. 1. Synthesis and antimicrobial activity of the enantiomers of 6-fluoro-7-(1-piperazinyl)-1-(2′-transphenyl-1′-cyclopropyl)-1,4-dihydro-4-oxoquinoline-3-carboxylic acid. *J. Med. Chem.* **29:**2044–2047.
104. **Mitscher, L. A., P. N. Sharma, D. T. W. Chu, L. L. Shen, and A. G. Pernet.** 1987. Chiral DNA gyrase inhibitors. 2. Asymmetric synthesis and biological activity of enantiomers of 9-fluoro-3-methyl-10-(4-methyl-1-piperazinyl)-7-oxo-2,3-dihydro-7H-pyrido[1,2,3,-de]-1,4-benzoxazine-6-carboxylic acid (ofloxacin). *J. Med. Chem.* **30:**2283–2286.
105. **Monk, J. P., and D. M. Campoli-Richards.** 1987. Ofloxacin. A review of its antibacterial activity, pharmacokinetic properties, and therapeutic use. *Drugs* **33:**346–391.
106. **Moody, J. A., D. N. Gerding, and L. R. Peterson.** 1987. Evaluation of ciprofloxacin's synergism with other agents by multiple in vitro methods. *Am. J. Med.* **82**(Suppl. 4A)**:**44–54.
107. **Moody, J. A., L. R. Peterson, and D. N. Gerding.** 1985. In vitro activity of ciprofloxacin combined with azlocillin. *Antimicrob. Agents Chemother.* **28:**849–850.
108. **Mouton, R. P., and S. L. Mulders.** 1987. Combined resistance to quinolones and beta-lactams after in vitro transfer on single drugs. *Chemotherapy* (Basel) **33:**189–196.
109. **Mulligan, M. E., P. J. Ruane, L. Johnston, P. Wong, J. P. Wheelock, K. MacDonald, J. F. Reinhardt, et al.** 1987. Ciprofloxacin for eradication of methicillin-resistant *Staphylococcus aureus* colonization. *Am. J. Med.* **82**(Suppl. 4A)**:**215–223.
110. **Munshi, M. H., K. Haider, M. M. Rahaman, D. A. Sack, Z. U. Ahmed, and M. G. Morshed.** 1987. Plasmid-mediated resistance to nalidixic acid in *Shigella dysenteriae* type 1. *Lancet* **ii:**419–421.

111. **Nakamura, S., S. Inoue, M. Simizu, S. Iyobe, and S. Mitsuhashi.** 1976. Inhibition of conjugal transfer of R plasmids by pipemidic acid and related compounds. *Antimicrob. Agents Chemother.* **10:**779–785.
112. **Neu, H. C.** 1987. Clinical use of the quinolones. *Lancet* **ii:**1319–1322.
113. **Neu, H. C., N.-X. Chin, and P. Labthavikul.** 1984. Antibacterial activity of coumermycin alone and in combination with other antibiotics. *Antimicrob. Agents Chemother.* **25:**678–689.
114. **O'Connor, M. B., and M. H. Malamy.** 1985. Mapping of DNA gyrase cleavage sites in vivo. Oxolinic acid induced cleavages in plasmid pBR322. *J. Mol. Biol.* **181:**545–550.
115. **Paton, J. H., and E. W. Williams.** 1987. Interaction between ciprofloxacin and vancomycin against staphylococci. *J. Antimicrob. Chemother.* **20:**251–254.
116. **Perronne, C. M., R. Malinverni, and M. P. Glauser.** 1987. Treatment of *Staphylococcus aureus* endocarditis in rats with coumermycin A1 and ciprofloxacin, alone and in combination. *Antimicrob. Agents Chemother.* **31:**539–543.
117. **Phillips, I., E. Culebras, F. Moreno, and F. Baquero.** 1987. Induction of the SOS response by new 4-quinolones. *J. Antimicrob. Chemother.* **20:**631–638.
118. **Piddock, L. J. V., J. M. Andrews, J. M. Diver, and R. Wise.** 1986. In vitro studies of S-25930 and S-25932, two new 4-quinolones. *Eur. J. Clin. Microbiol.* **5:**303–310.
119. **Piddock, L. J. V., and R. Wise.** 1987. Induction of the SOS response in *Escherichia coli* by 4-quinolone antimicrobial agents. *FEMS Microbiol. Lett.* **41:**289–294.
120. **Platt, D. J., and A. C. Black.** 1987. Plasmid ecology and the elimination of plasmids by 4-quinolones. *J. Antimicrob. Chemother.* **20:**137–142.
121. **Raeburn, J. A., J. R. W. Govan, W. M. McCrae, A. P. Greening, P. S. Collier, M. E. Hodson, and M. C. Goodchild.** 1987. Ciprofloxacin therapy in cystic fibrosis. *J. Antimicrob. Chemother.* **20:**295–296.
122. **Rau, D. C., M. Gellert, F. Thoma, and A. Maxwell.** 1987. Structure of the DNA gyrase-DNA complex as revealed by transient electric dichroism. *J. Mol. Biol.* **193:**555–569.
123. **Rella, M., and H. Dieter.** 1982. Resistance of *Pseudomonas aeruginosa* PAO to nalidixic acid and low levels of β-lactam antibiotics: mapping of chromosomal genes. *Antimicrob. Agents Chemother.* **22:**242–249.
124. **Robillard, N. J., and A. L. Scarpa.** 1988. Genetic and physiological characterization of ciprofloxacin resistance in *Pseudomonas aeruginosa* PAO. *Antimicrob. Agents Chemother.* **32:**535–539.
125. **Roche, Y., M.-A. Gougerot-Pocidalo, M. Fay, D. Etienne, N. Forest, and J.-J. Pocidalo.** 1987. Comparative effects of quinolones on human mononuclear leucocyte functions. *J. Antimicrob. Chemother.* **19:**781–790.
126. **Saito, A., K. Sawatari, Y. Fukuda, M. Nagasawa, H. Koga, A. Tomonaga, H. Nakazato, K. Fujita, Y. Shigeno, Y. Suzuyama, K. Yamaguchi, K. Izumikawa, and K. Hara.** 1985. Susceptibility of *Legionella pneumophila* to ofloxacin in vitro and in experimental *Legionella* pneumonia in guinea pigs. *Antimicrob. Agents Chemother.* **28:**15–20.
127. **Sanders, C. C., W. E. Sanders, R. V. Goering, and V. Werner.** 1984. Selection of multiple antibiotic resistance by quinolones, β-lactams, and aminoglycosides with special reference to cross-resistance between unrelated drug classes. *Antimicrob. Agents Chemother.* **26:**797–801.
128. **Sanders, C. C., and C. Watanakunakorn.** 1986. Emergence of resistance to β-lactams, aminoglycosides, and quinolones during combination therapy for infection due to *Serratia marcescens*. *J. Infect. Dis.* **153:**617–619.
129. **Sato, K., Y. Inoue, T. Fujii, H. Aoyama, M. Inoue, and S. Mitsuhashi.** 1986. Purification and properties of DNA gyrase from a fluoroquinolone-resistant strain of *Escherichia coli*. *Antimicrob. Agents Chemother.* **30:**777–780.

130. **Sato, K., Y. Inoue, T. Fujii, H. Aoyama, and S. Mitsuhashi.** 1986. Antibacterial activity of ofloxacin and its mode of action. *Infection* **14**(Suppl. 4):S226–S230.
131. **Scully, B. E., M. Nakatomi, C. Ores, S. Davidson, and H. C. Neu.** 1987. Ciprofloxacin therapy in cystic fibrosis. *Am. J. Med.* **82**(Suppl. 4A):196–201.
132. **Scully, B. E., H. C. Neu, M. F. Parry, and W. Mandell.** 1986. Oral ciprofloxacin therapy of infections due to *Pseudomonas aeruginosa. Lancet* **i**:819–822.
133. **Setlow, J. K., E. Cabrera-Juárez, W. L. Albritton, D. Spikes, and A. Muschler.** 1985. Mutations affecting gyrase in *Haemophilus influenzae. J. Bacteriol.* **164**:525–534.
134. **Shalit, I., H. R. Stutman, M. I. Marks, S. A. Chartrand, and B. C. Hilman.** 1987. Randomized study of two dosage regimens of ciprofloxacin for treating chronic bronchopulmonary infection in patients with cystic fibrosis. *Am. J. Med.* **82**(Suppl. 4A):189–195.
135. **Shen, L. L., and A. G. Pernet.** 1985. Mechanism of inhibition of DNA gyrase by analogues of nalidixic acid: the target of the drugs is DNA. *Proc. Natl. Acad. Sci. USA* **82**:307–311.
136. **Simberkoff, M. S., and J. J. Rahal.** 1986. Bactericidal activity of ciprofloxacin against amikacin- and cefotaxime-resistant gram-negative bacilli and methicillin-resistant staphylococci. *Antimicrob. Agents Chemother.* **29**:1098–1100.
137. **Smith, J. T.** 1984. Awakening the slumbering potential of the 4-quinolone antibacterials. *Pharm. J.* **233**:299–305.
138. **Snyder, M., and K. Drlica.** 1979. DNA gyrase on the bacterial chromosome: DNA cleavage induced by oxolinic acid. *J. Mol. Biol.* **131**:287–302.
139. **Steck, T. R., and K. Drlica.** 1984. Bacterial chromosome segregation: evidence for DNA gyrase involvement in decatenation. *Cell* **36**:1081–1088.
140. **Sugino, A., and K. F. Bott.** 1980. *Bacillus subtilis* deoxyribonucleic acid gyrase. *J. Bacteriol.* **141**:1331–1339.
141. **Sugino, A., C. L. Peebles, K. N. Kreuzer, and N. R. Cozzarelli.** 1977. Mechanism of action of nalidixic acid: purification of *Escherichia coli nalA* gene product and its relationship to DNA gyrase and a novel nicking-closing enzyme. *Proc. Natl. Acad. Sci. USA* **74**:4767–4771.
142. **Swanberg, S. L., and J. C. Wang.** 1987. Cloning and sequencing of the *Escherichia coli gyrA* gene coding for the A subunit of DNA gyrase. *J. Mol. Biol.* **197**:729–736.
143. **Tabary, X., N. Moreau, C. Dureuil, and F. Le Goffic.** 1987. Effect of DNA gyrase inhibitors pefloxacin, five other quinolones, novobiocin, and clorobiocin on *Escherichia coli* **topoisomerase I.** *Antimicrob. Agents Chemother.* **31**:1925–1928.
144. **Tenney, J. H., R.W. Maack, and G. R. Chippendale.** 1983. Rapid selection of organisms with increasing resistance on subinhibitory concentrations of norfloxacin on agar. *Antimicrob. Agents Chemother.* **23**:188–189.
145. **Then, R. L., and P. Angehern.** 1986. Multiply resistant mutants of *Enterobacter cloacae* selected by β-lactam antibiotics. *Antimicrob. Agents Chemother.* **30**:684–688.
146. **Traub, W. H.** 1984. Intraphagocytic bactericidal activity of bacterial DNA gyrase inhibitors against *Serratia marcescens. Chemotherapy* (Basel) **30**:379–386.
147. **Traub, W. H.** 1985. Incomplete cross-resistance of nalidixic and pipemidic acid-resistant variants of *Serratia marcescens* against ciprofloxacin, enoxacin, and norfloxacin. *Chemotherapy* (Basel) **31**:34–39.
148. **Tsukamura, M., E. Nakamura, S. Yoshi, and H. Amano.** 1985. Therapeutic effect of a new antibacterial substance ofloxacin (DL8280) on pulmonary tuberculosis. *Am. Rev. Respir. Dis.* **131**:352–356.
149. **VanBogelen, R. A., P. M. Kelley, and F. C. Neidhardt.** 1987. Differential induction of heat shock, SOS, and oxidation stress regulons and accumulation of nucleotides in *Escherichia coli. J. Bacteriol.* **169**:26–32.

150. **Van der Auwera, P.** 1985. Interaction of gentamicin, dibekacin, netilmicin, and amikacin with various penicillins, cephalosporins, minocycline, and new fluoro-quinolones against Enterobacteriaceae and *Pseudomonas aeruginosa*. *J. Antimicrob. Chemother.* **16:**581–587.
151. **Van der Auwera, P., and P. Joly.** 1987. Comparative in-vitro activities of teicoplanin, vancomycin, coumermycin and ciprofloxacin, alone and in combination with rifampin or LM427, against *Staphylococcus aureus*. *J. Antimicrob. Chemother.* **19:**313–320.
152. **Van der Auwera, P., and J. Klasterky.** 1986. Bactericidal activity and killing rate of serum in volunteers receiving ciprofloxacin alone or in combination with vancomycin. *Antimicrob. Agents Chemother.* **30:**892–895.
153. **Van der Auwera, P., J. Klastersky, S. Lieppe, M. Husson, D. Lauzon, and A. P. Lopez.** 1986. Bactericidal activity and killing rate of serum from volunteers receiving pefloxacin alone and in combination with amikacin. *Antimicrob. Agents Chemother.* **29:**230–234.
154. **Van der Auwera, P., A. Vanermies, P. Grenier, and J. Klastersky.** 1987. Comparative in vitro activity of CI934, a new fluoroquinolone, alone and in combination with coumermycin, against gram-positive bacteria. *Drugs Exp. Clin. Res.* **13:**125–132.
155. **Vildé, J. L., E. Dournon, and P. Rajagopalan.** 1986. Inhibition of *Legionella pneumophila* multiplication within human macrophages by antimicrobial agents. *Antimicrob. Agents Chemother.* **30:**743–748.
156. **Wang, J. C.** 1971. Interaction between DNA and an *Escherichia coli* protein omega. *J. Mol. Biol.* **55:**523–533.
157. **Wang, J. C.** 1985. DNA topoisomerases. *Annu. Rev. Biochem.* **54:**655–697.
158. **Wang, J. C.** 1987. Recent studies of DNA topoisomerases. *Biochim. Biophys. Acta* **909:**1–9.
159. **Watson, J. D., and F. H. C. Crick.** 1953. The structure of DNA. *Cold Spring Harbor Symp. Quant. Biol.* **18:**123–131.
160. **Watson, J. D., N. H. Hopkins, J. W. Roberts, J. A. Steitz, and A. M. Weiner (ed.).** 1987. *Molecular Biology of the Gene*, 4th ed., vol. 1. *General Principles.* Benjamin/Cummings Publishing Co., Inc., Menlo Park, Calif.
161. **Weisser, J., and B. Wiedemann.** 1985. Elimination of plasmids by new 4-quinolones. *Antimicrob. Agents Chemother.* **28:**700–702.
162. **Whiting, J. L., N. Cheng, and A. W. Chow.** 1987. Interactions of ciprofloxacin with clindamycin, metronidazole, cefoxitin, cefotaxime, and mezlocillin against gram-positive and gram-negative anaerobic bacteria. Antimicrob. Agents Chemother. **31:**1379–1382.
163. **Wolfson, J. S., and D. C. Hooper.** 1985. The fluoroquinolones: structures, mechanisms of action and resistance, and spectra of activity in vitro. *Antimicrob. Agents Chemother.* **28:**581–586.
163a. **Wolfson, J. S., and D. C. Hooper.** 1988. Norfloxacin: a targeted fluoroquinolone antimicrobial agent. *Ann. Intern. Med.* **108:**238–251.
164. **Wolfson, J. S., D. C. Hooper, E. Y. Ng, K. S. Souza, G. L. McHugh, and M. N. Swartz.** 1987. Antagonism of wild-type and resistant *Escherichia coli* and its DNA gyrase by the tricyclic 4-quinolone analogs ofloxacin and S-25930 steroisomers. *Antimicrob. Agents Chemother.* **31:**1861–1863.
165. **Yamagishi, J.-I., Y. Furutani, S. Inoue, T. Ohue, S. Nakamura, and M. Shimizu.** 1981. New nalidixic acid resistance mutations related to deoxyribonucleic acid gyrase activity. *J. Bacteriol.* **148:**450–458.
166. **Yamagishi, J., H. Yoshida, M. Yamayoshi, and S. Nakamura.** 1986. Nalidixic acid-resistant mutations of the *gyrB* gene of *Escherichia coli*. *Mol. Gen. Genet.* **204:**367–373.
167. **Yoshida, H., T. Kojima, J. Yamagishi, and S. Nakamura.** 1988. Quinolone-resistant mutations of the *gyrA* gene of *Escherichia coli*. *Mol. Gen. Genet.* **211:**1–7.

168. **Zeiler, H.-J.** 1985. Evaluation of the in vitro bactericidal action of ciprofloxacin on cells of *Escherichia coli* in the logarithmic and stationary phases of growth. *Antimicrob. Agents Chemother.* **28**:524–527.

169. **Zweerink, M. M., and A. Edison.** 1986. Inhibition of *Micrococcus luteus* DNA gyrase by norfloxacin and 10 other quinolone carboxylic acids. *Antimicrob. Agents Chemother.* **29**:598–601.

Chapter 3

Quinolone Antimicrobial Agents: Activity In Vitro

G. M. Eliopoulos and C. T. Eliopoulos

Department of Medicine, New England Deaconess Hospital, and Harvard Medical School, Boston, Massachusetts 02215

Among the characteristics of the newer quinolone antimicrobial agents which have generated the intense interest, none is more compelling than the remarkable activities of these drugs in vitro (55, 216). The fluoroquinolones have demonstrated both extraordinary potency against many common bacterial pathogens and also extremely promising activity against a number of less common but therapeutically troublesome organisms. Included among the latter are multiply drug-resistant nosocomial isolates, various mycobacteria, *Legionella* species, and even rickettsiae.

The purpose of this chapter is to review the in vitro antimicrobial activities of representative members of this class, to discuss those technical factors which influence measurement of drug activity, and to consider the effects of drug combinations in vitro. Among the numerous compounds which have been synthesized and evaluated (48), a few have emerged as candidates for clinical development. These compounds will serve as a focus for this discussion, the primary intent of which is to consider the potential antimicrobial spectrum of the newer quinolones as a class, rather than to emphasize differences in activities among individual agents.

Antimicrobial Activity

Gram-negative bacteria

Enterobacteriaceae. The clinical utility of the older quinolone antimicrobial agents derived largely from their activities against members of the family *Enterobacteriaceae*. In comparison with these older agents, the fluoroquinolones have demonstrated greatly improved potency against this group of organisms (55, 216). Data presented in Tables 1 and 2 are repre-

Table 1. Susceptibilities of *Enterobacteriaceae* to earlier fluoroquinolone antimicrobial agents and nalidixic acid

Organism	MIC_{90} (MIC_{max}) (μg/ml) of[a,b]:						
	CPX	NFX	ENX	OFX	PFX	AMX	NAL
Escherichia coli	≤0.06 (2)	0.125 (4)	0.25 (16)	0.125 (2)	0.125 (8)	0.125 (8)	8 (>128)
Klebsiella spp.	0.125 (2)	0.5 (16)	1.0 (8)	0.5 (8)	0.5 (8)	0.5 (4)	8 (>128)
Enterobacter cloacae	0.125 (1)	0.25 (8)	0.5 (8)	0.5 (16)	0.5 (16)	0.5 (4)	8 (>128)
Enterobacter aerogenes	≤0.06 (0.5)	0.25 (4)	0.5 (2)	0.25 (8)	0.25 (0.5)	0.25 (2)	8 (32)
Serratia spp.	0.25 (2)	2 (8)	4 (8)	2 (8)	1.0 (8)	2 (8)	≥8 (>128)
Citrobacter spp.	0.125 (2)	0.5 (8)	0.5 (2)	0.5 (4)	0.5 (2)	0.25 (8)	16 (64)
Proteus mirabilis	≤0.06 (1)	0.25 (8)	0.5 (8)	0.5 (2)	0.25 (4)	0.5 (0.5)	8 (>128)
Proteus vulgaris	≤0.06 (1)	0.25 (4)	0.25 (4)	0.25 (2)	0.25 (2)		16 (≥128)
Morganella morganii	≤0.06 (0.25)	0.125 (2)	0.25 (2)	0.125 (2)	0.25 (4)	0.25 (1)	8 (≥128)
Providencia stuartii	2 (8)	2 (32)	2 (32)	2 (8)	4 (8)	1.0 (2)	32 (≥128)
Providencia rettgeri	0.25 (2)	1.0 (16)	1.0 (16)	1.0 (4)	0.5 (16)		16 (>128)

[a]Abbreviations: MIC_{max}, upper limit of MIC range; CPX, ciprofloxacin; NFX, norfloxacin; ENX, enoxacin; OFX, ofloxacin; PFX, pefloxacin; AMX, amifloxacin; NAL, nalidixic acid.

[b]Derived from the following references: CPX—4, 14, 15, 17, 24, 30, 33, 36, 40, 56, 57, 62, 82, 83, 89, 95, 108, 112, 116, 118, 119, 131, 133–135, 143, 147, 156, 163, 170, 171, 183, 203, 209, 213, 215; NFX—4, 14, 15, 17, 28, 30, 32, 33, 36, 40, 42, 56, 57, 62, 82, 83, 95, 97, 106, 108, 114, 116, 122, 131, 133–135, 143, 144, 146, 147, 149, 150, 156, 163, 172, 177, 185, 188, 209, 213–215; ENX—4, 15, 17, 29, 30, 32, 91, 93, 116, 118, 119, 122, 131, 144–147, 170, 172, 207, 209, 214; OFX—4, 15, 17, 28, 30, 38, 83, 95, 97, 116, 118, 133–135, 146, 147, 177, 203, 209, 214; PFX—4, 28, 36, 108, 116, 131, 203, 213, 218; AMX—4, 17, 74, 103, 105, 106, 158, 166; NAL—4, 14, 28, 36, 40, 42, 56, 82, 93, 108, 114, 116, 122, 131, 143, 144, 146, 156, 163, 177, 203, 213, 214.

Table 2. Susceptibilities of *Enterobacteriaceae* to newer fluoroquinolone antimicrobial agents

Organism	MIC$_{90}$ (μg/ml) of[a,b]:							
	DFX	A-56620	CI-934	FLX	S-29530	TMX	EN 272	NY-198
Escherichia coli	0.25	≤0.06	1.0	0.125	0.25	0.06	1.0	0.5
Klebsiella spp.	1.0	0.25	1.0	0.5	0.5	0.5	1.0	0.5
Enterobacter cloacae	0.25	≤0.06	0.5	0.5	0.5		2	2
Enterobacter aerogenes	0.25	0.125	0.25		1.0		2	
Serratia spp.	2	0.5	2	0.5	1.0	0.5	2	12.5
Citrobacter spp.	0.5	0.25	0.25	0.25	1.0	0.125	1.0	4
Proteus mirabilis	2	0.25	1.0	0.5	0.5		1.0	1.0
Proteus vulgaris	2	0.5	0.5	0.25	0.5		4	0.5
Morganella morganii	0.5	0.25	0.5	0.12	0.25		0.5	0.5
Providencia stuartii	4	1.0	4	1.0	16		4	
Providencia rettgeri	2	0.25	0.5	0.5	2		8	

[a] Abbreviations: DFX, difloxacin; FLX, fleroxacin; TMX, temafloxacin.

[b] Derived from the following references: DFX and A-56620—15, 57, 89, 116, 170, 185, 188, 209; CI-934—38, 116, 119, 134; FLX—30, 36, 95, 100, 133, 135, 209; S-29530—147, 156, 170; TMX—89; EN 272—146; NY-198—97.

sentative of published values for activities of the various newer agents in comparison with nalidixic acid. In view of the fact that such data are derived from several studies, each of which examines a limited number of agents against a potentially unique group of bacterial isolates, it is inappropriate to use these tables for critical comparison of the various compounds. Nevertheless, certain generalizations can be made. Although ciprofloxacin is the most active of the newer fluoroquinolones, differences in potencies among this group are less striking than those between the new quinolones and the earlier agents, including nalidixic acid, cinoxacin (11, 14, 103, 109, 131), pipemidic acid (4, 97, 131), and oxolinic acid (14, 40). This very potent activity of ciprofloxacin against *Enterobacteriaceae* is evident given MICs for 90% of isolates (MIC_{90}s) of ≤0.25 μg/ml against virtually all species. Although occasional clinical isolates are encountered which are substantially more resistant to one or more newer quinolones than are typical strains of its species, such isolates are still almost always susceptible to concentrations of the fluoroquinolones attainable in urine. Among the *Enterobacteriaceae*, strains of *Providencia stuartii*, *Providencia rettgeri*, and *Serratia marcescens* often tend to be less exquisitely susceptible to the new drugs than are other species.

Representatives of the latest group of fluoroquinolones to come under intensive investigation are included in Table 2. Comparison between difloxacin (A-56619) and a closely related aryl-fluoroquinolone, A-56620, illustrates the fact that in vitro potency is not the sole determinant of clinical promise. While the latter compound is consistently more active than the former against gram-negative organisms, higher peak levels in serum and the more prolonged elimination half-life of difloxacin confer relative advantages to that compound (15). Not shown in Table 2 are data pertaining to S-25932, a compound closely related to S-25930 in both structure and activity against *Enterobacteriaceae* (147, 156).

***Pseudomonas* spp. and other gram-negative organisms.** In contrast to nalidixic acid, several of the newer quinolones demonstrate substantial activity against *Pseudomonas aeruginosa* (Table 3). Activities of these drugs against other pseudomonads are more variable and depend upon the species tested. Many strains of *Pseudomonas maltophilia* and *Pseudomonas cepacia* are resistant to readily achievable (except perhaps in urine) concentrations of these agents. Among other species, reported MIC_{90}s of ciprofloxacin include the following: *Pseudomonas fluorescens-Pseudomonas putida*, 0.25 to 1.0 μg/ml (2, 16, 40); *Pseudomonas stutzeri*, 0.5 μg/ml (2); and *Pseudomonas putrefaciens*, 1.0 μg/ml (40). More detailed information on specific drug susceptibilities of the various species is limited by the tendency to group these organisms as "non-*P. aeruginosa*" pseudomonads in many studies.

Table 3. Susceptibilities of *Pseudomonas* species to fluoroquinolones

Drug	Range of MIC_{90}s[a] (μg/ml) against:		
	P. aeruginosa	*P. cepacia*	*P. maltophilia*
Ciprofloxacin	0.25–1.0	2–8	1.0–8
Norfloxacin	0.5–8	8–32	8–100
Enoxacin	2–8	2–32	4–32
Ofloxacin	2–8	4–32	2–32
Pefloxacin	2–8	8	2–4
Amifloxacin	2–16	16	4
Difloxacin	4–16		4
Fleroxacin	2–8		4–8
Temafloxacin	1.0		
S-29530	8–16		1.0
EN 272	100		50
NY-198	12.5	25	12.5

[a]For abbreviations and references, see Tables 1 and 2, footnotes *a* and *b*, as well as the following additional references: CPX—2, 16, 102, 121, 171, 183; NFX—2, 16, 49, 96, 117, 121, 123, 149, 150; ENX—16, 123; OFX—16, 123; PFX—2; AMX—2, 16, 199; DFX—184.

Among other nonfermenters, *Moraxella* spp. are generally susceptible to ciprofloxacin and fleroxacin (Ro23-6240) at concentrations of ≤0.5 μg/ml (2, 100) and to norfloxacin, amifloxacin, and pefloxacin at concentrations of ≤2 μg/ml (2). *Alcaligenes* spp. tend to be relatively resistant to the fluoroquinolones. MIC_{90}s may reach or exceed 32 μg/ml, although some strains may be considerably more susceptible to individual agents (2, 4, 95, 100, 102, 131). Against *Acinetobacter* spp., MIC_{90}s of ciprofloxacin, pefloxacin, or ofloxacin may reach 4 to 8 μg/ml (16, 107, 209). Ranges of reported inhibitory concentrations are very broad (133, 170, 190), however, with MIC_{90}s against *Acinetobacter calcoaceticus* subsp. *anitratus* of as low as 0.25 μg/ml (134, 168).

Strains of *Aeromonas* spp. are generally quite susceptible to the new drugs, with MIC_{90}s frequently of ≤0.06 μg/ml (4, 30, 36, 57, 83, 100, 116, 119, 134, 147, 170). This is not surprising in view of the excellent activities of some older quinolones against many isolates. Against these organisms, MIC_{90}s of nalidixic or pipemidic acid of as low as 0.12 μg/ml have been reported (4, 36, 116). Ciprofloxacin has also been shown to be highly active against *Brucella melitensis*, with a MIC_{90} of 0.5 μg/ml when tested against 95 isolates (22).

Enteropathogens. Among the gram-negative bacilli, it is useful to consider gastrointestinal pathogens as a special group because of the potential use of oral fluoroquinolones in the therapy of infectious diarrhea and other syndromes caused by these agents (Table 4). Goossens et al. (83) have examined activities of ciprofloxacin, norfloxacin, and ofloxacin against

Table 4. Susceptibilities of gastrointestinal tract pathogens to quinolone antimicrobial agents

Drug	Range of MIC_{90}s (μg/ml) against[a]:				
	Salmonella spp.	*Shigella* spp.	*Campylobacter jejuni*	*Yersinia enterocolitica*	*Vibrio* spp.
Nalidixic acid	4	2–8	8–256	2–4	0.25–1.0
Ciprofloxacin	≤0.06	≤0.06	≤0.06–1.0	≤0.06	≤0.06
Norfloxacin	≤0.06–0.25	≤0.06–0.5	0.5–8	≤0.06–0.5	≤0.06–0.25
Enoxacin	0.25	≤0.06–0.25	1.0–32	0.125–0.25	
Ofloxacin	≤0.06–0.125	≤0.06–0.5	1.0–2	≤0.06–0.125	
Pefloxacin	≤0.06	≤0.06–0.125	1.0–2	≤0.06–0.25	
Difloxacin	0.5–2	≤0.06–4	1.0–4		
Fleroxacin	≤0.06	≤0.06–0.125	1.0	≤0.06–0.125	

[a]Derived from the following references (see Tables 1 and 2, footnote *a*, for abbreviations): NAL—4, 28, 36, 40, 68, 82, 84, 117, 122, 123, 127, 137, 142–144, 146, 156, 163, 181, 182, 195, 203; CPX—4, 30, 33, 36, 40, 56, 68, 82, 83, 89, 95, 117, 129, 134, 137, 143, 147, 156, 163, 203, 209; NFX—4, 25, 28, 30, 36, 40, 41, 56, 57, 82, 83, 95, 97, 117, 122, 123, 127, 134, 142–144, 146, 147, 156, 163, 181, 182, 184, 188, 203, 209; ENX—4, 25, 29, 30, 32, 122, 123, 144, 146, 147, 195, 209; OFX—4, 28, 30, 37, 38, 83, 95, 123, 134, 146, 147, 203, 209; PFX—4, 28, 36, 41, 126, 203; DFX—57, 89, 184, 188, 209; FLX—30, 36, 95, 100, 209.

enteropathogenic, enterotoxigenic, and enteroinvasive *Escherichia coli*. Ninety percent of strains from each group were inhibited by the drugs at concentrations of $\leq$0.1 μg/ml. *Salmonella* spp., including *S. typhi* (83), *Shigella* spp., *Yersinia enterocolitica*, and *Vibrio* spp. are exquisitely susceptible to fluoroquinolone antimicrobial agents (Table 4). Among the latter are included strains of *V. cholerae* and *V. parahaemolyticus* (142). *Campylobacter jejuni*, while not as highly susceptible as other species mentioned above, is usually inhibited by several of the new compounds at achievable concentrations. Against this group of isolates, ciprofloxacin is the most active drug. Among the newest agents, temafloxacin (A-62254) appears to be particularly promising, with a MIC_{90} of 0.25 μg/ml for *Campylobacter jejuni* (89). Ciprofloxacin (MIC_{90}, 0.25 μg/ml) and norfloxacin (MIC_{90}, 1.0 μg/ml) have demonstrated activity against *Campylobacter pylori*, while pefloxacin (MIC_{90}, 8 μg/ml) is less active (126, 137, 181). However, each of these results was obtained with different test media and, therefore, cannot be directly compared. *Campylobacter fetus*, which is resistant to nalidixic acid, is inhibited by ciprofloxacin at 1.0 μg/ml (68).

Respiratory tract pathogens. Several species of gram-negative respiratory tract pathogens are very susceptible to the new quinolone antimicrobial agents (Table 5). With few exceptions, most studies have found several fluoroquinolones to inhibit *Haemophilus influenzae* and *Branhamella catarrhalis* at concentrations of $\leq$0.25 μg/ml. *Neisseria meningitidis* has likewise proven to be highly susceptible (MIC_{90}s, $\leq$0.06 μg/ml) to most of the quinolones tested. *Capnocytophaga* spp. are most susceptible to ciprofloxacin and ofloxacin (MIC_{90}s, 0.12 and 0.25 μg/ml), followed by pefloxacin (MIC_{90}, 0.5 μg/ml) and norfloxacin (3).

Reported inhibitory concentrations of the fluoroquinolones against *Legionella* spp. vary widely. For each of several drugs, 100-fold or greater differences in MICs have been noted. Composition of the growth medium appears to have a dramatic impact on susceptibility results with this class of antimicrobial agents. For example, against 21 strains of *Legionella* spp. tested by agar dilution in a charcoal-yeast extract medium, we obtained a ciprofloxacin MIC_{90} of 2 μg/ml. Against this same group of strains, the MIC_{90} determined in the starch-yeast extract medium described by Saito et al. (174) was $\leq$0.06 μg/ml. The presence of charcoal in growth media adversely affected the activities of several fluoroquinolones not only against *Legionella* spp. but also against a test strain of *Staphylococcus epidermidis* (59). The potential clinical validity of results obtained in the starch-based medium has been supported by evidence that ofloxacin at concentrations of as low as 0.05 μg/ml is lethal to *Legionella pneumophila* growing intracellularly within human monocytes and by successful use of this drug in the therapy of experimental *Legionella* infections (174).

Table 5. Activities of fluoroquinolones against gram-negative respiratory pathogens

Organism	Range of MIC_{90}s (μg/ml)[a,b]						
	CPX	NFX	ENX	OFX	PFX	FLX	DFX
Haemophilus influenzae	≤0.06	≤0.06–0.5	≤0.06–0.5	≤0.06–0.12	≤0.06	≤0.06–0.12	≤0.06–0.12
Branhamella catarrhalis	≤0.06–0.5		0.25	≤0.06–1.0	0.25	0.25	≤0.06–0.12
Neisseria meningitidis	≤0.06	≤0.06–0.5	≤0.06	≤0.06	≤0.06	≤0.06	≤0.06
Eikenella corrodens	≤0.06	≤0.06	0.125	≤0.06			≤0.06
Bordetella spp.	0.5	0.25–4			2		
Legionella spp.	0.02–2	≤0.12–25	0.08–8	≤0.03–3		0.25	≤0.06–1.0

[a]See Tables 1 and 2, footnote *a*, for abbreviations.
[b]Derived from the following references: CPX—2–4, 14, 15, 30, 33, 36, 40, 59, 60, 78, 81, 89, 90, 95, 116, 117, 131, 134, 135, 156, 159, 173, 209, 215; NFX—2–4, 14, 15, 30, 32, 36, 40, 59, 78, 95, 97, 116, 117, 120, 122, 123, 131, 134, 135, 144, 156, 159, 173, 177, 188, 209, 214, 215, 220; ENX—4, 15, 29, 30, 32, 59, 60, 78, 81, 90, 93, 116, 122, 131, 144, 159, 173, 209, 214; OFX—3, 4, 15, 78, 81, 95, 97, 116, 134, 135, 173, 174, 177, 209, 214; PFX—2–4, 36, 116, 123, 131; FLX—30, 35, 36, 100, 135, 209; DFX—15, 59, 78, 89, 116, 188, 209.

A study by Osada and Ogawa (153) showed that ofloxacin inhibited each of 48 strains of *Mycoplasma pneumoniae* at a concentration of 1.56 μg/ml. This drug was more active than norfloxacin (MIC, 12.5 to 25 μg/ml) against the same isolates. Nakamura et al. (144) found similar results with norfloxacin, while enoxacin inhibited 10 strains at 6.25 μg/ml.

Nocardia spp. (included here for convenience) appear to be relatively resistant to ciprofloxacin (MIC_{90}, 8 μg/ml) and substantially more resistant to several other quinolones (MIC_{90}, ≥64 μg/ml) (4, 81). Mycobacterial susceptibilities will be discussed below.

Gram-positive bacteria

Table 6 shows the susceptibilities of gram-positive bacteria to new quinolones.

Staphylococci. Unlike nalidixic acid and the other older agents (180), several of the new agents exhibit relatively good activity against staphylococci. Coagulase-negative species are inhibited at concentrations generally comparable to those active against *Staphylococcus aureus*; activity extends to methicillin-resistant as well as to methicillin-susceptible strains (1, 45, 63, 185, 187). Several studies indicate that strains of *Staphylococcus saprophyticus* are somewhat more resistant to the fluoroquinolones (by 1 or 2 twofold dilutions) than are other coagulase-negative isolates (4, 63, 113, 135, 156).

Streptococci. Activities of the fluoroquinolones against group A streptococci and pneumococci are inferior to those against staphylococci. Of the agents extensively evaluated to date, CI-934 and ciprofloxacin have shown the greatest activity, although temafloxacin appears promising (89). MICs of ciprofloxacin against penicillin-resistant strains of pneumococci overlap considerably those against penicillin-susceptible strains (57, 79). Activities of the drugs against group B streptococcal isolates are comparable to those against *Streptococcus pyogenes*. MICs of various quinolones against the streptococci come uncomfortably close to, or exceed, easily attainable concentrations of the agents in serum.

***Listeria monocytogenes* and enterococci**. Ciprofloxacin inhibits strains of *Listeria monocytogenes* at concentrations of between 0.5 and 4 μg/ml (30, 40, 55, 57, 205). Although bactericidal concentrations of the drug appear to be only twice those which inhibit growth (4), these concentrations are high enough to cast serious doubt on the utility of the quinolones against organisms so notable for their tropism for the central nervous system.

Of the drugs tested to date, CI-934 is the most active against enterococci. MIC_{90}s of ciprofloxacin against *Streptococcus faecalis* isolates range from 0.5 to 4 μg/ml. In some studies, strains of *Streptococcus faecium* are

Table 6. Susceptibilities of gram-positive bacteria to new quinolones

Organism	MIC$_{90}$ (μg/ml) of[a,b]:									
	CPX	CI-934	NFX	OFX	PFX	DFX	FLX	TMX	S-25930	NY-198
Staphylococcus aureus	0.5	0.125	2	0.5	0.5	0.5	0.5	0.125	0.25	4
Coagulase-negative staphylococci	0.5	0.125	2	0.5	1.0	0.5	1.0	0.25	0.25	2
Streptococcus pneumoniae	2	0.5	8	4	8	4	8	1.0	2	8
Streptococcus groups A, C, and G	2	0.5	8	2	8	4	16	0.5	2	8
Listeria monocytogenes	1.0	0.5	4	2	8	2	4		2	
Enterococci	2	0.5	8	4	4	4	4	2	1.0	12.5
Corynebacterium spp.	1.0	0.25	4	1.0		1.0			0.5	

[a]See Tables 1 and 2, footnote *a*, for abbreviations.
[b]Derived from the following references: CPX—1, 4, 14–16, 24, 30, 33, 36, 40, 56–58, 62, 66, 67, 70, 79, 89, 95, 108, 112, 113, 116, 117, 119, 131, 133–135, 143, 147, 156, 163, 167, 169, 183, 185, 187, 191, 192, 205, 209, 213, 215; NFX—1, 4, 14–16, 28, 30, 32, 33, 36, 40, 41, 56, 57, 62, 67, 70, 79, 95, 97, 108, 116, 117, 122, 123, 131, 133–135, 143, 144, 146, 147, 156, 163, 177, 184, 185, 188, 209, 213–215; CI-934—38, 45, 58, 116, 119, 134, 167, 169; OFX—4, 15, 16, 28, 30, 38, 67, 95, 97, 116, 123, 133–135, 146, 147, 177, 185, 209, 214; PFX—4, 28, 36, 41, 63, 108, 116, 131, 208, 213; FLX—30, 36, 95, 133, 135, 209; DFX—15, 57, 89, 116, 169, 184, 185, 188, 209; TMX—89; S-25930—147, 156, 169; NY-198—97.

two- to fourfold more resistant to the quinolones than are isolates of *Streptococcus faecalis* (15, 57, 147). In other studies, the two species demonstrate comparable susceptibilities to various compounds (40, 45).

Anaerobic bacteria

Activities of the quinolones, as a class, against anaerobic bacteria have been disappointing (Table 7). For example, while reported MIC_{90}s of ciprofloxacin against *Bacteroides fragilis* group isolates have ranged from 0.12 to >128 μg/ml, most studies have yielded MIC_{90}s in the range of 4 to 32 μg/ml. These data suggest that a substantial proportion of clinical isolates will be resistant to concentrations of the fluoroquinolones which can be easily attained in serum. Strains of *Bacteroides melaninogenicus* appear to be no more susceptible to these agents than are the other *Bacteroides* species (108, 117). *Clostridium difficile* is generally resistant to the quinolones. However, concentrations of drug detectable in stool following oral administration of several quinolone compounds are manyfold in excess of inhibitory concentrations against *Clostridium difficile* (72). To what extent these agents could be used therapeutically for treatment of *Clostridium difficile*-associated colitis is uncertain. Among the more extensively studied fluoroquinolones, none offers any clear advantage over ciprofloxacin in activity against anaerobic organisms. The limited data available suggest that temafloxacin may be particularly active against this group of organisms (89). Ciprofloxacin inhibits 90% of *Propionibacterium* spp. and *Eubacterium* spp. at concentrations of between 2 and 16 μg/ml (10, 65, 77, 115, 160, 194, 212).

Genital pathogens

Bacterial pathogens. Various quinolones have shown excellent in vitro activity against *Neisseria gonorrhoeae* (14, 38, 54, 89, 108, 110, 116, 118, 120, 123, 128, 146–148, 153, 155, 179, 209, 214, 215). For ciprofloxacin, MICs of as low as 0.002 μg/ml (against almost 400 isolates) have been reported (110). Strains producing β-lactamase and those which possess chromosomally mediated penicillin resistance are also susceptible to quinolones (23). *Haemophilus ducreyi*, the etiologic agent of chancroid, is also susceptible to this group of drugs (108, 128, 130, 176, 193, 210). Difloxacin, ofloxacin, and ciprofloxacin are the most active quinolones against *Gardnerella vaginalis*, with MIC_{90}s of as low as 2 μg/ml (116, 118). Reported MIC ranges for these drugs extend to >16 μg/ml (13, 23, 108).

Mycoplasmas and ureaplasmas. Strains of *Mycoplasma hominis* are inhibited and killed by theoretically attainable concentrations (≤2 μg/ml) of ciprofloxacin (60, 164, 200) and difloxacin (200). *Ureaplasma urealyticum* has demonstrated variable, but generally poor, susceptibility to these

Table 7. Expected susceptibilities of anaerobic bacteria to fluoroquinolones

Organism	MIC_{90} (μg/ml) of[a,b]:							
	CPX	NFX	PFX	OFX	DFX	FLX	TMX	NAL
Bacteroides fragilis	8	32	32	8	4	8	2	>128
Bacteroides spp.	16	64	16	8	4	8	4	>128
Fusobacterium spp.	8	32	32	4	4			>128
Clostridium spp.	8	32	64	8	4	16	0.5	>128
Clostridium difficile	8	64	128	16	8			>128
Gram-positive cocci	2	16	8	4	4		2	>128
Actinomyces spp.	8	32			4			128
Mobiluncus spp.	2	16	16	2	4			>128

[a]See Tables 1 and 2, footnote *a*, for abbreviations.
[b]Derived from the following references: NAL—10, 36, 46, 77, 82, 108, 116, 117, 122, 156, 160, 163, 177, 182, 203, 214; CPX—10, 15, 21, 30, 33, 35, 36, 46, 65, 77, 82, 89, 95, 108, 116, 117, 134, 135, 147, 151, 156, 160, 163, 194, 203, 211, 212, 215; NFX—10, 15, 30, 32, 33, 35, 36, 46, 65, 77, 82, 95, 97, 108, 116, 117, 122, 134, 135, 147, 151, 156, 163, 177, 182, 203, 214, 215; PFX—36, 46, 108, 116, 203; OFX—30, 38, 46, 65, 95, 97, 116, 134, 135, 147, 151, 177, 203, 214; DFX—10, 15, 35, 65, 89, 116; FLX—30, 36, 95, 135; TMX—89.

agents. MIC_{90}s of ciprofloxacin have ranged from 2 to 32 μg/ml (6, 130, 164, 200). MIC_{90}s of enoxacin or amifloxacin may reach 64 μg/ml (6, 71).

***Chlamydia trachomatis*.** Several studies have examined the susceptibilities of *Chlamydia trachomatis* strains to the fluoroquinolones. Ciprofloxacin has inhibited isolates at concentrations of ≤2 μg/ml (6, 92, 101, 130, 164, 200). Activities of ofloxacin in the same range have been reported (6, 7, 23, 94, 130, 189). Against these organisms, difloxacin has demonstrated excellent activity, inhibiting isolates at concentrations of approximately 0.25 μg/ml (23, 130, 200). These values can be compared with inhibitory concentrations of doxycycline or tetracycline against such isolates, ≤0.06 or 0.3 μg/ml, respectively (6, 7). While bactericidal concentrations of tetracycline are substantially higher than inhibitory concentrations (7), ofloxacin appears to exert a lethal effect against *Chlamydia trachomatis* at concentrations near the MIC (7, 23). The complexities of assessing antichlamydial drug effects have been discussed by How et al. (101).

Rickettsiae

Ciprofloxacin was shown to inhibit *Rickettsia conorii,* the agent of Boutonneuse fever, at concentrations of as low as 0.25 μg/ml (161). These assays were performed by plaque inhibition techniques on Vero cell monolayers. The drug was rickettsiacidal at concentrations near the MIC. At concentrations of 1.0 μg/g, ciprofloxacin also enhanced the survival of infected embryonated hen eggs. Pefloxacin at 0.5 to 1.0 μg/ml also inhibited *Rickettsia conorii,* as well as *Rickettsia rickettsii* (the agent of Rocky Mountain spotted fever), in cell culture (162). This drug was likewise effective in ovo.

Difloxacin at a concentration of 2 μg/ml was effective against *Coxiella burnetii,* as measured by a reduction in the percentage of infected cells in tissue culture to less than 10% by 48 h of exposure to the drug (217). Ciprofloxacin was also effective in this study. Unfortunately, concentrations of 5 μg of this drug per ml were used, which exceeds readily attainable levels in serum. The rapid decline in the fraction of infected cells upon treatment suggested to these authors a rickettsiacidal effect of the quinolones.

Mycobacteria

Several of the new fluoroquinolones have demonstrated activity against pathogenic mycobacteria. *Mycobacterium tuberculosis* isolates can be inhibited by ciprofloxacin at concentrations of as low as 0.25 μg/ml (64, 75). Five in vitro studies have each derived MIC_{90}s of ≤1.6 μg/ml for ciprofloxacin; the total number of strains examined in these reports exceeded 300 (19, 44, 64, 75). However, for this drug, MICs of as high as 6.25 μg/ml have been reported by others (39, 202). Ofloxacin (19, 44, 64), CI-934

(19), pefloxacin (44), enoxacin (19), and norfloxacin (19) have also been reported to inhibit 90% of isolates at concentrations of ≤2.0 μg/ml.

The rapid growers of the *Mycobacterium chelonei-Mycobacterium fortuitum* complex have shown variable susceptibilities to the quinolones. Using a 1% proportional agar method, Gay et al. (75) found that ciprofloxacin inhibited isolates of this group at concentrations of between 0.25 and 16 μg/ml. The MIC_{90} against *Mycobacterium fortuitum* (0.25 μg/ml) was lower than that against *Mycobacterium chelonei* (8 μg/ml). Activities of norfloxacin paralleled those of ciprofloxacin, with MIC_{90}s of 2 μg/ml (range, 0.25 to >16) and >16 μg/ml (range, 0.5 to >16), respectively. Other studies have confirmed the greater activity of ciprofloxacin against *Mycobacterium fortuitum* than against *Mycrobacterium chelonei* (39, 99). Fleroxacin has also demonstrated better activity against the former (MIC_{90}, 0.5 μg/ml) than against the latter (MIC_{90}, >32 μg/ml) group (99). This drug was found to be bactericidal against susceptible mycobacteria at concentrations of 2 μg/ml (>99.9% killing within 48 h) and was effective against intracellular organisms within mouse peritoneal macrophages (52). Determination of ciprofloxacin MICs against the rapid growers is method dependent. For these atypical mycobacteria, MICs obtained with the BACTEC radiometric system were significantly lower than those obtained with a macrodilution broth method (202). Which results prove to be the best predictions of clinical response remain to be determined.

Ciprofloxacin has demonstrated activity against *Mycobacterium xenopi* (MIC_{90}, 1.56 μg/ml for 36 strains) (39). Higher concentrations of the drug are required to inhibit *Mycobacterium kansasii*. For the latter species, MICs of ciprofloxacin ranged from 1.0 to 4 μg/ml, with a MIC_{90} of 4 μg/ml in one study (75). Strains of the *Mycobacterium avium-Mycobacterium intracellulare* complex tend to be relatively resistant to the fluoroquinolones. While individual isolates have been susceptible to ciprofloxacin at concentrations of as low as 0.25 μg/ml (64), several studies have compiled MIC_{90}s of ≥8 μg/ml (39, 75, 202). MICs determined by a radiometric method (MIC_{90}, 12.5 μg/ml) have been lower than those obtained by macrodilution broth techniques (MIC_{90}, >100 μg/ml) (202). For ofloxacin, a MIC_{90} of 8 μg/ml against this group of organisms has been reported (64).

Influence of Methodology on Assessment of Quinolone Antimicrobial Activity

Media and techniques

There is generally good correlation between measurements of quinolone antimicrobial activities by agar and microdilution broth techniques. Results obtained by macrodilution broth methods sometimes yield slightly higher MICs (80, 111, 152). Studies comparing activities of ciprofloxacin (111), norfloxacin (123), CI-934 (134), amifloxacin (103), and S-25930 (147) in several commonly employed standard laboratory media have demonstrated similar results (usually within 1 twofold dilution) in each. An exception to this generalization was found in one study in which the activity of ciprofloxacin against anerobic organisms was significantly greater in Wilkens-Chalgren agar as compared with brucella agar (21).

Activities of these drugs have also been examined in media supplemented with human serum in an attempt to simulate more closely conditions in vivo. Forty or fifty percent serum has no effect on the activities of ciprofloxacin (20, 33, 215), norfloxacin (188), enoxacin (20, 144, 214), ofloxacin (28, 123), amifloxacin (41), or CI-934 (38, 134). In contrast, addition of 50% serum reduced the activity of difloxacin to a variable degree, depending upon the species tested (10, 98, 188). Against one *Acinetobacter* strain, the MIC of difloxacin in 50% serum-supplemented Mueller-Hinton broth was 2 μg/ml, while in unsupplemented broth the MIC was 0.03 μg/ml (188). This represents a 64-fold decrease in drug activity in the presence of serum. Most differences between results obtained in serum-supplemented versus unsupplemented media are of smaller magnitude. Addition of serum reduced the activity of the highly (approximately 95%) protein-bound drug S-25930 by up to eightfold, while the activity of S-25932 was less affected (156). Activity of temafloxacin was adversely affected by serum against some bacterial isolates, although for most strains serum supplementation yielded MICs of only one- to twofold higher than those in unsupplemented media (89).

Activity in urine and effect of pH and magnesium

In view of the potential use of the new quinolones for the treatment of urinary tract infections, tests of in vitro activity have also been carried out in urine-supplemented media. In urine, activities of the new drugs are typically reduced by 8- to 64-fold or more compared with MICs obtained in standard media (32, 33, 85, 98, 123, 134, 188). For several new quinolones, reduced potency in urine could be explained in part by diminished activity at low pH values commonly prevailing in that medium. For example, activities of ciprofloxacin or norfloxacin against *E. coli* and *S. aureus* are eightfold higher at pH 6.8 than at pH 4.8 (18). Decreased activities in standard media adjusted to acidic pH levels have also been noted with enoxacin (27,144), amifloxacin (105), ofloxacin (123), CI-934 (134), fleroxacin (30), and EN 272 (146). In contrast, the activity of difloxacin is not adversely affected to a significant degree at lower pHs (10, 98, 188), while activities of S-25930 and S-25932 may actually be greater at pH 5.5 than at pH 7.5 (147).

It is clear, however, that the effect of urine pH alone does not fully account for the diminished potencies of the new quinolones in this medium. Several studies comparing activities of these drugs in urine-supplemented media and in standard media adjusted to acidic pHs prevailing in the urine reveal superior activity in the latter, when comparisons at any given pH value are made (30, 33, 123, 147). Evidence now points to high concentrations of magnesium ions in urine as the major cause of diminished activity of these compounds in that medium. Urinary concentrations of the cation are commonly in the range of 8 to 10 mM, whereas levels in unsupplemented Mueller-Hinton broth are approximately 0.3 mM. Supplementation of standard laboratory media with magnesium to levels attained in human urine results in 4- to 16-fold increases in MICs of CI-934 (134), 4- to 8-fold increases in MICs of difloxacin (98), and 2- to 8-fold increases in MICs of fleroxacin (30). However, addition of even lower levels of the cation to standard media may result in statistically significant decreases in activities of various quinolone antimicrobial agents (4, 20). Despite the combined effects of high cation concentrations and acidic pHs which prevail in normal urine, inhibitory concentrations of the new quinolones still generally remain considerably below concentrations of these drugs achievable within the urinary tract.

Because of the potential applications of fluoroquinolones as oral agents in the outpatient therapy of peritonitis associated with chronic ambulatory peritoneal dialysis, the in vitro activity of ciprofloxacin in dialysis medium has been examined. One group compared the activity of this quinolone in Mueller-Hinton agar and in Mueller-Hinton agar reconstituted with pooled peritoneal dialysis effluent (86). Activities of the drug against *S. aureus* measured in the two media were not significantly different. Although the mean MIC and MIC_{50} against coagulase-negative staphylococci were higher in dialysate-containing medium than in standard agar, 90% of isolates were still inhibited at concentrations of $\leq$1.0 μg/ml. Ciprofloxacin also retains potent (bactericidal) activity against *E. coli* and *P. aeruginosa* in fresh dialysate, pH 5.5, albeit at rates somewhat slower than those in dialysate buffered to pH 7.4 (136).

Effect of inoculum size on activity

Compared with many antimicrobial agents, the in vitro activities of the fluoroquinolones are only modestly affected by changes in inoculum size. Most studies which have compared activities of these drugs at inocula of 10^3 to 10^5 CFU/ml with those at inocula of 10^6 to 10^7 CFU/ml have detected differences in activities against most isolates of fourfold or less (18, 29, 30, 41, 103, 123, 134, 152, 177, 188, 209). A minority of isolates are affected to greater degrees. Examples of such include the following: (i) 10-fold-higher

MICs of norfloxacin against *Klebsiella pneumoniae* and *S. marcescens* at 10^7 versus 10^3 CFU (148); (ii) 8-fold-higher MICs of amifloxacin against *P. aeruginosa* (and >64-fold increase in MBC) at 10^8 versus 10^6 CFU/ml (105); (iii) 10-fold-higher MICs of ciprofloxacin against *S. aureus* at 10^7 versus 10^5 CFU (111); and (iv) 10- to 100-fold-higher MICs of CI-934 (10^7 versus 10^3 CFU) and ciprofloxacin (10^7 versus 10^5 CFU) against enterococci at higher inocula (38, 111).

Bactericidal Activity

MBCs and SBTs

Determination of accurate MBCs is subject to a number of methodologic difficulties, the most troublesome of which is the transfer onto antimicrobial agent-free agar of sample volumes large enough to detect surviving colonies, yet not so large as to result in substantial antimicrobial agent carry-over. Despite these difficulties, data pertaining to MBCs of the fluoroquinolones are remarkably consistent. For the most part, MBCs of these drugs against a wide range of bacteria are generally within 1 or 2 dilutions of the MICs at standard inocula (ciprofloxacin [43, 186, 215], norfloxacin [104, 148, 152, 197], enoxacin [29, 33, 122, 214], ofloxacin [123, 177], amifloxacin [106], CI-934 [134], pefloxacin [4], fleroxacin [30, 209], difloxacin [188], and S-25930 and S-25932 [156]). For example, Chartrand et al. (29) found that the MBCs of enoxacin (defined at only a 99% killing level) were within 1 dilution of the MIC against 97% of 495 clinical bacterial isolates. However, occasionally strains are encountered with MBCs which are 10-fold or more above the MIC of one or more quinolone antimicrobial agents (30, 105, 106). Some studies have suggested greater discrepancies between the MBC and MIC at high bacterial inocula (28, 105), while others have not noted a significant inoculum effect (30). It is not clear to what extent these observations are representative of real phenomena or merely reflect the greater sensitivity inherent in sampling from denser bacterial suspensions. The new fluoroquinolones are also generally bactericidal in urine, although, like MICs, bactericidal effects are substantially reduced in that medium (124).

Another measure of bactericidal activity is the serum bactericidal titer (SBT). Studies of SBTs, whether carried out with volunteer subjects or actual patients, confirm the bactericidal activities of the fluoroquinolones at actual achievable concentrations in serum. For example, against eight isolates of *Salmonella typhi*, peak SBTs obtained after oral administration of ciprofloxacin (500 mg) or ofloxacin (200 mg) ranged from 1:100 to 1:400 and from 1:20 to 1:80, respectively (201). Peak SBTs achievable with ciprofloxacin are higher against *Enterobacteriaceae* (1:128 to 1:256) than against *P. aeruginosa* (1:4), *S. aureus* (1:2), or *Streptococcus faecalis* (<1:2) (125).

Bactericidal activity by time-kill techniques

Time-kill curve methods offer advantages over MBC determinations in that both the rate and the absolute magnitude of killing can be established by the former. Such studies, however, are more time-consuming. As a result, only a few drug concentrations and a limited number of strains are usually examined in any one study. Like MBC determinations, time-kill studies must be evaluated with an understanding of potential methodologic difficulties which can influence the validity of test results. Studies published to date have not consistently specified lower limits of detection applicable to their techniques and have not always addressed issues of drug carry-over which could potentially lead to spuriously high rates of killing. In view of the extreme potency of many newer fluoroquinolones, the latter problem may be particularly troublesome with this group of agents. Further complicating comparisons between studies, various studies use different arbitrary sampling times, and some use log-phase bacterial inocula while others use stationary-phase bacterial inocula. The latter may be particularly important, as illustrated by the work of Zeiler (221) who found that ciprofloxacin was radidly bactericidal against two strains of *E. coli* when log-phase inocula were used, whereas for nongrowing bacteria susceptibility to the lethal effects of this drug was strain dependent.

Nevertheless, to the extent that they are comparable, data from various sources are consistent in reporting excellent bactericidal activities of the fluoroquinolones against most gram-negative bacteria. Against *Enterobacteriaceae*, concentrations of these drugs of as low as two to four times the MIC produce reductions of viable cell concentrations of 3 to 5 $\log_{10}$ within 2 h of exposure (33, 41, 104, 122, 147, 188). Onset of bactericidal activity against highly susceptible organisms has been documented as early as 15 min after exposure to ciprofloxacin (26). With *P. aeruginosa*, comparable 3- to 5-$\log_{10}$ reductions in the number of viable cells have been reported within 2 h of exposure to fluoroquinolones at concentrations of two to four times the MIC (32, 41).

Bactericidal activities of these drugs against staphylococci appear to be somewhat strain dependent, method dependent, or both. In one study, ciprofloxacin at four times the MIC reduced the number of viable bacteria of one *S. aureus* strain by 5 $\log_{10}$ within 2 h; against another isolate, the drug decreased viable cells by $<2 \log_{10}$ in the same time period (33). There is a general sense that rates of killing of staphylococci by the newer quinolones are slower than killing rates observed against gram-negative bacteria (26, 32,

57). Nevertheless, by 24 h of incubation, most strains undergo 3- to 4-$\log_{10}$ killing in the presence of achievable levels of the new drugs in serum (32, 63, 186). In one study in which the activity of pefloxacin was compared with those of other antimicrobial agents against several species of staphylococci, bactericidal activity of the quinolone was equivalent to that of gentamicin and superior to those of oxacillin and vancomycin (63). At concentrations near the MIC, after initial killing some regrowth can occur at 24 h. Colonies recovered after regrowth may demonstrate reduced susceptibility to the drug (70). Bactericidal effects of the fluoroquinolones against staphylococci can also extend to viable intracellular organisms within phagocytic cells (53, 138).

The quinolone CI-934, which is one of the most active fluoroquinolones against gram-positive bacteria, has been shown to exert bactericidal activity (approximately 3-$\log_{10}$ killing) against 20 strains of *Streptococcus faecalis* at drug concentrations of as low as 1.0 μg/ml (219). At concentrations equal to four times the MIC, ciprofloxacin led to an average of 99.4% killing at 24 h of incubation against 11 enterococcal isolates (191). This level of bactericidal activity was comparable to those observed with ampicillin and daptomycin (LY 146032).

Concentrations of norfloxacin or ciprofloxacin which are achievable in the urine have been shown to exert bactericidal activity against enterococci and staphylococci when tested in a pooled urine medium. However, killing of these organisms takes place at a slower rate than observed with *Enterobacteriaceae* and *P. aeruginosa* (124, 125).

Postantibiotic effect

The period of time which elapses between transient exposure of a bacterial culture to an antimicrobial agent and the point that growth resumes among surviving bacteria is a measure of the postantibiotic effect of the antimicrobial agent (73). This phenomenon may be important because concentrations of antimicrobial agents in serum or tissue sometimes fall below inhibitory concentrations for part of the interval between drug doses. The longer the period of postantibiotic effect, the lower the probability that growth of infecting organisms will resume during that fraction of the interdose interval that antimicrobial levels fall below the MIC. The duration of the postantibiotic effect depends on the nature of the microorganism, on characteristics of the drug, and on both the concentration of antimicrobial agent employed and the duration of exposure.

For staphylococci and *P. aeruginosa*, exposure to inhibitory concentrations of ciprofloxacin for 1 to 2 h results in a postantibiotic effect of approximately 2 h (34, 73, 187). With *E. coli*, a postantibiotic effect of more than 4 h was noted after exposure to that drug (34). Similar results have been obtained with S-25930 (147). Against *Streptococcus faecalis,* Chin and Neu (34) found no measurable postantibiotic effect after exposure to ciprofloxacin.

Resistance to Quinolones

Use of the older quinolone antimicrobial agents for treatment of urinary tract infections was complicated by a significant incidence of emergence of drug-resistant strains during therapy. As a result, in vitro evaluation of newly developed quinolone antimicrobial agents has included intensive study of bacterial resistance to these agents (55). For this reason, the issue of resistance to quinolones which is discussed in detail in chapter 2 will be briefly mentioned here.

Extensive evidence amassed to date suggests that the frequencies of spontaneous mutants resistant to the new fluoroquinolones are low in comparison with frequencies of resistance to nalidixic acid. Most studies examining the emergence of resistant clones when large bacterial inocula are plated onto media containing new quinolones at concentrations of eight times the MIC have reported resistance frequencies of $<10^{-9}$ for a representative variety of bacteria (15, 28, 30, 122, 123, 134, 178). Occasional strains have exhibited somewhat higher frequencies of resistance. Isolates of *E. coli, Klebsiella pneumoniae, Enterobacter cloacae, Providencia stuartii, P. aeruginosa,* and *S. aureus* have demonstrated resistance to one or more fluoroquinolones at frequencies in the range of 10^{-7} to 10^{-8} (30, 33, 51, 57, 123).

Strains resistant to fluoroquinolones can be derived from initially susceptible isolates by serial passage in media containing incremental concentrations of drug (57, 98, 132, 136, 178, 196, 209). Resulting colonies may have MICs which are 2 to 3 orders of magnitude above those of the parent isolates (209). Selection of resistance to one quinolone by this method confers relative cross resistance to other drugs of the class (12, 98, 132, 165). Relative cross resistance among the various fluoroquinolones also occurs with naturally occurring resistant isolates. If bacterial isolates are stratified according to levels of susceptibility or resistance to nalidixic acid, the group fully resistant to nalidixic acid also demonstrates decreased susceptibility to fluoroquinolones in comparison with strains of the same species which are fully susceptible to the older drug. Organisms with intermediate levels of resistance to nalidixic acid are inhibited by intermediate MICs of the newer quinolones (15, 144, 157, 198). However, while fluoroquinolone MICs may be higher against nalidixic acid-resistant organisms than against susceptible ones, MICs of various fluoroquinolones are likely to fall within the range judged "susceptible" for clinical uses even for strains fully resistant to the older drug.

Attention has also been called to the apparent association between relative resistance to the fluoroquinolones and to other classes of antimicrobial agents. For example, some aminoglycoside-resistant strains of *P. aeruginosa* and *S. marcescens* tend to require higher MICs or MBCs or both of fluoroquinolones than do aminoglycoside-susceptible strains (17, 69, 183). However, such differences in fluoroquinolone activities tend to be quite small, usually only 1 dilution step. Among *S. marcescens*, cefotaxime resistance has correlated with higher MICs or MBCs or both of ciprofloxacin (183). For *Klebsiella* isolates, in vitro selection of norfloxacin-resistant mutants can also result in greater resistance to cefotaxime. Similarly, selection for cefotaxime resistance can confer cross resistance to norfloxacin (as well as to tetracycline and chloramphenicol) (175). Either mode of selection for antimicrobial resistance resulted in loss or diminution of a 39,000-molecular-weight outer membrane protein as compared with parent strains.

Fortunately, while drug resistance has been reported during therapy of infections with new quinolone antimicrobial agents, this has been associated with treatment failure only in a minority of cases (5, 87, 139). At present, primary resistance to the more potent fluoroquinolones appears to be quite uncommon among routine clinical isolates in the United States. Nevertheless, possibilities of relative cross resistance to quinolone antimicrobial agents among isolates selected for resistance to other drug classes suggest that particular caution should be exercised in the use of newer quinolones in hospital settings where multiply drug-resistant isolates are common (183).

Antimicrobial Combinations

There has been considerable interest in the study of new quinolones in combination with other antimicrobial agents. One reason for this interest arises from the fact that the site of action of these drugs differs from those of other clinically useful antimicrobial agents, suggesting the possibility that synergistic activity may occur. Mechanisms of synergism, when it occurs, have not been systematically studied. In addition, the new quinolones are frequently active in vitro against organisms resistant to many other drugs and to which further drug resistance might emerge rapidly (e.g., mycobacteria or methicillin-resistant staphylococci). Thus, the possibility exists that combining quinolones with other drugs may result in suppression of the emergence during treatment of isolates resistant to the quinolones.

Gram-negative bacteria

Combinations with aminoglycosides. As might be expected, results of combination studies have not yielded consistent results given differences in methods employed and composition of strain collections. For example, using agar dilution checkerboard techniques and stringent definitions for patterns of interaction, Haller (88) found evidence of true synergism or antagonism between ciprofloxacin and an aminoglycoside in less than 1% of interactions against >200 strains of gram-negative bacteria. Lack of significant synergistic or antagonistic activity against *Enterobacteriaceae* has also been noted when enoxacin (8), norfloxacin (148), CI-934 (134), or difloxacin (or A-56620) (188) has been combined with an aminoglycoside in checkerboard titrations. On the other hand, in one study, synergistic inhibitory activity between gentamicin and enoxacin or norfloxacin was detected against 17.5 and 15%, respectively, of *Enterobacteriaceae*, with bactericidal interactions against 17.5 and 20% (204). Rates of synergistic interactions between fluoroquinolones and aminoglycosides of as high as 57% (against *S. marcescens*) have been reported (140), although such high rates appear to be distinctly unusual. Although quinolone-aminoglycoside bactericidal synergism has been convincingly demonstrated against some strains of *Enterobacteriaceae* by time-kill methods (26), interactions more frequently noted by such studies involve the suppression of resistance emerging to either of the antimicrobial agents employed (88).

Synergism between fluoroquinolones and aminoglycosides by checkerboard titrations has been noted against 0 to 30% of *P. aeruginosa* isolates (43, 61). One study documented enhanced killing at clinically relevant concentrations against 11 of 26 (42%) multiply drug-resistant *P. aeruginosa* isolates when ciprofloxacin was combined with amikacin by time-kill methods (76).

Combinations with beta-lactams. Numerous studies have examined combinations of quinolones with beta-lactam antibiotics against gram-negative bacteria. Several of these report little evidence of meaningful synergistic interactions against *Enterobacteriaceae*, with <10% of strains so affected (8, 31, 50, 123, 134, 188). Higher rates of synergistic interactions between quinolones and beta-lactams have been reported against *Acinetobacter* spp., with ciprofloxacin plus azlocillin synergistically inhibiting 3 of 10 isolates (141). Against *P. aeruginosa*, synergism between ciprofloxacin and azlocillin has been noted against 10 to 56% of isolates when checkerboard techniques are used (31, 43, 141). Results in this range have also been seen with enoxacin-piperacillin (17% synergy) and enoxacin-cefsulodin (28.5% synergy) combinations against *P. aeruginosa* (8). Inhibitory synergism between ciprofloxacin and imipenem was observed against 36% of imipenem-susceptible strains of this species but in <10% of imipenem-resistant isolates (24). By time-kill methods, ciprofloxacin-imipenem synergism was found against 42% of *P. aeruginosa* strains studied (76), but in that report the presence or absence of enhanced killing (synergism after 3 or 5 h of incubation) could not be correlated with susceptibilities to the individual agents. Others have also been unable to predict the presence or absence of synergistic interactions between ciprofloxacin and a beta-lactam against *P. aeruginosa* based on susceptibility or resistance to the latter drug (31, 141). Combinations of fluoroquinolones with ceftazidime have resulted in synergistic interactions against *P. aeruginosa* only infrequently (43, 47, 61).

Anaerobes

When ciprofloxacin was combined with antianaerobic antimicrobial agents, synergism was infrequently seen against 41 isolates of anaerobic bacteria (212). At clinically relevant drug concentrations, synergism was noted when the quinolone was combined with cefotaxime (16% of isolates), mezlocillin (9%), clindamycin (9%), and cefoxitin (2%). Only the first two of these produced synergistic inhibition of *Bacteroides* spp. or *Fusobacterium* spp. when combined with ciprofloxacin.

Staphylococci

Studies of combinations against staphylococci have generally examined only small numbers of strains. Among the more interesting combinations is that of ciprofloxacin with coumermycin. The latter acts at the B subunit of the DNA gyrase enzyme. At the lowest concentrations of drugs tested in combination (0.1 μg/ml), five of five strains were synergistically inhibited, but at higher drug concentrations, only indifference was noted (206).

Combinations of quinolones with penicillins, clindamycin, or vancomycin have generally resulted in few synergistic interactions (31, 123, 154, 186). On the other hand, combination with rifampin has led to antagonism of the bactericidal effects of fluoroquinolones against staphylococci (63, 206). Bactericidal activities of quinolones can also be antagonized by chloramphenicol (221).

Discussion of animal models to assess interactions between quinolones and other antimicrobial agents is beyond the scope of this chapter. Suffice it to say that in vitro synergism has been associated with enhanced in vivo efficacy in some (31), but not all (9, 88, 140), animal models.

Conclusions

In comparison with the older quinolone antimicrobial agents, the newer fluoroquinolone compounds are more potent against those gram-negative isolates susceptible to both groups of drugs, often by several orders of magnitude. In addition, the new agents extend the antimicrobial spectrum of the quinolones to include *P. aeruginosa* and a number of gram-positive organisms. Strains of *Pseudomonas maltophilia* and some streptococci tend to be more resistant to this class of agents. Activities of these drugs against anaerobic bacteria are generally weak. Activities of several drugs against *Legionella* spp., rickettsiae, and certain mycobacteria appear to be very promising and deserve further study.

Measurement of the in vitro activity of any antimicrobial compound is but one step in its assessment for possible clinical application. In some cases, favorable pharmacokinetic or toxicologic properties of one drug may outweigh the superior in vitro potency of a related agent. These factors, and eventually a record of clinical efficacy, determine the ultimate role of a newer antimicrobial agent.

Literature Cited

1. **Aldridge, K. E., A. Janney, and C. V. Sanders.** 1985. Comparison of the activities of coumermycin, ciprofloxacin, teichoplanin, and other non-β-lactam antibiotics against clinical isolates of methicillin-resistant *Staphylococcus aureus* from various geographical locations. *Antimicrob. Agents Chemother.* **28:**634–638.
2. **Appelbaum, P. C., S. K. Spangler, and L. Sollenberger.** 1986. Susceptibility of nonfermentative gram-negative bacteria to ciprofloxacin, norfloxacin, amifloxacin, pefloxacin and cefpirome. *J. Antimicrob. Chemother.* **18:**675–679.
3. **Arlet, G., M.-J. Sanson-LePors, I. M. Cosin, M. Ortenberg, and Y. Perol.** 1987. In vitro susceptibility of 96 *Capnocytophaga* strains, including a β-lactamase producer, to new β-lactam antibiotics and six quinolones. *Antimicrob. Agents Chemother.* **31:**1283–1284.
4. **Auckenthaler, R., M. Michea-Hamzehpour, and J. C. Pechere.** 1986. In vitro activity of newer quinolones against aerobic bacteria. *J. Antimicrob. Chemother.* **17**(Suppl. B)**:**29–39.
5. **Azadian, B. S., J. W. A. Bendig, and D. M. Samson.** 1986. Emergence of ciprofloxacin-resistant *Pseudomonas aeruginosa* after combined therapy with ciprofloxacin and amikacin. *J. Antimicrob. Chemother.* **18:**771.
6. **Aznar, J., M. C. Caballero, M. C. Lozano, C. DeMiguel, J. C. Palomares, and E. J. Perea.** 1985. Activities of new quinolone derivatives against genital pathogens. *Antimicrob. Agents Chemother.* **27:**76–78.
7. **Bailey, J. M. G., C. Heppelston, and S. J. Richmond.** 1984. Comparison of the in vitro activities of ofloxacin and tetracycline against *Chlamydia trachomatis* as assessed by indirect immunofluorescense. *Antimicrob. Agents Chemother.* **26:**13–16.
8. **Baltch, A. L., C. Bassey, G. Fanciullo, and R. P. Smith.** 1987. In vitro antimicrobial activity of enoxacin in combination with eight other antibiotics against *Pseudomonas aeruginosa*, Enterobacteriaceae, and *Staphylococcus aureus*. *J. Antimicrob. Chemother.* **19:**45–48.
9. **Bamberger, D. M., L. R. Peterson, D. N. Gerding, J. A. Moody, and C. E. Fasching.** 1986. Ciprofloxacin, azlocillin, ceftizoxime and amikacin alone and in combination against gram-negative bacilli in an infected chamber model. *J. Antimicrob. Chemother.* **18:**51–63.
10. **Bansal, M. B., and H. Thadepalli.** 1987. Activity of difloxacin (A-56619) and A-56620 against clinical anaerobic bacteria in vitro. *Antimicrob. Agents Chemother.* **31:**619–621.
11. **Barbhaiya, R. H., A. U. Gerber, W. A. Craig, and P. G. Welling.** 1982. Influence of uri-

nary pH on the pharmacokinetics of cinoxacin in humans and on antibacterial activity in vitro. *Antimicrob. Agents Chemother.* **21**:472–480.

12. **Barry, A. L., and R. N. Jones.** 1984. Cross-resistance among cinoxacin, ciprofloxacin, DJ-6783, enoxacin, nalidixic acid, norfloxacin, and oxolinic acid after in vitro selection of resistant populations. *Antimicrob. Agents Chemother.* **25**:775–777.
13. **Barry, A. L., R. N. Jones, C. Thornsberry, L. W. Ayers, T. L. Gavan, and E. H. Gerlach.** 1986. In vitro activity of the aryl-fluoro-quinolones A-56619 and A-56620 and evaluation of disk susceptibility tests. *Eur. J. Clin. Microbiol.* **5**:18–22.
14. **Barry, A. L., R. N. Jones, C. Thornsberry, L. W. Ayers, E. H. Gerlach, and H. M. Sommers.** 1984. Antibacterial activities of ciprofloxacin, norfloxacin, oxolinic acid, cinoxocin, and nalidixic acid. *Antimicrob. Agents Chemother.* **25**:633–637.
15. **Barry, A. L., C. Thornsberry, and R. N. Jones.** 1986. In vitro evaluation of A-56619 and A-56620, two new quinolones. *Antimicrob. Agents Chemother.* **29**:40–43.
16. **Bassey, C. M., A. L. Baltch, and R. P. Smith.** 1986. Comparative antimicrobial activity of enoxacin, ciprofloxacin, amifloxacin, norfloxacin and ofloxacin against 177 bacterial isolates. *J. Antimicrob. Chemother.* **17**:623–628.
17. **Bassey, C. M., A. L. Baltch, R. P. Smith, and P. E. Conley.** 1984. Comparative in vitro activities of enoxacin (CI-919, AT-2266) and eleven antipseudomonal agents against aminoglycoside-susceptible and -resistant *Pseudomonas aeruginosa* strains. *Antimicrob. Agents Chemother.* **26**:417–418.
18. **Bauernfeind, A., and C. Petermuller.** 1983. In vitro activity of ciprofloxacin, norfloxacin and nalidixic acid. *Eur. J. Clin. Microbiol.* **2**:111–115.
19. **Berlin, O. G. W., L. S. Young, and D. A. Bruckner.** 1987. In vitro activity of six fluorinated quinolones against *Mycobacterium tuberculosis. J. Antimicrob. Chemother.* **19**:611–615.
20. **Blaser, J., M. N. Dudley, D. Gilbert, and S. H. Zinner.** 1986. Influence of medium and method on the in vitro susceptibility of *Pseudomonas aeruginosa* and other bacteria to ciprofloxacin and enoxacin. *Antimicrob. Agents Chemother.* **29**:927–929.
21. **Borobio, M. V., and E. J. Perea.** 1984. Effect of inoculum, pH, and medium on the activity of ciprofloxacin against anaerobic bacteria. *Antimicrob. Agents Chemother.* **25**:342–343.
22. **Bosch, J., J. Linares, M. J. Lopez de Goicoechea, J. Ariza, M. C. Cisnal, and R. Martin.** 1986. In vitro activity of ciprofloxacin, ceftriaxone and five other antimicrobial agents against 95 strains of *Brucella melitensis. J. Antimicrob. Chemother.* **17**:459–461.
23. **Bowie, W. R., C. E. Shaw, D. G. W. Chan, J. Boyd, and W. A. Black.** 1986. In vitro activity of difloxacin hydrochloride (A-56619), A-56620, and cefixime (CL284,635; FK027) against selected genital pathogens. *Antimicrob. Agents Chemother.* **30**:590–593.
24. **Bustamante, C. I., G. L. Drusano, R. C. Wharton, and J. C. Wade.** 1987. Synergism of the combinations of imipenem plus ciprofloxacin and imipenem plus amikacin against *Pseudomonas aeruginosa* and other bacterial pathogens. *Antimicrob. Agents Chemother.* **31**:632–634.
25. **Carlson, J. R., S. A. Thornton, H. L. DuPont, A. H. West, and J. J. Mathewson.** 1983. Comparative in vitro activities of ten antimicrobial agents against bacterial enteropathogens. *Antimicrob. Agents Chemother.* **24**:509–513.
26. **Chalkley, L. J., and H. J. Koornhof.** 1985. Antimicrobial activity of ciprofloxacin against *Pseudomonas aeruginosa, Escherichia coli,* and *Staphylococcus aureus* determined by the killing curve method: antibiotic comparisons and synergistic interactions. *Antimicrob. Agents Chemother.* **28**:331–342.
27. **Chan, C. H. T., A. Cowlishaw, A. Eley, G. Slater, and D. Greenwood.** 1985. Laboratory evaluation of enoxacin: comparison with norfloxacin and nalidixic acid. *J. Antimicrob. Chemother.* **15**:45–52.

28. **Chantot, J. F., and A. Bryskier.** 1985. Antibacterial activity of ofloxacin and other 4-quinolone derivatives: in vitro and in vivo comparison. *J. Antimicrob. Chemother.* **16:**475–484.
29. **Chartrand, S. A., R. K. Scribner, A. H. Weber, D. F. Welch, and M. I. Marks.** 1983. In vitro activity of CI-919 (AT-2266), an oral antipseudomonal compound. *Antimicrob. Agents Chemother.* **23:**658–663.
30. **Chin, N.-X., D. C. Brittain, and H. C. Neu.** 1986. In vitro activity of Ro 23-6240, a new fluorinated 4-quinolone. *Antimicrob. Agents Chemother.* **29:**675–680.
31. **Chin, N.-X., K. Jules, and H. C. Neu.** 1986. Synergy of ciprofloxacin and azlocillin in vitro and in a neutropenic mouse model of infection. *Eur. J. Clin. Microbiol.* **5:**23–28.
32. **Chin, N.-X., and H. C. Neu.** 1983. In vitro activity of enoxacin, a quinolone carboxylic acid, compared with those of norfloxacin, new β-lactams, aminoglycosides, and trimethoprim. *Antimicrob. Agents Chemother.* **24:**754–763.
33. **Chin, N.-X., and H. C. Neu.** 1984. Ciprofloxacin, a quinolone caraboxylic acid compound active against aerobic and anaerobic bacteria. *Antimicrob. Agents Chemother.* **25:**319–326.
34. **Chin, N.-X., and H. C. Neu.** 1987. Post-antibiotic suppressive effect of ciprofloxacin against gram-positive and gram-negative bacteria. *Am. J. Med.* **82**(Suppl. A)**:**58–62.
35. **Chow, A. W., N. Cheng, and K. H. Bartlett.** 1985. In vitro susceptibility of *Clostridium difficile* to new β-lactam and quinolone antibiotics. *Antimicrob. Agents Chemother.* **28:**842–844.
36. **Clarke, A. M., and S. J. V. Zemcov.** 1987. In vitro activity of the new 4-quinolone compound Ro 23-6240. *Eur. J. Clin. Microbiol.* **6:**161–164.
37. **Clarke, A. M., S. J. V. Zemcov, and M. E. Campbell.** 1985. In vitro activity of pefloxacin compared to enoxacin, norfloxacin, gentamicin and new β-lactams. *J. Antimicrob. Chemother.* **15:**39–44.
38. **Cohen, M. A., T. J. Griffin, P. A. Bien, C. L. Heifetz, and J. M. Domagala.** 1985. In vitro activity of CI-934, a quinolone carboxylic acid active against gram-positive and -negative bacteria. *Antimicrob. Agents Chemother.* **28:**766–772.
39. **Collins, C. H., and A. H. C. Uttley.** 1985. In vitro susceptibility of mycobacteria to ciprofloxacin. *J. Antimicrob. Chemother.* **16:**575–580.
40. **Cornaglia, G., R. Pompei, B. Dainelli, and G. Satta.** 1987. In vitro activity of ciprofloxacin against aerobic bacteria isolated in a Southern European hospital *Antimicrob. Agents Chemother.* **31:**1651–1655.
41. **Cornett, J. B., R. B. Wagner, R. A. Dobson, M. R. Wentland, and D. M. Bailey.** 1985. In vitro and in vivo antibacterial activities of the fluoroquinolone WIN 49375 (amifloxacin). *Antimicrob. Agents Chemother.* **27:**4–10.
42. **Corrado, M. L., C. E. Cherubin, and M. Shulman.** 1983. The comparative activity of norfloxacin with other antimicrobial agents against gram-positive and gram-negative bacteria. *J. Antimicrob. Chemother.* **11:**369–376.
43. **Davies, G. S. R., and J. Cohen.** 1985. In vitro study of the activity of ciprofloxacin alone and in combination against strains of *Pseudomonas aeruginosa* with multiple antibiotic resistance. *J. Antimicrob. Chemother.* **16:**713–717.
44. **Davies, S., P. D. Sparham, and R. C. Spencer.** 1987. Comparative in vitro activity of five fluoroquinolones against mycobacteria. *J. Antimicrob. Chemother.* **19:**605–609.
45. **Debbia, E., A. Pesce, and G. C. Schito.** 1987. In vitro activity of CI-934 against gram-positive aerobes and anaerobes. *J. Antimicrob. Chemother.* **19:**135–136.
46. **Delmee, M., and V. Avesani.** 1986. Comparative in vitro activity of seven quinolones against 100 clinical isolates of *Clostridium difficile*. *Antimicrob. Agents Chemother.* **29:**374–375.

47. **Desplaces, N., L. Gutmann, J. Carlet, J. Guibert, and J. F. Acar.** 1986. The new quinolones and their combinations with other agents for therapy of severe infections. *J. Antimicrob. Chemother.* **17**(Suppl. A):25–39.
48. **Domagala, J. M., L. D. Hanna, C. L. Heifetz, M. P. Hutt, T. F. Mich, J. P. Sanchez, and M. Solomon.** 1986. New structure-activity relationships of the quinolone antibacterials using the target enzyme. The development and application of a DNA gyrase assay. *J. Med. Chem.* **29**:394–404.
49. **Downs, J., V. T. Andriole, and J. L. Ryan.** 1982. In vitro activity of MK-0366 against clinical urinary pathogens including gentamicin-resistant *Pseudomonas aeruginosa*. *Antimicrob. Agents Chemother.* **21**:670–672.
50. **Drugeon, H. B., J. Caillon, M. E. Juvin, and J. L. Pirault.** 1987. Dynamics of ceftazidime-pefloxacin interaction shown by a new killing curve-chequer-board method. *J. Antimicrob. Chemother.* **19**:197–203.
51. **Duckworth, G. J., and J. D. Williams.** 1984. Frequency of appearance of resistant variants to norfloxacin and nalidixic acid. *J. Antimicrob. Chemother.* **13**(Suppl. B):33–38.
52. **Easmon, C., and L. Verity.** 1987. Effect of Ro23-6240 on sensitive and resistant intracellular mycobacteria. *Eur. J. Clin. Microbiol.* **6**:165–166.
53. **Easmon, C. S. F., J. P. Crane, and A. Blowers.** 1986. Effect of ciprofloxacin on intracellular organisms: in vitro and in vivo studies. *J. Antimicrob. Chemother.* **18**(Suppl. D):43–48.
54. **Easmon, C. S. F., N. Woodford, and C. A. Ison.** 1987. The activity of the 4 quinolone Ro 23-6240 and the cephalosporins Ro 15-8074 and Ro 19-5247 against penicillin sensitive and resistant gonococci. *J. Antimicrob. Chemother.* **19**:761–765.
55. **Eliopoulos, G. M.** 1986. In vitro activity of new quinolone antimicrobial agents, p. 219–221. In L. Leive (ed.), *Microbiology—1986.* American Society for Microbiology, Washington, D.C.
56. **Eliopoulos, G. M., A. Gardella, and R. C. Moellering, Jr.** 1986. In vitro activity of ciprofloxacin, a new carboxyquinoline antimicrobial agent. *Antimicrob. Agents Chemother.* **25**:331–335.
57. **Eliopoulos, G. M., A. E. Moellering, E. Reiszner, and R. C. Moellering, Jr.** 1985. In vitro activities of the quinolone antimicrobial agents A-56619 and A-56620. *Antimicrob. Agents Chemother.* **28**:514–520.
58. **Eliopoulos, G. M., E. Reiszner, G. M. Caputo, and R. C. Moellering, Jr.** 1986. In vitro activity of CI-934, a new quinolone antimicrobial, against gram-positive bacteria. *Diagn. Microbiol. Infect. Dis.* **5**:341–344.
59. **Eliopoulos, G. M., E. Reiszner, M. J. Ferraro, C. Wennersten, and R. C. Moellering, Jr.** 1988. Effect of growth medium on the activity of fluoroquinolones and other antimicrobials against *Legionella* species. *Rev. Infect. Dis.* **10**(Suppl. 1):S56.
60. **Fallon, R. J., and W. M. Brown.** 1987. Sensitivity of *Legionellaceae,* meningococci and mycoplasmas to the quinolones ciprofloxacin and enoxacin. *Chemotherapia* **6**:466–467.
61. **Farrag, N. N., J. W. A. Bendig, C. Talboys, and B. S. Azadian.** 1986. In vitro study of the activity of ciprofloxacin combined with amikacin or ceftazidime against *Pseudomonas aeruginosa*. *J. Antimicrob. Chemother.* **18**:770.
62. **Fass, R. J.** 1983. In vitro activity of ciprofloxacin (Bay 09867). *Antimicrob. Agents Chemother.* **24**:568–574.
63. **Fass, R. J., and V. L. Helsel.** 1987. In vitro antistaphylococcal activity of pefloxacin alone and in combination with other antistaphylococcal drugs. *Antimicrob. Agents Chemother.* **31**:1457–1460.
64. **Fenlon, C. H., and M. H. Cynamon.** 1986. Comparative in vitro activities of ciprofloxacin and other 4-quinolones against *Mycobacterium tuberculosis* and *Mycobacterium intracellulare*. *Antimicrob. Agents Chemother.* **29**:386–388.

65. **Fernandes, P. B., N. Shipkowitz, R. R. Bower, K. P. Jarvis, J. Weisz, and D. T. W. Chu.** 1986. In vitro and in vivo potency of five new fluoroquinolones against anaerobic bacteria. *J. Antimicrob. Chemother.* **18:**693–701.
66. **Fernandez-Guerrero, M., M. S. Rouse, N. K. Henry, J. E. Geraci, and W. R. Wilson.** 1987. In vitro and in vivo activity of ciprofloxacin against enterococci isolated from patients with infective endocarditis. *Antimicrob. Agents Chemother.* **31:**430–433.
67. **Fernandez-Roblas, R., S. Prieto, M. Santamaria, C. Ponte, and F. Soriano.** 1987. Activity of nine antimicrobial agents against *Corynebacterium* group D2 strains isolated from clinical specimens and skin. *Antimicrob. Agents Chemother.* **31:**821–822.
68. **Fliegelman, R. M., R. M. Petrak, L. J. Goodman, J. Segreti, G. M. Trenholme, and R. L. Kaplan.** 1985. Comparative in vitro activities of twelve antimicrobial agents against *Campylobacter* species. *Antimicrob. Agents Chemother.* **27:**429–430.
69. **Forward, K. R., G. K. M. Harding, G. J. Gray, B. A. Urias, and A. P. Ronald.** 1983. Comparative activities of norfloxacin and fifteen other antipseudomonal agents against gentamicin-susceptible and -resistant *Pseudomonas aeruginosa* strains. *Antimicrob. Agents Chemother.* **24:**602–604.
70. **Foster, J. K., J. R. Lentino, R. Strodtman, and C. DiVincenzo.** 1986. Comparison of in vitro activity of quinolone antibiotics and vancomycin against gentamicin- and methicillin-resistant *Staphylococcus aureus* by time-kill kinetic studies. *Antimicrob. Agents Chemother.* **30:**823–827.
71. **Freeman, S., J. Wormuth, and J. Cornett.** 1985. In vitro activity of amifloxacin against *Chlamydia trachomatis* and *Ureaplasma urealyticum*. *Eur. J. Clin. Microbiol.* **4:**515–516.
72. **Fridmodt-Moller, P. C., K. M.-E. Jensen, and P. O. Madsen.** 1983. Antibacterial activity of norfloxacin in the gastrointestinal tracts of rats. *Antimicrob. Agents Chemother.* **24:**560–563.
73. **Fuursted, K.** 1987. Post-antibiotic effect of ciprofloxacin on *Pseudomonas aeruginosa*. *Eur. J. Clin. Microbiol.* **6:**271–274.
74. **Garcia, I., G. P. Bodey, V. Fainstein, D. H. Ho, and B. LeBlanc.** 1984. In vitro activity of WIN 49375 compared with those of other antibiotics in isolates from cancer patients. *Antimicrob. Agents Chemother.* **26:**421–423.
75. **Gay, J. D., D. R. DeYoung, and G. D. Roberts.** 1984. In vitro activity of norfloxacin and ciprofloxacin against *Mycobacterium tuberculosis, M. avium* complex, *M. chelonei, M. fortuitum,* and *M. kansasii*. *Antimicrob. Agents Chemother.* **26:**94–96.
76. **Giamarellou, H., and G. Petrikkos.** 1987. Ciprofloxacin interactions with imipenem and amikacin against multiresistant *Pseudomonas aeruginosa*. *Antimicrob. Agents Chemother.* **31:**959–961.
77. **Goldstein, E. J. C., and D. M. Citron.** 1985. Comparative activity of the quinolones against anaerobic bacteria isolated at community hospitals. *Antimicrob. Agents Chemother.* **27:**657–659.
78. **Goldstein, E. J. C., D. M. Citron, A. E. Vagvolgyi, and M. E. Gombert.** 1986. Susceptibility of *Eikenella corrodens* to newer and older quinolones. *Antimicrob. Agents Chemother.* **30:**172–173.
79. **Gombert, M. E., and T. M. Aulicino.** 1984. Susceptibility of multiply antibiotic-resistant pneumococci to the new quinoline antibiotics, nalidixic acid, coumermycin, and novobiocin. *Antimicrob. Agents Chemother.* **26:**933–934.
80. **Gombert, M. E., and T. M. Aulicino.** 1985. Comparison of agar dilution, microtitre broth dilution and tube macrodilution susceptibility testing of ciprofloxacin against several pathogens at two different inocula. *J. Antimicrob. Chemother.* **16:**709–712.
81. **Gombert, M. E., T. M. Aulicino, L. duBouchet, and L. R. Berkowitz.** 1987. Susceptibility of *Nocardia asteroides* to new quinolones and β-lactams. *Antimicrob. Agents*

Chemother. **31:**2013–2014.
82. **Goodman, L. J., R. M. Fliegelman, G. M. Trenholme, and R. L. Kaplan.** 1984. Comparative in vitro activity of ciprofloxacin against *Campylobacter* spp. and other bacterial enteric pathogens. *Antimicrob. Agents Chemother.* **25:**504–506.
83. **Goossens, H., P. DeMol, H. Coignau, J. Levy, O. Grados, G. Ghysels, H. A. Innocent, and J.-P. Butzler.** 1985. Comparative in vitro activities of aztreonam, ciprofloxacin, norfloxacin, ofloxacin, HR 810 (a new cephalosporin), RU28965 (a new macrolide), and other agents against enteropathogens. *Antimicrob. Agents Chemother.* **27:**388–392.
84. **Gordon, R. C., L. I. Stevens, C. E. Edmiston, Jr., and K. Mohan.** 1976. Comparative in vitro studies of cinoxacin, nalidixic acid, and oxolinic acid. *Antimicrob. Agents Chemother.* **10:**918–920.
85. **Greenwood, D., M. Osman, J. Goodwin, W. A. Cowlishaw, and R. Slack.** 1984. Norfloxacin activity against urinary tract pathogens and factors influencing the emergence of resistance. *J. Antimicrob. Chemother.* **13:**315–323.
86. **Guay, D., R. Klicker, T. Pence, and P. Peterson.** 1986. In vitro antistaphylococcal activity of teichoplanin and ciprofloxacin in peritoneal dialysis effluent. *Eur. J. Clin. Microbiol.* **5:**661–663.
87. **Guimaraes, M. A., and P. Noone.** 1986. The comparative in vitro activity of norfloxacin, ciprofloxacin, enoxacin, and nalidixic acid against 423 strains of gram-negative rods and staphylococci isolated from infected hospitalized patients. *J. Antimicrob. Chemother.* **17:**63–67.
88. **Haller, I.** 1985. Comprehensive evaluation of ciprofloxacin-aminoglycoside combinations against *Enterobacteriaceae* and *Pseudomonas aeruginosa* strains. *Antimicrob. Agents Chemother.* **28:**663–666.
89. **Hardy, D. J., R. N. Swanson, D. M. Hensey, N. R. Ramer, R. R. Bower, C. W. Hanson, D. T. W. Chu, and P. B. Fernandes.** 1987. Comparative antibacterial activities of temafloxacin hydrochloride (A-62254) and two reference fluoroquinolones. *Antimicrob. Agents Chemother.* **31:**1768–1774.
90. **Havlicheck, D., L. Saravolatz, and D. Pohlod.** 1987. Effect of quinolones and other antimicrobial agents on cell-associated *Legionella pneumophila*. *Antimicrob. Agents Chemother.* **31:**1529–1534.
91. **Hawkey, P. M., and C. A. Hawkey.** 1984. Comparative in vitro activity of quinolone carboxylic acids against *Proteeae*. *J. Antimicrob. Chemother.* **14:**485–489.
92. **Heesson, F. W. A., and H. L. Muytjens.** 1984. In vitro activities of ciprofloxacin, norfloxacin, pipemidic acid, cinoxacin, and nalidixic acid against *Chlamydia trachomatis*. *Antimicrob. Agents Chemother.* **25:**123–124.
93. **Henry, D., A. G. Skidmore, J. Ngui-Yen, A. Smith, and J. A. Smith.** 1985. In vitro activities of enoxacin, ticarcillin plus clavulanic acid, aztreonam, piperacillin, and imipenem and comparison with commonly used antimicrobial agents. *Antimicrob. Agents Chemother.* **28:**259–264.
94. **Heppleston, C., S. Richmond, and J. Bailey.** 1985. Antichlamydial activity of quinolone carboxylic acids. *J. Antimicrob. Chemother.* **15:**645–647.
95. **Hirai, K., H. Aoyama, M. Hosaka, Y. Oomori, Y. Niwata, S. Suzue, and T. Irikura.** 1986. In vitro and in vivo antibacterial activity of AM-833, a new quinolone derivative. *Antimicrob. Agents Chemother.* **29:**1059–1066.
96. **Hirai, K., A. Ito, Y. Abe, S. Suzue, T. Irikura, M. Inoue, and S. Mitsuhashi.** 1981. Comparative activities of AM-715 and pipemidic and nalidixic acids against experimentally induced systemic and urinary tract infections. *Antimicrob. Agents Chemother.* **19:**188–189.
97. **Hirose, T., E. Okezaki, H. Kato, Y. Ito, M. Inoue, and S. Mitsuhashi.** 1987. In vitro and

in vivo activity of NY-198, a new difluorinated quinolone. *Antimicrob. Agents Chemother.* **31**:854–859.

98. **Hirschhorn, L., and H. C. Neu.** 1986. Factors influencing the in vitro activity of two new aryl-fluoroquinolone antimicrobial agents, difloxacin (A-56619) and A-56620. *Antimicrob. Agents Chemother.* **30**:143–146.
99. **Hohl, P., M. Salfinger, and F. M. Kafader.** 1987. In vitro activity of the new quinolone Ro23-6240 (AM-833) and the new cephalosporins Ro15-8074 and Ro19-5247 (T-2525) against *Mycobacterium fortuitum* and *Mycobacterium chelonae. Eur. J. Clin. Microbiol.* **6**:487–488.
100. **Hohl, P., A. VonGraevenitz, and J. Zollinger-Iten.** 1987. Fleroxacin (Ro23-6240): activity in vitro against 355 enteropathogenic and non-fermentative gram-negative bacilli and *Legionella pneumophila. J. Antimicrob. Chemother.* **20**:373–378.
101. **How, S. J., D. Hobson, C. A. Hart, and E. Quayle.** 1985. A comparison of the in vitro activity of antimicrobials against *Chlamydia trachomatis* examined by Giemsa and fluorescent antibody stain. *J. Antimicrob. Chemother.* **15**:399–404.
102. **Husson, M. O., D. Izard, L. Bouillet, and H. Leclerc.** 1985. Comparative in vitro activity of ciprofloxacin against non-fermenters. *J. Antimicrob. Chemother.* **15**:457–462.
103. **Iravani, A., G. S. Welty, B. R. Newton, and G. A. Richard.** 1985. Effects of changes in pH, medium, and inoculum size on the in vitro activity of amifloxacin against urinary isolates of *Staphylococcus saprophyticus* and *Escherichia coli. Antimicrob. Agents Chemother.* **27**:449–451.
104. **Ito, A., K. Hirai, M. Inoue, H. Koga, S. Suzue, T. Irikura, and S. Mitsuhashi.** 1980. In vitro antibacterial activity of AM-715, a new nalidixic acid analog. *Antimicrob. Agents Chemother.* **17**:103–108.
105. **Jacobus, N. V., F. P. Tally, and M. Barza.** 1984. Antimicrobial spectrum of WIN 49375. *Antimicrob. Agents Chemother.* **26**:104–107.
106. **John, J. F., Jr., and T. A. Twitty.** 1984. Amifloxacin activity against well-defined gentamicin-resistant, gram-negative bacteria. *Antimicrob. Agents Chemother.* **26**:781–784.
107. **Joly-Guillou, M. L., and E. Bergogne-Berezin.** 1986. In vitro activity of antimicrobial agents against *Acinetobacter calcoaceticus. Drugs Exp. Clin. Res.* **12**:949–952.
108. **Jones, B. M., I. Geary, M. E. Lee, and B. I. Duerden.** 1986. Activity of pefloxacin and thirteen other antimicrobial agents in vitro against isolates from hospital and genitourinary infections. *J. Antimicrob. Chemother.* **17**:739–746.
109. **Jones, R. N., and P. C. Fuchs.** 1976. In vitro antimicrobial activity of cinoxacin against 2,968 clinical bacterial isolates. *Antimicrob. Agents Chemother.* **10**:146–149.
110. **Kaukoranta-Tolvanen, S. S. E., and O. V. J. Renkonen.** 1987. In vitro susceptibility of *Neisseria gonorrhoeae* to Ro23-6240 and ciprofloxacin. *Eur. J. Clin. Microbiol.* **6**:315–317.
111. **Kayser, F. H., and J. Novak.** 1987. In vitro activity of ciprofloxacin against gram-positive bacteria. *Am. J. Med.* **82**(Suppl. 4A):33–39.
112. **Kelley, S. G., M. A. Bertram, and L. S. Young.** 1986. Activity of ciprofloxacin against resistant clinical isolates. *J. Antimicrob. Chemother.* **17**:281–286.
113. **Kent, A. F., and D. C. E. Speller.** 1985. Sensitivity of urinary staphylococci to ciprofloxacin and acrosoxacin. *Eur. J. Clin. Microbiol.* **4**:139–140.
114. **Khan, M. Y., R. P. Gruninger, S. M. Nelson, and R. E. Klicker.** 1982. Comparative in vitro activity of norfloxacin (MK-0366) and ten other oral antimicrobial agents against urinary bacterial isolates. *Antimicrob. Agents Chemother.* **21**:848–851.
115. **King, A., and I. Phillips.** 1986. The comparative in vitro activity of pefloxacin. *J. Antimicrob. Chemother.* **17**(Suppl. B):1–10.

116. **King, A., and I. Phillips.** 1986. The comparative in vitro activity of eight newer quinolones and nalidixic acid. *J. Antimicrob. Chemother.* **18**(Suppl. D):1–20.
117. **King, A., K. Shannon, and I. Phillips.** 1984. The in vitro activity of ciprofloxacin compared with that of norfloxacin and nalidixic acid. *J. Antimicrob. Chemother.* **13:**325–331.
118. **King, A., K. Shannon, and I. Phillips.** 1985. The in vitro activities of enoxacin and ofloxacin compared with that of ciprofloxacin. *J. Antimicrob. Chemother.* **15:**551–558.
119. **King, A., K. Shannon, J. Slegg, and I. Phillips.** 1986. The in vitro activity of CI-934 compared with that of other new 4-quinolones and nalidixic acid. *J. Antimicrob. Chemother.* **18:**163–169.
120. **King, A., C. Warren, K. Shannon, and I. Phillips.** 1982. In vitro antibacterial activity of norfloxacin (MK-0366). *Antimicrob. Agents Chemother.* **21:**604–607.
121. **Klinger, J. D., and S. C. Aronoff.** 1985. In vitro activity of ciprofloxacin and other antibacterial agents against *Pseudomonas aeruginosa* and *Pseudomonas cepacia* from cystic fibrosis patients. *J. Antimicrob. Chemother.* **15:**679–684.
122. **Kouno, K., M. Inoue, and S. Mitsuhashi.** 1983. In vitro and in vivo antibacterial activity of AT-2266. *Antimicrob. Agents Chemother.* **24:**78–84.
123. **Kumada, T., and H. C. Neu.** 1985. In vitro activity of ofloxacin, a quinolone carboxylic acid compared to other quinolones and other antimicrobial agents. *J. Antimicrob. Chemother.* **16:**563–574.
124. **Lacey, R. W., V. L. Lord, and G. L. Howson.** 1984. Bactericidal effects of norfloxacin towards bacteria in urine. *J. Antimicrob. Chemother.* **13**(Suppl. B):49–54.
125. **Lagast, H., M. Husson, and J. Klastersky.** 1985. Bactericidal activity of ciprofloxacin in serum and urine against *Escherichia coli, Pseudomonas aeruginosa, Klebsiella pneumoniae, Staphylococcus aureus* and *Streptococcus faecalis. J. Antimicrob. Chemother.* **16:**341–347.
126. **Lambert, T., F. Megraud, G. Gerbaud, and P. Courvalin.** 1986. Susceptibility of *Campylobacter pyloridis* to 20 antimicrobial agents. *Antimicrob. Agents Chemother.* **30:**510–511.
127. **Lariviera, L. A., C. L. Gaudreau, and F. F. Turgeon.** 1986. Susceptibility of clinical isolates of *Campylobacter jejuni* to twenty-five antimicrobial agents. *J. Antimicrob. Chemother.* **18:**681–685.
128. **Le Saux, N. M., L. A. Slaney, F. A. Plummer, A. R. Ronald, and R. C. Brunham.** 1987. In vitro activity of ceftriaxone, cefetamet (Ro 15-8074), ceftetrame (Ro 19-5247; T-2588), and fleroxacin (Ro 23-6240; AM-833) versus *Neisseria gonorrhoeae* and *Haemophilus ducreyi. Antimicrob. Agents Chemother.* **31:**1153–1154.
129. **Lewis, A. M., and B. Chattopadhyay.** 1987. Comparative in vitro activity of thirteen antiobiotics against faecal isolates of *Yersinia* spp. *J. Antimicrob. Chemother.* **19:**406–407.
130. **Liebowitz, L. D., J. Saunders, G. Fehler, R. C. Ballard, and H. J. Koornhof.** 1986. In vitro activity of A-56619 (difloxacin), A-56620, and other new quinolone antimicrobial agents against genital pathogens. *Antimicrob. Agents Chemother.* **30:**948–950.
131. **Ligtvoet, E. E. J., and T. Wickerhoff-Minoggio.** 1985. In vitro activity of pefloxacin compared with six other quinolones. *J. Antimicrob. Chemother.* **16:**485–490.
132. **Limb, D. I., D. J. W. Dabbs, and R. C. Spencer.** 1987. In vitro selection of bacteria resistant to the 4-quinolone agents. *J. Antimicrob. Chemother.* **19:**65–71.
133. **Machka, K., and I. Braveny.** 1987. Comparative in vitro activity of Ro 23-6240 (fleroxacin), a new 4-quinolone derivative. *Eur. J. Clin. Microbiol.* **6:**482–485.
134. **Mandell, W., and H. C. Neu.** 1986. In vitro activity of CI-934, a new quinolone, compared with that of other quinolones and other antimicrobial agents. *Antimicrob. Agents Chemother.* **29:**852–857.
135. **Manek, N., J. M. Andrews, and R. Wise.** 1986. In vitro activity of Ro 23-6240, a new

difluoroquinolone derivative, compared with that of other antimicrobial agents. *Antimicrob. Agents Chemother.* **30:**330–332.
136. **McCormick, E. M., and R. M. Echols.** 1987. Effect of peritoneal dialysis fluid and pH on bactericidal activity of ciprofloxacin. *Antimicrob. Agents Chemother.* **31:**657–659.
137. **McNulty, C. A. M., J. Dent, and R. Wise.** 1985. Susceptibility of clinical isolates of *Campylobacter pyloridis* to 11 antimicrobial agents. *Antimicrob. Agents Chemother.* **28:**837–838.
138. **Milatovic, D.** 1986. Intraphagocytic activity of ciprofloxacin and CI-934. *Eur. J. Clin. Microbiol.* **5:**659–660.
139. **Milatovic, D., and I. Braveny.** 1987. Development of resistance during antibiotic therapy. *Eur. J. Clin. Microbiol.* **6:**234–244.
140. **Moody, J. A., D. N. Gerding, and L. R. Peterson.** 1987. Evaluation of ciprofloxacin's synergism with other agents by multiple in vitro methods. *Am. J. Med.* **82**(Suppl. 4A)**:**44–54.
141. **Moody, J. A., L. R. Peterson, and D. N. Gerding.** 1985. In vitro activity of ciprofloxacin combined with azlocillin. *Antimicrob. Agents Chemother.* **28:**849–850.
142. **Morris, J. G., Jr., J. H. Tenney, and G. L. Drusano.** 1985. In vitro susceptibility of pathogenic *Vibrio* species to norfloxacin and six other antimicrobial agents. *Antimicrob. Agents Chemother.* **28:**442–445.
143. **Muytjens, H. L., J. van Der Ros-van de Repe, and G. van Veldhuizen.** 1983. Comparative activities of ciprofloxacin (Bay o 9867), norfloxacin, pipemidic acid, and nalidixic acid. *Antimicrob. Agents Chemother.* **24:**302–304.
144. **Nakamura, S., A. Minami, H. Katae, S. Inoue, J. Yamagishi, Y. Takase, and M. Shimizu.** 1983. In vitro antibacterial properties of AT-2266, a new pyridonecarboxylic acid. *Antimicrob. Agents Chemother.* **23:**641–648.
145. **Nakamura, S., K. Nakata, H. Katae, A. Minami, S. Kashimoto, J. Yamagishi, Y. Takase, and M. Shimizu.** 1983. Activity of AT-2266 compared with those of norfloxacin, pipemidic acid, nalidixic acid, and gentamicin against various experimental infections in mice. *Antimicrob. Agents Chemother.* **23:**742–749.
146. **Neu, H. C.** 1985. The in vitro activity of EN 272, a quinolone-7-carboxylic acid, in comparison with other quinolones. *J. Antimicrob. Chemother.* **16:**43–48.
147. **Neu, H. C., and N.-X. Chin.** 1987. In vitro activity of two new quinolone antimicrobial agents, S-25930 and S-25932 compared with that of other agents. *J. Antimicrob. Chemother.* **19:**175–185.
148. **Neu, H. C., and P. Labthavikul.** 1982. In vitro activity of norfloxacin, a quinolinecarboxylic acid, compared with that of β-lactams, aminoglycosides, and trimethoprim. *Antimicrob. Agents Chemother.* **22:**23–27.
149. **Newsom, S. W. B.** 1984. The antimicrobial spectrum of norfloxacin. *J. Antimicrob. Chemother.* **13**(Suppl. B)**:**25–31.
150. **Newsom, S. W. B., J. Matthews, M. Amphlett, and R. W. Warren.** 1982. Norfloxacin and the antibacterial γ pyridone β-carboxylic acids. *J. Antimicrob. Chemother.* **10:**25–30.
151. **Newsom, S. W. B., J. Matthews, and A. M. Rampling.** 1985. Susceptibility of *Clostridium difficile* strains to new antibiotics: quinolones, efrotomycin, teicoplanin and imipenem. *J. Antimicrob. Chemother.* **15:**648–649.
152. **Norrby, S. R., and M. Jonsson.** 1983. Antibacterial activity of norfloxacin. *Antimicrob. Agents Chemother.* **23:**15–18.
153. **Osada, Y., and H. Ogawa.** 1983. Antimycoplasmal activity of ofloxacin (DL-8280). *Antimicrob. Agents Chemother.* **23:**509–511.
154. **Paton, J. H., and E. W. Williams.** 1987. Interaction between ciprofloxacin and vancomycin against staphylococci. *J. Antimicrob. Chemother.* **20:**251–254.
155. **Peeters, M., E. VanDyck, and P. Piot.** 1984. In vitro activities of the spectinomycin ana-

log U-63366 and four quinolone derivatives against *Neisseria gonorrhoeae. Antimicrob. Agents Chemother.* **26:**608–609.
156. **Piddock, L. J. V., J. M. Andrews, J. M. Diver, and R. Wise.** 1986. In vitro studies of S-25930 and S-25932, two new 4-quinolones. *Eur. J. Clin. Microbiol.* **5:**303–310.
157. **Piddock, L. J. V., J. M. Diver, and R. Wise.** 1986. Cross-resistance of nalidixic acid resistant *Enterobacteriaceae* to new quinolones and other antimicrobials. *Eur. J. Clin. Microbiol.* **5:**411–415.
158. **Pohlod, D. J., and L. D. Saravolatz.** 1984. In vitro susceptibilities of 393 recent clinical isolates to WIN 49375, cefotaxime, tobramycin, and piperacillin. *Antimicrob. Agents Chemother.* **25:**377–379.
159. **Pohlod, D. J., and L. D. Saravolatz.** 1986. Activity of quinolones against Legionellaceae. *J. Antimicrob. Chemother.* **17:**540–541.
160. **Prabhala, R. H., B. Rao, R. Marshall, M. B. Bansal, and H. Thadepalli.** 1984. In vitro susceptibility of anaerobic bacteria to ciprofloxacin (Bay o 9867). *Antimicrob. Agents Chemother.* **26:**785–786.
161. **Raoult, D., P. Rousselier, V. Galicher, R. Perez, and J. Tamalet.** 1986. In vitro susceptibility of *Rickettsia conorii* to ciprofloxacin as determined by suppressing lethality in chicken embryos and by plaque assay. *Antimicrob. Agents Chemother.* **29:**424–425.
162. **Raoult, D., P. Rousselier, G. Vestris, V. Galicher, R. Perez, and J. Tamalet.** 1987. Susceptibility of *Rickettsia conorii* and *R. rickettsii* to pefloxacin, in vitro and in ovo. *J. Antimicrob. Chemother.* **19:**303–305.
163. **Reeves, D. S., M. J. Bywater, H. A. Holt, and L. O. White.** 1984. In vitro studies with ciprofloxacin, a new 4-quinolone compound. *J. Antimicrob. Chemother.* **13:**333–346.
164. **Ridgway, G. L., G. Mumtaz, F. G. Gabriel, and J. D. Oriel.** 1984. The activity of ciprofloxacin and other 4-quinolones against *Chlamydia trachomatis* and *Mycoplasma* in vitro. *Eur. J. Clin. Microbiol.* **3:**344–346.
165. **Rohlfing, S. R., J. E. Landmesser, J. F. Gerster, S. E. Pecore, and R. M. Stern.** 1985. Differentiation of fluorinated quinolone antibacterials with *Neisseria gonorrhoeae* isolates. *J. Antimicrob. Chemother.* **15:**539–544.
166. **Rolston, K. V. I., M. E. Alvarez, K.-C. Hsu, and G. P. Bodey.** 1986. In vitro activity of cefpirome (HR-810), WIN-49375, BMY-28142 and other antibiotics against nosocomially important isolates from cancer patients. *J. Antimicrob. Chemother.* **17:**453–457.
167. **Rolston, K. V. I., and G. P. Bodey.** 1986. Activity of CI-934 against gram-positive bacteria. *J. Antimicrob. Chemother.* **18:**768–769.
168. **Rolston, K. V. I., and G. P. Bodey.** 1986. In vitro susceptibility of *Acinetobacter* species to various antimicrobial agents. *Antimicrob. Agents Chemother.* **30:**769–770.
169. **Rolston, K. V. I., and G. P. Bodey.** 1987. In vitro activity of various new antimicrobial agents against Group JK Corynebacteria. *Eur. J. Clin. Microbiol.* **6:**491–492.
170. **Rolston, K. V. I., D. H. Ho, B. LeBlanc, and G. P. Bodey.** 1987. In vitro evaluation of S-25930 and S-25932, two new quinolones, against aerobic gram-negative isolates from cancer patients. *Antimicrob. Agents Chemother.* **31:**102–103.
171. **Rudin, J. F., C. W. Norden, and E. M. Shinners.** 1984. In vitro activity of ciprofloxacin against aerobic gram-negative bacteria. *Antimicrob. Agents Chemother.* **26:**597–598.
172. **Rudnik, J. T., S. J. Cavalieri, and E. M. Britt.** 1984. In vitro activities of enoxacin and 17 other antimicrobial agents against multiply resistant, gram-negative bacteria. *Antimicrob. Agents Chemother.* **26:**97–100.
173. **Saito, A., H. Koga, H. Shigeno, K. Watanabe, K. Mori, S. Kohno, Y. Shigeno, Y. Suzuyama, K. Yamaguchi, M. Hirota, and K. Hara.** 1986. The antimicrobial activity of ciprofloxacin against *Legionella* species and the treatment of experimental *Legionella*

pneumonia in guinea pigs. *J. Antimicrob. Chemother.* **18:**251–260.

174. **Saito, A., K. Sawatari, Y. Fukuda, M. Nagasawa, H. Koga, A. Tomonaga, H. Nakazato, K. Fujita, Y. Shigeno, Y. Suzuyama, K. Yamaguchi, K. Izumikawa, and K. Hara.** 1985. Susceptibility of *Legionella pneumophila* to ofloxacin in vitro and in experimental *Legionella* pneumonia in guinea pigs. *Antimicrob. Agents Chemother.* **28:**15–20.
175. **Sanders, C. C., W. E. Sanders, Jr., R. V. Goering, and V. Werner.** 1984. Selection of multiple antibiotic resistance by quinolones, β-lactams, and aminoglycosides with special reference to cross-resistance between unrelated drug classes. *Antimicrob. Agents Chemother.* **26:**797–801.
176. **Sanson-Le Pors, M. J., I. M. Casin, M.-C. Thebault, G. Arlet, and Y. Perol.** 1986. In vitro activities of U-63366, a spectinomycin analog; roxithromycin (RU 28965), a new macrolide antibiotic; and five quinolone derivatives against *Haemophilus ducreyi*. *Antimicrob. Agents Chemother.* **30:**512–513.
177. **Sato, K., Y. Matsuura, M. Inoue, T. Une, Y. Osada, M. Ogawa, and S. Mitsuhashi.** 1982. In vitro and in vivo activity of DL-8280, a new oxazine derivative. *Antimicrob. Agents Chemother.* **22:**548–553.
178. **Scribner, R. K., D. F. Welch, and M. I. Marks.** 1985. Low frequency of bacterial resistance to enoxacin in vitro and in experimental pneumonia. *J. Antimicrob. Chemother.* **16:**597–603.
179. **Seth, A. D.** 1981. Sensitivity of gonococci to rosoxacin compared with that of penicillin, cefuroxime and tetracycline. *J. Antimicrob. Chemother.* **7:**331–334.
180. **Shimizu, M., Y. Takase, S. Nakamura, H. Katae, A. Minami, K. Nakata, S. Inoue, M. Ishiyama, and Y. Kubo.** 1975. Pipemidic acid, a new antibacterial agent active against *Pseudomonas aeruginosa*: in vitro properties. *Antimicrob. Agents Chemother.* **8:**132–138.
181. **Shungu, D. L., D. R. Nalin, R. H. Gilman, H. H. Gadebusch, A. T. Cerami, C. Gill, and B. Weissberger.** 1987. Comparative susceptibilities of *Campylobacter pylori* to norfloxacin and other agents. *Antimicrob. Agents Chemother.* **31:**949–950.
182. **Shungu, D. L., E. Weinberg, and H. H. Gadebusch.** 1983. In vitro antibacterial activity of norfloxacin (MK-0366, AM-715) and other agents against gastrointestinal tract pathogens. *Antimicrob. Agents Chemother.* **23:**86–90.
183. **Simberkoff, M. S., and J. J. Rahal, Jr.** 1986. Bactericidal activity of ciprofloxacin against amikacin- and cefotaxime-resistant gram-negative bacilli and methicillin-resistant staphylococci. *Antimicrob. Agents Chemother.* **29:**1098–1100.
184. **Smith, B. R., J. L. LeFrock, J. B. Donato, W. S. Joseph, and S. J. Weber.** 1986. In vitro activity of A-56619 and A-56620, two new aryl-fluoroquinolone antimicrobial agents. *Antimicrob. Agents Chemother.* **29:**355–358.
185. **Smith, S. M.** 1986. In vitro comparison of A-56619, A-56620, amifloxacin, ciprofloxacin, enoxacin, norfloxacin, and ofloxacin against methicillin-resistant *Staphylococcus aureus*. *Antimicrob. Agents Chemother.* **29:**325–326.
186. **Smith, S. M., and R. H. K. Eng.** 1985. Activity of ciprofloxacin against methicillin-resistant *Stephylococcus aureus*. *Antimicrob. Agents Chemother.* **27:**688–691.
187. **Smith, S. M., R. H. K. Eng, and E. Berman.** 1986. The effect of ciprofloxacin on methicillin-resistant *Staphylococcus aureus*. *J. Antimicrob. Chemother.* **17:**287–295.
188. **Stamm, J. M., C. W. Hanson, D. T. W. Chu, R. Bailer, C. Vojtko, and P. B. Fernandes.** 1986. In vitro evaluation of A-56619 (difloxacin) and A-56620: new aryl-fluoroquinolones. *Antimicrob. Agents Chemother.* **29:**193–200.
189. **Stamm, W. E., and R. Suchland.** 1986. Antimicrobial activity of U-70138F (paldimycin), roxithromycin (RU 965), and ofloxacin (ORF 18489) against *Chlamydia trachomatis* in cell culture. *Antimicrob. Agents Chemother.* **30:**806–807.
190. **Stiver, H. G., K. H. Bartlett, and A. W. Chow.** 1986. Comparison of susceptibility of

gentamicin-resistant and -susceptible *"Acinetobacter anitratus"* to 15 alternative antibiotics. *Antimicrob. Agents Chemother.* **30:**624–625.

191. **Stratton, C. W., C. Liu, H. B. Ratner, and L. S. Weeks.** 1987. Bactericidal activity of deptomycin (LY146032) compared with those of ciprofloxacin, vancomycin, and ampicillin against enterococci as determined by kill-kinetic studies. *Antimicrob. Agents Chemother.* **31:**1014–1016.
192. **Stratton, C. W., C. Liu, and L. S. Weeks.** 1987. Activity of LY146032 compared with that of methicillin, cefazolin, cefamandole, cefuroxime, ciprofloxacin, and vancomycin against staphylococci as determined by kill-kinetic studies. *Antimicrob. Agents Chemother.* **31:**1210–1215.
193. **Sturm, A. W.** 1987. Comparison of antimicrobial susceptibility patterns of fifty-seven strains of *Haemophilus ducreyi* isolated in Amsterdam from 1978 to 1985. *J. Antimicrob. Chemother.* **19:**187–191.
194. **Sutter, V. L., Y. Y. Kwok, and J. Bulkacz.** 1985. Comparative activity of ciprofloxacin against anaerobic bacteria. *Antimicrob. Agents Chemother.* **27:**427–428.
195. **Taylor, D. E., L.-K. Ng, and H. Lior.** 1985. Susceptibility of *Campylobacter* species to nalidixic acid, enoxacin, and other DNA gyrase inhibitors. *Antimicrob. Agents Chemother.* **28:**708–710.
196. **Tenney, J. H., R. W. Maack, and G. R. Chippendale.** 1983. Rapid selection of organisms with increasing resistance on subinhibitory concentrations of norfloxacin in agar. *Antimicrob. Agents Chemother.* **23:**188–189.
197. **Tenney, J. H., and J. W. Warren.** 1983. Bactericidal activity of norfloxacin and nine other urinary tract antibiotics against gram-negative bacilli causing bacteriuria in chronically catheterized patients. *J. Antimicrob. Chemother.* **11:**287–290.
198. **Thabaut, A., and J. L. Durosoir.** 1983. Comparative in vitro antibacterial activity of pefloxacin (1589 RB), nalidixic acid, pipemidic acid and flumaquin. *Drugs Exp. Clin. Res.* **9:**229–234.
199. **Thompson, K. D., J. P. O'Keefe, and W. A. Tatarowicz.** 1984. In vitro comparison of amifloxacin and six other antibiotics against aminoglycoside-resistant *Pseudomonas aeruginosa. Antimicrob. Agents Chemother.* **26:**275–276.
200. **Tjiam, K. H., J. H. T. Wagenvoort, B. van Klingeren, P. Piot, E. Stolz, and M. F. Michel.** 1986. In vitro activity of the two new 4-quinolones A-56619 and A-56620 against *Neisseria gonorrhoeae, Chlamydia trachomatis, Mycoplasma hominis, Ureaplasma urealyticum* and *Gardnerella vaginalis. Eur. J. Clin. Microbiol.* **5:**498–501.
201. **Trautmann, M., B. Krause, D. Birnbaum, J. Wagner, and V. Lenk.** 1986. Serum bactericidal activity of two newer quinolones against *Salmonella typhi* compared with standard therapeutic regimens. *Eur. J. Clin. Microbiol.* **5:**297–302.
202. **Trimble, K. A., R. B. Clark, W. E. Sanders, Jr., J. W. Frankel, R. Cacciatore, and H. Valdez.** 1987. Activity of ciprofloxacin against *Mycobacteria* in vitro: comparison of BACTEC and macrobroth dilution methods. *J. Antimicrob. Chemother.* **19:**617–622.
203. **Van Caekenberghe, D. L., and S. R. Pattyn.** 1984. In vitro activity of ciprofloxacin compared with those of other new fluorinated piperazinyl-substituted quinoline derivatives. *Antimicrob. Agents Chemother.* **25:**518–521.
204. **Van der Auwera, P.** 1985. Interaction of gentamicin, dibekacin, netilmicin and amikacin with various penicillins, cephalosporins, minocycline and new fluoroquinolones against Enterobacteriaceae and *Pseudomonas aeruginosa. J. Antimicrob. Chemother.* **16:**581–587.
205. **Van der Auwera, P., P. Grenier, and J. Klastersky.** 1987. In vitro activity of LY 146032 against *Staphylococcus aureus, Listeria monocytogenes, Corynebacterium* JK, and *Bacillus* spp., in comparison with various antibiotics. *J. Antimicrob. Chemother.* **20:**209–212.
206. **Van der Auwera, P., and P. Joly.** 1987. Comparative in vitro activities of teicoplanin,

vancomycin, coumermycin and ciprofloxacin, alone and in combination with rifampicin or LM427, against *Staphylococcus aureus. J. Antimicrob. Chemother.* **19:**313–320.

207. **VanRoof, R., J. M. Hubrechts, E. Roebben, M. Claeys, and H. J. Nyssen.** 1986. Antibacterial activity of enoxacin: comparison with aminoglycosides, β-lactams, and other antimicrobial agents. *J. Antimicrob. Chemother.* **17:**297–302.
208. **Verbist, L.** 1986. In vitro activity of pefloxacin against microorganisms multiply resistant to β-lactam antibiotics and aminoglycosides. *J. Antimicrob. Chemother.* **17**(Suppl. B)**:**11–17.
209. **Verbist, L.** 1987. Comparative in vitro activity of Ro 23-6240, a new trifluorinated quinolone. *J. Antimicrob. Chemother.* **20:**363–372.
210. **Wall, R. A., D. C. W. Mabey, C. S. S. Bello, and D. Felmingham.** 1985. The comparative in vitro activity of twelve 4-quinolone antimicrobials against *Haemophilus ducreyi. J. Antimicrob. Chemother.* **16:**165–168.
211. **Watt, B., and F. V. Brown.** 1986. Is ciprofloxacin active against clinically important anaerobes? *J. Antimicrob. Chemother.* **17:**605–613.
212. **Whiting, J. L., N. Cheng, and A. W. Chow.** 1987. Interactions of ciprofloxacin with clindamycin, metronidazole, cefoxitin, cefotaxime, and mezlocillin against gram-positive and gram-negative anaerobic bacteria. *Antimicrob. Agents Chemother.* **31:**1379–1382.
213. **Willems, F. T. C., J. B. J. Boerema, and T. R. K. M. v Summeren.** 1986. The in vitro comparative activity of quinolones against bacteria from urinary tract infections in general practice. *J. Antimicrob. Chemother.* **17:**69–73.
214. **Wise, R., J. M. Andrews, and G. Danks.** 1984. In vitro activity of enoxacin (CI-919), a new quinoline derivative, compared with that of other antimicrobial agents. *J. Antimicrob. Chemother.* **13:**237–244.
215. **Wise, R., J. M. Andrews, and L. J. Edwards.** 1983. In vitro activity of Bay 09867, a new quinoline derivative, compared with those of other antimicrobial agents. *Antimicrob. Agents Chemother.* **23:**559–564.
216. **Wolfson, J. S., and D. C. Hooper.** 1985. The fluoroquinolones: structures, mechanisms of action and resistance, and spectra of activity in vitro. *Antimicrob. Agents Chemother.* **28:**581–586.
217. **Yeaman, M. R., L. A. Mitscher, and O. G. Baca.** 1987. In vitro susceptibility of *Coxiella burnetii* to antibiotics, including several quinolones. *Antimicrob. Agents Chemother.* **31:**1079–1084.
218. **Yourassowsky, E., M. P. Van der Linden, F. Crokaert, and Y. Glupczynski.** 1986. In vitro activity of pefloxacin compared to other antibiotics. *J. Antimicrob. Chemother.* **17**(Suppl. B)**:**19–28.
219. **Yourassowsky, E., M. P. Van der Linden, M. J. Lismont, F. Crokaert, and Y. Glupczynski.** 1986. Rate of bactericidal activity for *Streptococcus faecalis* of a new quinolone, CI-934, compared with that of amoxicillin. *Antimicrob. Agents Chemother.* **30:**258–259.
220. **Zackrisson, G., J. E. Brorson, B. Bjornegard, and B. Trollfors.** 1985. Susceptibility of *Bordetella pertussis* to doxycycline, cinoxacin, nalidixic acid, norfloxacin, imipenem, mecillinam and rifampicin. *J. Antimicrob. Chemother.* **15:**629–632.
221. **Zeiler, H. J.** 1985. Evaluation of the in vitro bactericidal action of ciprofloxacin on cells of *Escherichia coli* in the logarithmic and stationary phases of growth. *Antimicrob. Agents Chemother.* **28:**524–527.

Chapter 4

Pharmacokinetics of the Quinolone Antimicrobial Agents

George L. Drusano

Division of Infectious Diseases, University of Maryland School of Medicine, Baltimore, Maryland 21201

Introduction

The past decade has seen breakthroughs in medicinal chemistry resulting in a new series of compounds, the 6-fluoroquinolone carboxylic acids, which exhibit markedly improved activity against major hospital-acquired pathogens, including *Pseudomonas aeruginosa*. In this chapter the pharmacokinetic properties of the new fluoroquinolones are examined. Because the number of fluoroquinolones is now substantial and continuing to grow, a classification system would be useful. One such system used here is classification of these agents by their predominant clearance pathway: renal (ofloxacin), hepatic (pefloxacin and difloxacin), or both (norfloxacin, ciprofloxacin, enoxacin, and fleroxacin). Interactions of these agents with other drugs are discussed in chapter 14.

Fluoroquinolones Excreted Predominantly by the Kidneys

Ofloxacin

Ofloxacin (see Fig. 1 in chapter 2) is the prototype of a renally cleared fluoroquinolone and is available in both oral and intravenous dosing forms (71).

Verho et al. (58) evaluated the oral pharmacokinetics of ofloxacin in an open randomized crossover study for single doses of 100, 300, and 600 mg. Maximal concentrations in serum were 1 μg/ml for the 100-mg dose, 3.4 μg/ml for the 300-mg dose, and 6.9 μg/ml for the 600-mg dose (Fig. 1). Absorption was rapid, with maximal concentrations occurring at or before 1 h.

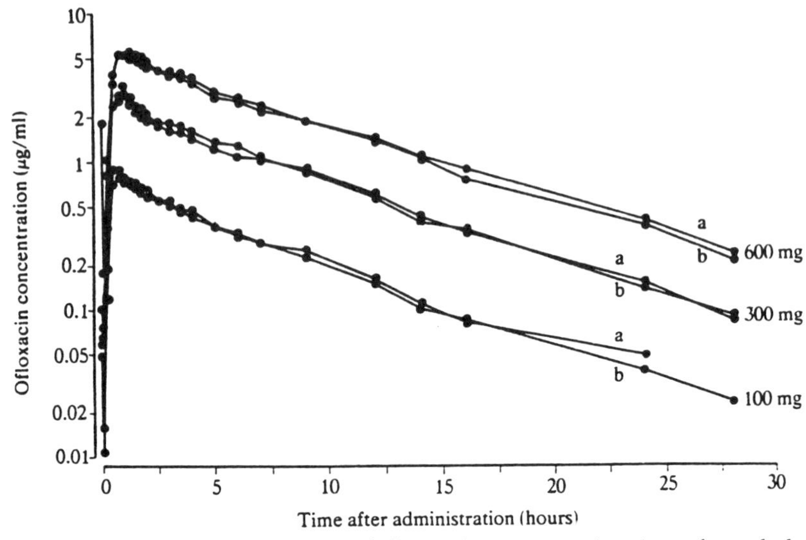

Figure 1. Median concentrations of ofloxacin in serum at various times after oral administration of different single doses. a, Microbiologic essay; b, high-pressure liquid chromatography assay. (From reference 58.)

Terminal elimination half-lives varied between 5.6 and 6.4 h. Urinary recovery varied between 75 and 81%. Lockley et al. (43) evaluated a 600-mg dose of ofloxacin and found similar results, except for higher maximal concentrations in serum. The area under the curve (AUC) of the concentration in serum plotted versus time, a measure of bioavailability, did not differ between the two studies. Lockley et al. (43) also studied the penetration of ofloxacin into inflammatory blister fluid. High peak concentrations of ofloxacin in blister fluid were obtained (5.2 ± 0.9 μg/ml), with maximal concentrations at 5.3 ± 0.8 h. The terminal half-life was longer in blister fluid (8.0 ± 1.7 h) than in serum, and penetration, as judged by the ratios of the AUC for blister fluid to the AUC for serum, was in excess of unity (125%). Similar results were obtained for ofloxacin (300-mg dose) penetration into noninflammatory blister fluid (36).

The fluoroquinolones will find substantial use in the outpatient setting. Consequently, it becomes important to examine the effect of concurrent food administration on drug pharmacokinetics. Leroy et al. (42) and Kalager et al. (36) have examined the effect of food on the absorption of ofloxacin and reported similar results. For a 200-mg dose (42) maximum concentrations were 0.83 ± 0.31 (mean ± standard deviation) h in fasting subjects and 1.85 ± 1.15 h in fed subjects. AUCs were, however, not statistically significantly different between groups. Half-lives in the study by Leroy et al. (42) were 7.86 ± 1.81 h in fasting subjects and 8.00 ± 1.71 h in nonfasting subjects. Urinary recovery was 74.8 ± 11.6% in fasting subjects and 76.2 ± 17.4% in fed subjects.

As renal elimination accounts for the majority of ofloxacin clearance, it is clinically important to detail the effect of altered renal function on drug disposition. Fillastre et al. (23) examined the influence of various degrees of renal impairment on the disposition of a single 200-mg oral dose of ofloxacin (Table 1). Renal impairment had a major impact on ofloxacin disposition. Although peak concentrations were relatively unaffected, AUCs increased from 13.8 ± 3.12 μg·h/ml for subjects with normal renal function to 86.05 ± 49.72 μg·h/ml for subjects with severely impaired renal function. Most of the change resulted from an altered terminal elimination half-life, which increased from 7.86 ± 1.81 to 37.16 ± 23.26 h. Renal clearance and urinary recovery declined as a function of a decreasing glomerular filtration rate (GFR).

Lode et al. (44) examined the pharmacokinetics of ofloxacin after both parenteral and oral administration in single doses of 25, 50, 100, and 200 mg (Table 2). AUCs for serum increased with the dose in a linear fashion over the range examined, paralleling the results in the oral dose-ranging trials. Volumes of distribution were large, ranging from 1.2 to 1.5 liters/kg. Systemic clearances of 0.20 to 0.24 liters/h per kg were seen, with renal clearances accounting for 75 to 80% of the total serum clearance. Protein binding averaged 25 ± 2.5%, with no dependence on the drug concentration noted within the range of 0.57 to 2.1 μg/ml.

Ofloxacin is less biotransformed than other new fluoroquinolones, with desmethylofloxacin and ofloxacin *N*-oxide being the two major metabolites formed (Table 3). There were no differences between oral and parenteral administration, indicating no first-pass hepatic metabolism after oral administration.

Because both oral and parenteral dosing forms of ofloxacin are available, Lode et al. (44) determined the absolute bioavailability of this agent. For a 200-mg dose, the ratio of the AUC after oral administration to the AUC after intravenous administration was 0.96 for a three-compartment analysis, indicating a virtually complete absorption of the drug when given orally. Despite the excellent absorption of ofloxacin, concentrations in stool are high, up to 327 ± 274 μg/g after administration of 200 mg orally twice daily for 4 days (51).

Fluoroquinolones Cleared Predominantly by the Liver

Pefloxacin

Pefloxacin is a fluoroquinolone which is primarily cleared by the liver. The structure of pefloxacin (see Fig. 1 in chapter 2) differs from that of

Table 1. Ofloxacin pharmacokinetic data for subjects with normal and impaired renal function after a single oral dose of 200 mg[a] (from reference 23)

Subjects[b]	C_{max} (μg/ml)	T_{max} (h)	AUC (μg·h/ml)	V/F (liters/kg)	$t_{1/2}$ (h)	ml/min per 1.73 m²	
						CL/F	CL_R
≥80	2.24 ± 0.90	0.83 ± 0.31	13.18 ± 3.12	2.53 ± 0.78	7.86 ± 1.81	241.4 ± 53.8	196.5 ± 42.9
≥40, ≤80	2.18 ± 0.53	1.75 ± 1.66	32.33 ± 4.18	2.14 ± 0.50	15.00 ± 4.40	109.4 ± 18.3	60.6 ± 9.3
≥20, <39	1.84 ± 0.32	2.36 ± 1.60	47.48 ± 14.00	2.15 ± 0.31	25.37 ± 8.56	70.8 ± 15.9	30.9 ± 7.7
<20	1.71 ± 0.62	1.60 ± 1.34	65.98 ± 15.00	2.02 ± 0.26	34.84 ± 15.46	50.6 ± 14.5	13.7 ± 6.2
Hemodialysis	1.97 ± 0.56	3.20 ± 2.59	86.05 ± 49.72	2.02 ± 0.37	37.16 ± 23.26	49.2 ± 21.6	

[a]Abbreviations: C_{max}, maximum concentration of drug in serum; T_{max}, time to maximum concentration of drug in serum; V/F, volume of distribution divided by bioavailability; $t_{1/2}$, half-life; CL/F, total serum clearance divided by bioavailability; CL_R, renal clearance.
[b]GFR in milliliters per minute per 1.73 m².

Table 2. Pharmacokinetic parameters for ofloxacin in 10 healthy volunteers (normalized to a mean body weight of 70 kg)[a] (from reference 44)

Dose (mg)	Route of adminis-tration	Mean ± SD							
		C_{max} (mg/liter)	T_{max} (min)	V_{area} (liters)	$t_{1/2\beta}$ (min)	AUC_{tot} (mg·h/liter)	CL_{tot} (ml/min)	CL_R (ml/min)	Urinary excretion (% of dose/24 h)
25	i.v.			93.9 ± 8.7	231 ± 34.5	1.5 ± 0.18	286 ± 32.6	245 ± 29.4	82.2 ± 4.5
50	i.v.			103.0 ± 31.8	260 ± 34.6	3.1 ± 0.58	276 ± 68.2	276 ± 7.0	81.1 ± 13.5
100	i.v.			89.8 ± 14.7	267 ± 36.8	7.3 ± 1.2	235 ± 37.0	185 ± 41.2	73.1 ± 10.1
200	i.v.			86.0 ± 10.4	256 ± 19.8	14.4 ± 1.8	234 ± 27.8	190 ± 27.3	77.0 ± 8.2
200	Oral	2.19 ± 0.43	76.8 ± 39.2	111.0 ± 26.0	334 ± 98.9	14.6 ± 2.7		197 ± 44.3	73.6 ± 7.3
400	Oral	3.51 ± 0.7	114.9 ± 38.7	102.0 ± 16.3	294 ± 45.2	28.0 ± 4.9		202 ± 35.0	73.3 ± 6.9

[a]Abbreviations: V_{area}, volume of distribution (per 70 kg of body weight); $t_{1/2\beta}$, half-life at β phase; AUC_{tot}, total AUC; CL_{tot}, total clearance; i.v., intravenous. For other abbreviations see Table 1, footnote *a*.

Table 3. Mean renal elimination of ofloxacin, desmethylofloxacin, and ofloxacin *N*-oxide after 100 mg orally and intravenously (i.v.) (from reference 44)

Dose (mg)	Route of administration	Mean excretion (% of dose/24 h) ± SD		
		Ofloxacin	Desmethyl-ofloxacin	Ofloxacin *N*-oxide
200	Oral	73.6 ± 7.3	3.0 ± 0.80	1.0 ± 0.20
200	i.v.	77.0 ± 8.2	3.2 ± 0.61	1.1 ± 0.18

norfloxacin only in the addition of a methyl group to position 4′ of the piperazinyl substituent at position 7. This minor difference in structure markedly alters the half-life and renal handling of pefloxacin, changing it from a compound secreted by the renal tubule to one for which there is net renal tubular reabsorption. Consequently, hepatic extraction and biotransformation dominate the clearance process.

Barre et al. (3) examined the pharmacokinetics of pefloxacin when administered both orally and intravenously in single doses to volunteers (Table 4). For oral administration (Fig. 2) peak concentrations approximated 1.5 μg/ml for the 200-mg dose, 3.2 μg/ml for the 400-mg dose, 5.5 μg/ml for the 600-mg dose, and slightly less than 7 μg/ml for the 800-mg dose. Peak concentrations increased roughly proportionately with the dose. AUCs were examined for several doses, and no evidence for a saturable clearance pathway was found, although only small numbers of volunteers were examined (three per dosing level). Absorption was rapid at all doses, with peak concentrations occurring at 1 to 2 h. Terminal elimination half-lives ranged from 10.5 ± 2 h (400-mg dose) to 12.6 ± 2.5 h (800-mg dose). Urinary recoveries ranged from 11.1 to 17% of the administered oral dose. Renal clearances ranged from 12.9 to 21.9 ml/min across doses. Total clearances ranged from 111 ml/min (600-mg dose) to 135 ml/min (200-mg dose). Thus, nonrenal clearance accounted for the vast majority of total drug clearance.

Frydman et al. (25) examined the multiple-dose pharmacokinetics of pefloxacin in 12 normal subjects who received 400 mg of pefloxacin either orally or intravenously. The pharmacokinetic parameter values observed after the first dose were virtually identical when compared with the studies of Barre et al. (3). After the last dose (dose 18), however, a major change in pharmacokinetic parameter values was found. Total serum clearance decreased statistically significantly from 123 ml/min for the first dose to 87 ml/min on the last dose and was associated with a significant increase in half-life from 12 to 14.8 h. These data suggested that multiple dosing saturated a nonrenal clearance pathway of pefloxacin.

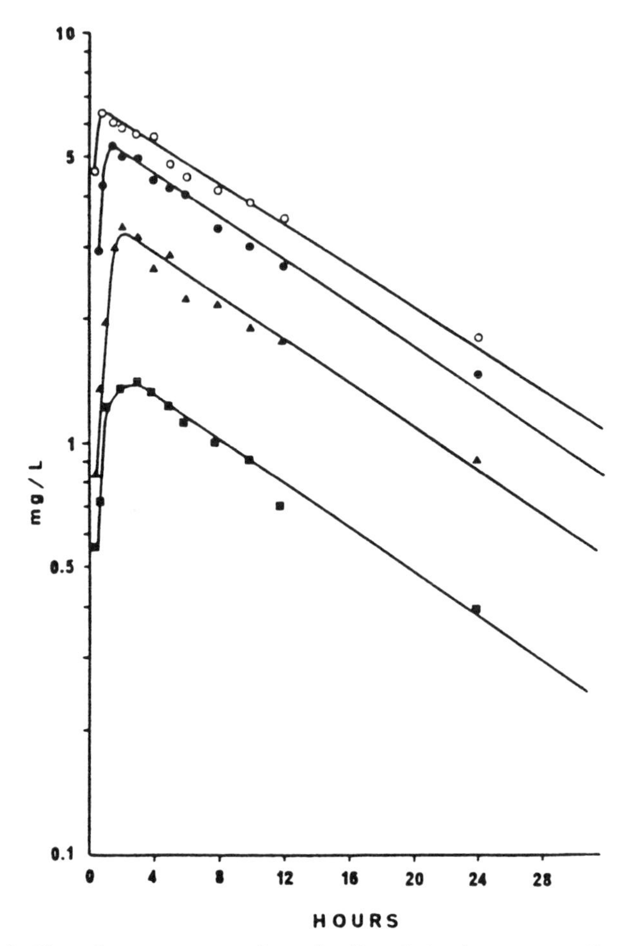

Figure 2. Plots of mean concentrations of pefloxacin in plasma versus time obtained in three subjects for each orally administered dose. The lines were generated from the best-fit pharmacokinetic parameters. Symbols: ■, 200 mg; ▲, 400 mg; ●, 600 mg; ○, 800 mg. (From reference 3.)

The disposition of pefloxacin when given intravenously was also examined in dose-ranging (3) and multiple-dosing (25) studies. The dose-ranging study revealed a terminal half-life ranging from 9.7 h (200-mg dose) to 13.8 h (800-mg dose). Total serum clearances were similar to those seen in the oral dose-ranging study, ranging from 105 ml/min (800-mg dose) to 156 ml/min (200-mg dose). Renal recoveries varied from 11 to 14% of the administered dose.

AUCs were the same as those observed in the oral studies, indicating that pefloxacin absorption was complete and uninfluenced by the dose. Al-

Table 4. Pharmacokinetic parameters obtained after intravenous infusion and oral administration of pefloxacin (from reference 3)

Parameter [a] (unit) obtained after:	Mean ± SD at following dose (mg):			
	200	400	600	800
Intravenous infusion				
V_1 (liters)	23.2 ± 19.1	46.9 ± 8.7	68.2 ± 20.0	54.6 ± 21.8
V_B (liters)	100.9 ± 40.9	64.8 ± 4.1	83.5 ± 14.7	65.3 ± 30.9
AUC (mg·h/liter)	23.4 ± 9.3	54.3 ± 4.1	82.2 ± 22.5	130.9 ± 23.7
CL (ml/min)	156.4 ± 53.7	123.1 ± 9.8	27.2 ± 31.0	104.8 ± 19.3
U_∞ (mg)	27.4 ± 13.1		76.2 ± 37.1	84.8 ± 19.3
U_∞ (%)	13.7 ± 6.5		12.5 ± 6.6	10.8 ± 2.1
CL_R (ml/min)	19.6 ± 5.6		15.2 ± 6.6	12.0 ± 0.3
Oral administration				
Lag time (h)	0.32 ± 0.14	0.48 ± 0.23	0.26 ± 0.11	0.28 ± 0.22
$t_{1/2\beta}$ (h)	11.7 ± 3.6	10.5 ± 2.0	11.3 ± 1.1	12.6 ± 2.5
V (liters)	132.2 ± 26.9	112.0 ± 20.5	109.9 ± 17.7	137.9 ± 2.1
AUC (mg·h/liter)	25.7 ± 6.3	54.5 ± 11.2	87.9 ± 8.9	105.0 ± 20.7
CL_T (ml/min)	135.3 ± 35.5	125.7 ± 25.7	111.3 ± 7.0	130.6 ± 28.6
U_∞ (mg)	23.6 ± 8.0	44.3 ± 17.3	84.1 ± 23.5	136.4 ± 26.7
U_∞ (%)	11.8 ± 4.0	11.1 ± 4.3	14.0 ± 3.9	17.0 ± 3.3
CL_R (ml/min)	15.3 ± 3.8	12.9 ± 2.6	14.7 ± 3.6	21.9 ± 4.8

[a]Abbreviations: V_1, volume of distribution in the central compartment; V_B, volume of distribution of drug in the body; CL, clearance; U_∞ (mg), amount of drug excreted into the urine from time zero to infinity; U_∞ (%), amount of drug excreted into the urine as a percentage of the administered dose from time zero to infinity; CL_T, total clearance. For other abbreviations see Tables 1 and 2, footnote *a*.

though the mean AUC increased more than in proportion to the increasing dose, the increase was not statistically significant. Although it was concluded that pefloxacin behaved in an approximately dose-linear fashion, significant increases in serum half-life and decreases in serum clearance were observed when first and last doses were compared (Fig. 3). Thus, studies with intravenous pefloxacin, as those with oral pefloxacin, suggest the existence of a saturable nonrenal clearance pathway.

Montay et al. (45) examined the profile of metabolites in plasma, urine, and bile after the administration of pefloxacin to normal volunteers. The major metabolites in serum were pefloxacin *N*-oxide and *N*-desmethylpefloxacin (norfloxacin). The parent compound, pefloxacin *N*-oxide, norfloxacin, oxonorfloxacin, oxopefloxacin, and traces of pefloxacin glucuronide were recovered in urine (Table 5). Over 72 h after a 800-mg dose, the total urinary recovery of the parent compound plus metabolites accounted for 58.9 ± 3.1% of the administered dose. Pefloxacin concentrations in bile were 10 to 20 μg/ml at 2 to 12 h after the administration of 800 mg of pefloxacin.

For either oral or intravenous dosing of pefloxacin, a two- or threefold accumulation of the two major metabolites, pefloxacin *N*-oxide and

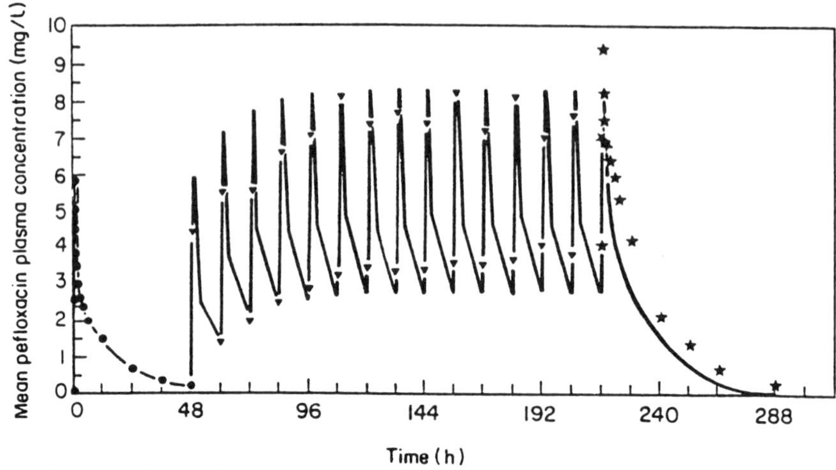

Figure 3. Mean pefloxacin concentrations in plasma predicted by superimposition (continuous line) and measured (●, ▼, ★) during repeated intravenous administration (400 mg twice a day). Symbols: ●, day 1; ▼, days 3 to 9; ★, day 10. (From reference 25.)

norfloxacin, was found after multiple doses, but the absolute concentrations of the metabolites were low and unlikely to be clinically important. Frydman et al. (25) reported that the terminal half-lives determined for norfloxacin and the *N*-oxide metabolite of pefloxacin were 9.75 and 12 h, respectively. The half-life of norfloxacin, however, is likely artifactually prolonged because of the relatively slow rate of formation of norfloxacin from pefloxacin.

As with other quinolones, pefloxacin penetrates well into extravascular spaces. Wise et al. (66) examined the concentrations of pefloxacin in serum and blister fluid after 400 mg was given as an intravenous infusion over 1 h. The blisters were induced by an inflammatory method (cantharides plaster

Table 5. Urinary recovery of pefloxacin and main metabolites after oral administration[a] (from reference 45)

Compound	Mean recovery (% of dose [b]) ± SE
Pefloxacin	9.3 ± 1
Norfloxacin	20.2 ± 1
Pefloxacin glucuronide	tr
Pefloxacin *N*-oxide	23.2 ± 1.6
Oxonorfloxacin	5.4 ± 0.5
Oxopefloxacin	0.75 ± 0.05
Total	58.9 ± 3.1

[a]The 0- to 72-h urine was used.
[b]Eight-tenths gram per person (n = 5).

application). The penetration, as calculated by the ratio of the AUC for blister fluid to the AUC for serum, was 70%.

The penetration of pefloxacin into the cerebrospinal fluid (CSF) of humans, in both the presence and the absence of meningeal inflammation, has been studied by several groups. Dow et al. (15) examined nine subjects with hydrocephalus and external ventricular drains, but without meningeal inflammation. CSF and plasma samples were obtained from 1 to 48 h after the infusion of 400 mg of pefloxacin (Fig. 4). Maximum pefloxacin concentrations averaged 8.54 ± 1.53 μg/ml of plasma and 2.97 ± 0.32 μg/ml of CSF. Peak concentrations in CSF occurred at 5.44 ± 0.42 h after drug administration. The ratio of drug in CSF to that in plasma was relatively stable between 6 and 24 h after the dose (range, 0.57 ± 0.08 to 0.64 ± 0.15).

Wolff et al. (70) examined the penetration of pefloxacin into CSF of 15 patients with meningitis or ventriculitis. Three doses of pefloxacin (either 7.5 or 15 mg/kg) were administered at 12-h intervals to 11 patients intravenously and to 4 patients orally. Concentrations of pefloxacin in CSF measured 2 h after the third intravenous dose and 4 h after the third oral dose ranged from 2.4 to 9 μg/ml in patients receiving 7.5 mg of pefloxacin per kg

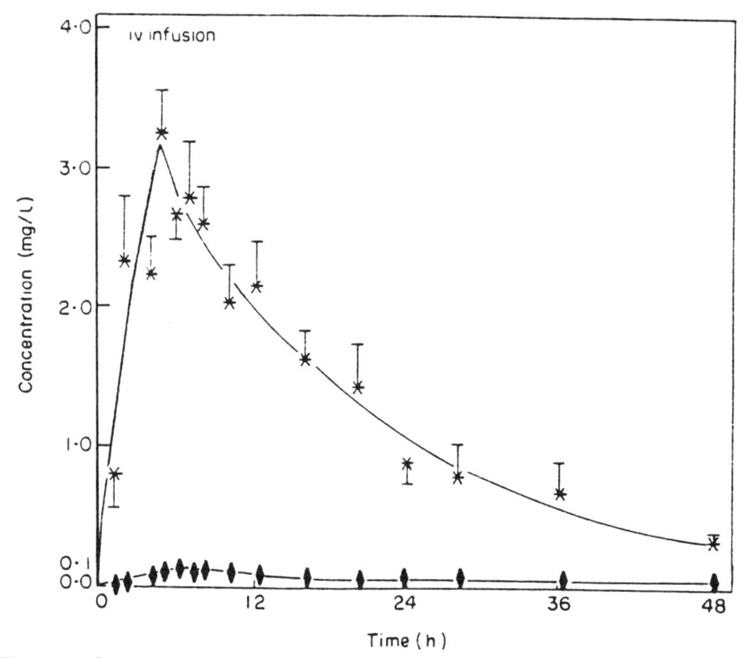

Figure 4. Concentrations (mean + standard error of the mean) of pefloxacin (*) and *N*-desmethylpefloxacin (◆) in CSF of humans ($n = 9$) after a single 1-h intravenous (iv) infusion of 400 mg of pefloxacin. (From reference 15.)

and from 6.5 to 13 μg/ml in patients receiving 15 mg/kg. These concentrations of pefloxacin in CSF are in excess of the MICs for 90% of the strains of important nonstreptococcal pathogens.

Montay et al. (46) examined the disposition of pefloxacin after a single intravenous infusion of 8 mg/kg in 15 male patients with various degrees of renal failure. No difference in the distribution or elimination of the drug between patients with mild or severe renal impairment was observed, and pharmacokinetic parameter values were similar to those reported for healthy subjects. No accumulation of the *N*-desmethyl metabolite (norfloxacin) occurred.

Hemodialysis (for 4 h) resulted in little removal of pefloxacin from the blood, with an extraction ratio of drug across the dialyzer of 0.28 ± 0.06 and a dialyzer clearance of 62 ± 9 ml/min. Redosing of pefloxacin after hemodialysis may therefore be unnecessary. Thus, overall, renal failure has a minimal effect on the pharmacokinetics of pefloxacin, but studies in which the accumulation of metabolites of pefloxacin in patients with renal failure is evaluated are needed.

Because the major route of total serum clearance of pefloxacin is by hepatic biotransformation, it is important to examine the impact of hepatic cirrhosis on pefloxacin pharmacokinetics. In one study (13) (Table 6) the terminal elimination half-life was significantly longer (35.1 ± 19 h) in 16 subjects with histologically proved hepatic cirrhosis than in normal subjects (11 ± 2.64 h). The total plasma clearance was markedly decreased from 8.19 ± 2.8 liters/h per 1.73 m^2 in normal subjects to 2.66 ± 1.85 liters/h per 1.73 m^2 in cirrhotic subjects. Urinary excretion of unchanged pefloxacin was higher in the volunteers with cirrhosis than in normal volunteers, while the excretion of *N*-desmethylpefloxacin (norfloxacin) was lower. A significant correlation between the nonrenal clearance and the prothrombin time was seen.

Table 6. Kinetic parameters of pefloxacin in 12 healthy subjects receiving a single dose of 400 mg intravenously and in 16 patients with cirrhosis receiving a single dose of 8 mg/kg intravenously (from reference 13)

Parameter [a] (unit)	Mean ± SD	
	Subjects	Patients
$t_{1/2}$ (h)	11.00 ± 2.64	35.10 ± 19.00[b]
V_{area} (liters/kg)	1.88 ± 0.37	1.54 ± 0.26[c]
CL (liters/h·1.73 m^2)	8.19 ± 2.80	2.66 ± 1.85[b]

[a]For abbreviations see Tables 1, 2, and 4, footnote *a*.
[b]$P < 0.001$.
[c]$P < 0.02$.

Although the clearance of pefloxacin is affected by alterations in hepatic function, the extent to which pefloxacin dosing should be reduced to prevent drug accumulation in patients with hepatic impairment is uncertain. Further investigation into the degree of accumulation with multiple doses and the correlation of clearance with other measures of hepatic function is warranted.

Difloxacin

Difloxacin is a 6-fluoroquinolone with a *p*-fluorophenyl substituent at position 1 (Fig. 5; see Fig. 1 in chapter 2). Granneman et al. (31) examined the pharmacokinetics of this compound after single doses of 200, 400, and 600 mg (Table 7). Peak concentrations in serum occurred approximately 4 h after administration and increased linearly with the dose (2.17, 4.09, and 6.12 μg/ml, respectively) (Fig. 6), as did AUCs. The terminal elimination half-life was in excess of 24 h, a value much longer than that of other quinolones discussed in this chapter.

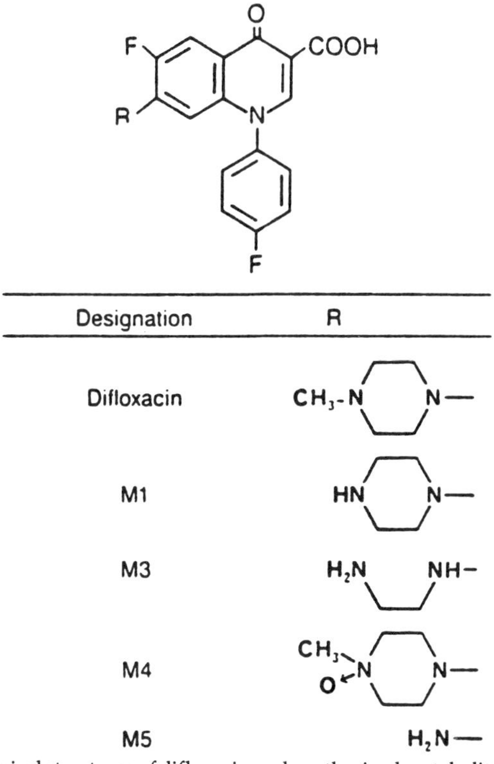

Designation	R
Difloxacin	CH_3-N N—
M1	HN N—
M3	H_2N NH-
M4	CH_3 O←N N—
M5	H_2N—

Figure 5. Chemical structures of difloxacin and synthesized metabolites. (From reference 31.)

Table 7. Pharmacokinetic parameters of difloxacin[a] (from reference 31)

Dose (mg)	Mean ± SD								
	T_{max} (h)	C_{max}/D (µg/ml)	AUC_{0-48}/D (µg·h/ml)	$X_{u0-\infty}$ (% of D)	CL_R (ml/min)	CL/F (ml/min)	Lag (h)	V/F (liters)	$t_{1/2}$ (h)
200 (n = 6)	3.9 ± 2.1	1.09 ± 0.14	26.6 ± 3.1	11.4 ± 2.3	5.6 ± 1.0	50.1 ± 7.0	0.4 ± 0.3	89.7 ± 8.1	20.6 ± 1.4
400 (n = 6)	5.2 ± 2.9	1.02 ± 0.15	28.1 ± 2.4	9.6 ± 2.5	4.3 ± 1.4	41.3 ± 4.5	0.6 ± 0.8	97.5 ± 9.8	27.1 ± 3.3
600 (n = 11)	4.7 ± 1.3	1.02 ± 0.11	28.3 ± 2.6	10.0 ± 2.0	4.1 ± 0.9	39.3 ± 5.5	0.6 ± 0.8	98.5 ± 9.9	28.8 ± 4.9
Mean	4.5 ± 1.9	1.04 ± 0.12	27.8 ± 2.7	10.2 ± 2.2	4.4 ± 1.3	42.7 ± 7.3	0.5 ± 0.7	96.0 ± 9.6	25.7 ± 4.9

[a]Abbreviations: C_{max}/D, maximum drug concentration in plasma; AUC_{0-48}/D, area under the 0- to 48-h drug concentration curve in plasma. For other abbreviations see Tables 1 and 4, footnote *a*. T_{max}, C_{max}/D, and CL_R are model independent; C_{max}/D and AUC/D are normalized per 100 mg administered. X_u-required extrapolation to infinite time, using $t_{1/2}$, CL/F, lag, V/F, and $t_{1/2}$, is based on NONLIN regression by using a one-compartment model.

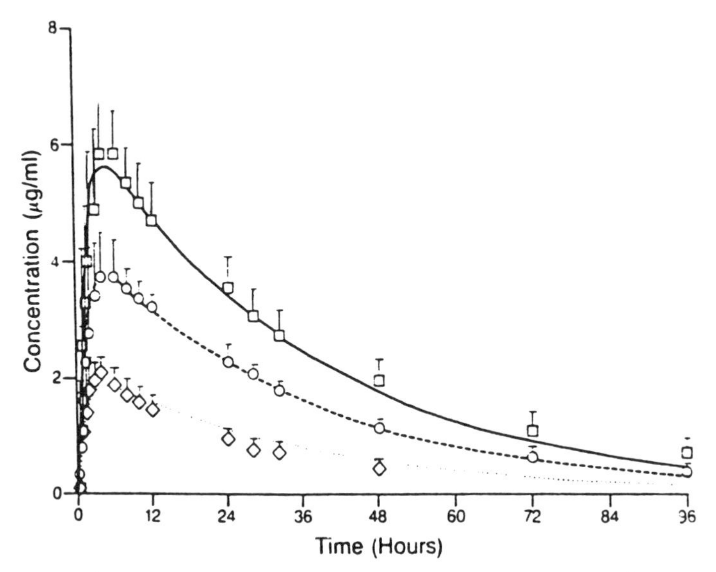

Figure 6. Difloxacin concentrations in plasma after oral doses of 200 (◇), 400 (○), and 600 (□) mg. (From reference 31.)

Renal recoveries of the intact parent compound were low (10.2 ± 2.2% of the administered dose). Because serum protein binding of difloxacin was also low (42% by ultrafiltration), net renal tubular absorption appeared to be occurring. In urine, additional metabolites were recovered, including the major metabolite, the glucuronide ester of difloxacin, and five oxidative metabolites (Fig. 5 and 7).

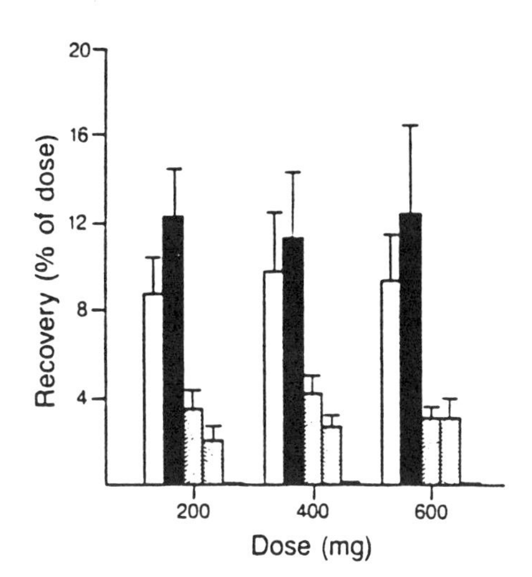

Figure 7. Urinary recoveries of (left to right) difloxacin, difloxacin glucuronide, and oxidative metabolites M1, M4, and M3 after oral doses of 200, 400, and 600 mg in humans. (From reference 31.)

Fluoroquinolones Cleared by Both Renal and Hepatic Mechanisms

Norfloxacin

Norfloxacin, the first fluoroquinolone available for clinical use in the United States, is available only as an oral formulation.

Swanson et al. (56) examined the disposition of norfloxacin after sequentially increasing oral doses from 200 to 1,600 mg (Fig. 8). Peak concentrations in serum increased linearly with increasing doses from 0.75 ± 0.2 μg/ml (200-mg dose) to 3.87 ± 1.3 μg/ml (1,600-mg dose). The time to reach the peak concentration also increased with increasing doses. Serum protein binding was low (14%).

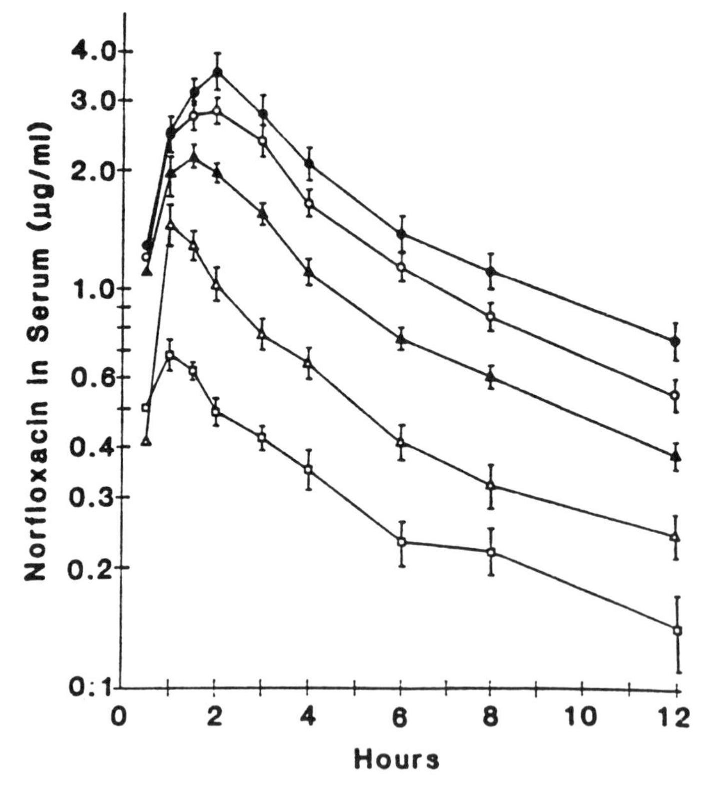

Figure 8. Mean concentrations of norfloxacin in serum. Norfloxacin was administered orally at weekly intervals as single doses of 200 (□), 400 (△), 800 (▲), 1,200 (○), and 1,600 (●) mg. Bars indicate the standard error of the mean. (From reference 56.)

Renal clearances were similar for each dose and ranged from 272 ± 96 to 298 ± 40 ml/min (Table 8). These renal clearances are in excess of GFRs and thus imply net renal tubular secretion of norfloxacin.

Edlund et al. (21) examined the multiple-dose pharmacokinetics of norfloxacin in a study in which 200 mg was administered every 12 h for 7 days. At the steady state the maximum concentration and the time to the maximum concentration were similar to those found by Swanson et al. (56). The half-life was 4.19 ± 0.57 h, and the AUC from 0 to 12 h at the steady state (which approximates AUCs extrapolated from zero to infinity for the first dose) was 3.17 ± 0.40 μg·h/ml. Concentrations of norfloxacin in feces at between 3 and 7 days were stable at approximately 950 μg/g of feces. A mean of 28% (range, 8.3 to 53%) of a 400-mg dose of norfloxacin was recovered as the intact parent compound in the feces over 48 h in another study (12).

The coadministration of norfloxacin with food delays the time of attainment of peak concentrations in serum, but recovery in urine over 8 h was little changed (from 28 to 20%) (37), suggesting that the coadministration of food is unlikely to alter the efficacy of norfloxacin in the treatment of urinary tract infections.

Hepatic biotransformation of norfloxacin occurs to some extent. All six metabolites described have modifications in the piperazine ring. The oxo and ethylenediamine derivatives are the two major metabolites recovered in urine, although cumulative recoveries in urine are low relative to norfloxacin itself (20% for the oxo form and 4% for the ethylenediamine form). These metabolites have not been detected in serum even after a 1,600-mg dose (49).

Adhami et al. (1) examined the penetration of norfloxacin into inflammatory blister fluid after a dose of 400 mg. Peak concentrations in blister fluid averaged 1.01 ± 0.27 μg/ml, with peak concentrations in serum averaging 1.45 ± 0.09 μg/ml. Serum half-lives and blister fluid half-lives were

Table 8. Pharmacokinetic parameters for norfloxacin in 14 healthy men[a] (from reference 56)

Oral dose (mg)	Mean ± SD				
	AUC_{0-12} (μg·h/ml)	C_{max} (μg/ml)	T_{max} (h)	$t_{1/2}^{6-12}$ (h)	CL_R^{0-12} (ml/min)
200	3.56 ± 1.13	0.75 ± 0.20	1.1 ± 0.4	7.3 ± 2.9	272 ± 96
400	6.26 ± 2.05	1.58 ± 0.60	1.3 ± 0.4	7.4 ± 2.5	292 ± 76
800	11.4 ± 2.12	2.41 ± 0.43	1.5 ± 0.4	6.2 ± 1.8	273 ± 37
1,200	16.1 ± 4.00	3.15 ± 0.76	1.8 ± 0.8	5.7 ± 1.2	288 ± 56
1,600	19.7 ± 6.07	3.87 ± 1.27	1.9 ± 0.6	6.8 ± 1.4	298 ± 40

[a]Abbreviations: AUC_{0-12}, AUC from 0 to 12 h after drug administration; $t_{1/2}^{6-12}$, half-life with 6- to 12-h data; CL_R^{0-12}, drug excreted/serum AUC (0- to 12-h data). For other abbreviations see Table 1, footnote a.

similar (3.50 ± 0.79 and 3.25 ± 0.52 h, respectively). AUCs from zero to infinity in blister fluid and serum were also similar (5.74 ± 1.6 and 5.4 ± 1.7 μg·h/ml, respectively), indicating an excellent penetration of norfloxacin into the interstitial fluid.

Because norfloxacin is cleared by both renal and hepatic mechanisms, major changes in the profile of the concentration in serum plotted versus time with dysfunction of a single organ system are not expected. Hughes et al. (34) examined the pharmacokinetics of norfloxacin in patients with impaired renal function. Half-lives increased from 3.14 h in the normal group (GFR, $>$80 ml/min) to 8.87 h in those with severe renal impairment (GFR, $<$10 ml/min). AUCs increased from 6.55 to 18.32 μg·h/ml in the patients with the most severe renal dysfunction. Fillastre et al. (22) reported that the mean half-life of norfloxacin was 7.9 h in hemodialyzed patients between dialysis sessions and not influenced by hemodialysis. Both Hughes et al. (34) and Fillastre et al. (22) recommend altering the administration of norfloxacin for a GFR of 20 to 30 ml/min or less, either by halving the dose or by increasing the dosing interval twofold.

Eandi et al. (20) examined the alteration of norfloxacin disposition in patients with liver dysfunction compared with normal volunteers. No change in either serum half-life or AUCs was detected.

Ciprofloxacin

Ciprofloxacin, like norfloxacin, is cleared by balanced renal and nonrenal mechanisms. This compound is available in both oral and parenteral formulations. While its structure (see Fig. 1 in chapter 2) is similar to the other aforementioned quinolones, it differs in that a cyclopropane ring replaces the ethyl group attached at N-1.

Tartaglione et al. (57) performed a dose-ranging study for single doses (250 to 1,000 mg) of the oral formulation (Fig. 9 and Table 9). Peak concentrations were 0.76, 1.60, 2.54, and 3.38 μg/ml for the 250-, 500-, 750-, and 1,000-mg doses, respectively. Absorption was rapid but lengthened from 1.1 h at 250 mg to 1.8 h at 1,000 mg. Peak concentrations and AUCs increased proportionally with the dose. Terminal half-lives averaged 4.1 h after the 250- and 500-mg doses, but were 6.9 and 6.3 h for the higher doses. Renal clearances were relatively reproducible, ranging from 316 ml/min (1,000 mg) to 477 ml/min (500 mg), and in excess of GFR, indicating a net renal tubular secretion. Consistent with this presumption are the findings of Wingender et al. (W. Wingender, D. Beerman, D. Foerster, K.-H. Graefe, and P. Schacht, *4th Mediterr. Congr. Chemother.*, abstr. no. 621, 1984) that probenecid reduces the renal clearance of ciprofloxacin by 46%. Urinary recoveries ranged from 28.5% (1,000 mg) to 44.0% (500 mg).

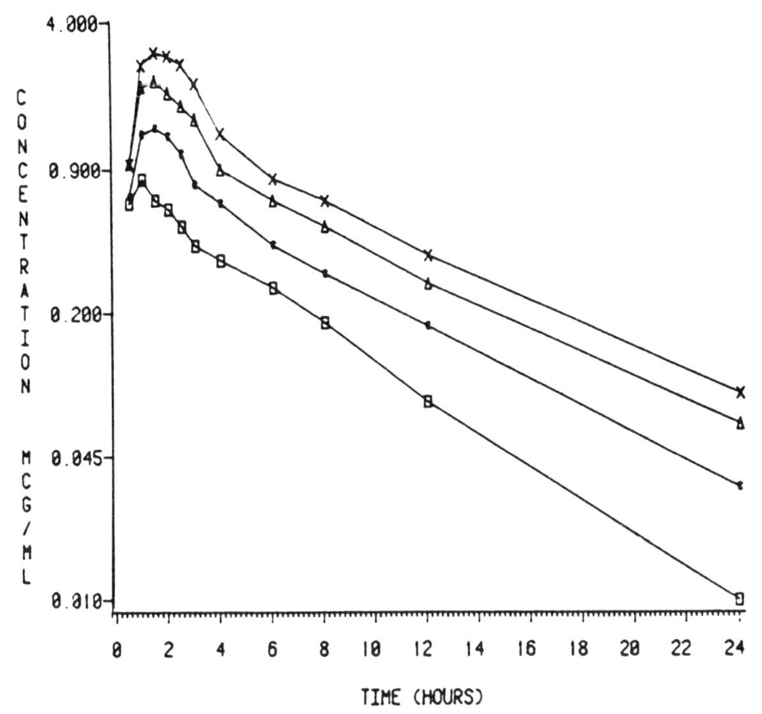

Figure 9. Semilogarithmic plot of mean ciprofloxacin concentrations in serum after single doses. Symbols: □, 250-mg dose; ●, 500-mg dose; △, 750-mg dose; ×, 1,000-mg dose. (From reference 57.)

The pharmacokinetics of orally administered ciprofloxacin have been studied by several investigators. Aronoff et al. (2) examined 250 mg administered every 12 h for 13 doses. Pharmacokinetic parameters determined on days 1 and 4 were consistent with those determined during single-dose trials. By day 7 the terminal half-life and apparent serum clearance had both declined significantly, but these changes were small and not likely to be clinically important. LeBel et al. (40) also found changes in the half-life and in the nonrenal component of clearance after multiple 500-mg doses of ciprofloxacin. Similar findings have been reported by other investigators for doses between 250 and 750 mg and using microbiologic assays (9, 30). An additional study in which a 500-mg dose and a high-pressure liquid chromatography assay were used, however, found no alteration in AUC from days 1 to 5 of drug administration (5).

Concentrations of drug in the urine with multiple dosing were quite high for all of the studies cited above. With 250 mg of ciprofloxacin given every 12 h (2), 6- to 12-h urine collections averaged 45 ± 25 to 69 ± 45 μg/

Table 9. Mean values of some pharmacokinetic variables reported for healthy volunteers after single oral doses of ciprofloxacin[a] (from reference 10)

Dose (mg)	C_{max} (mg/liter)	T_{max} (h)	$t_{1/2}$ (h)	AUC (mg/liter·h)	Reference
50	0.28	0.58	3.40	1.00	33
100	0.49	0.82	4.10	1.90	
250	1.45	1.00	3.97	6.37	8
500	2.56	1.33	4.15	11.10	
500	2.91	1.25	4.82	12.7	14
750	2.65	1.10	4.75	12.20	33
1,000	3.38	1.80	6.30	16.60	57

[a]For abbreviations see Table 1, footnote *a*.

ml. In studies with a single 500-mg dose (14), the lowest concentrations in urine in the 12- to 24-h collection was 8 μg/ml.

The absolute oral bioavailability of ciprofloxacin is good. In studies with healthy volunteers for doses from 50 to 750 mg, bioavailability ranged from 46 to 84% (6, 17, 33, 51). In some studies (6, 51), a variation in bioavailability was greater with higher doses. As with norfloxacin, the pharmacokinetic parameter values of ciprofloxacin are little affected by administration of the drug with food, except for an increase in the time to achieve peak concentrations in serum (41).

The intravenous administration of ciprofloxacin has also been studied. Drusano et al. (16) examined single doses of 100 and 200 mg administered as constant-rate infusions over 30 min (Fig. 10). The total serum clearance averaged 23.0 ± 9.1 liters/h per 1.73 m^2 for the 100-mg dose and 23.7 ± 5.1 liters/h per 1.73 m^2 for the 200-mg dose. Clearance was somewhat greater (28.9 ± 2.7 liters/h per 1.73 m^2), however, when 100 mg was given as a 15-min infusion followed by a 25-mg/h constant infusion for 4 h. Half-lives were 4.7 to 4.8 h, and renal clearance accounted for 65 to 67% of the total clearance. The volumes of distribution were large (2.27 ± 0.85 liters/kg for the 100-mg dose and 2.44 ± 0.51 liters/kg for the 200-mg dose).

Other investigators (6, 19, 33) have examined single doses of intravenously administered ciprofloxacin in this range and have found results with serum clearance similar to the larger value found in the study by Drusano et al. (Table 10). Similar half-lives, volumes of distribution, and dose linearity have also been observed in each of these studies.

Gonzalez et al. (28, 29) performed two multiple-dose and dose-ranging studies with intravenously administered ciprofloxacin, using a microbiologic assay. At doses ranging from 25 to 200 mg given to volunteers every 12 h for 1 week, drug accumulation consistent with the half-life and dosing frequency was seen at each dose. Half-lives were consistently in the range of

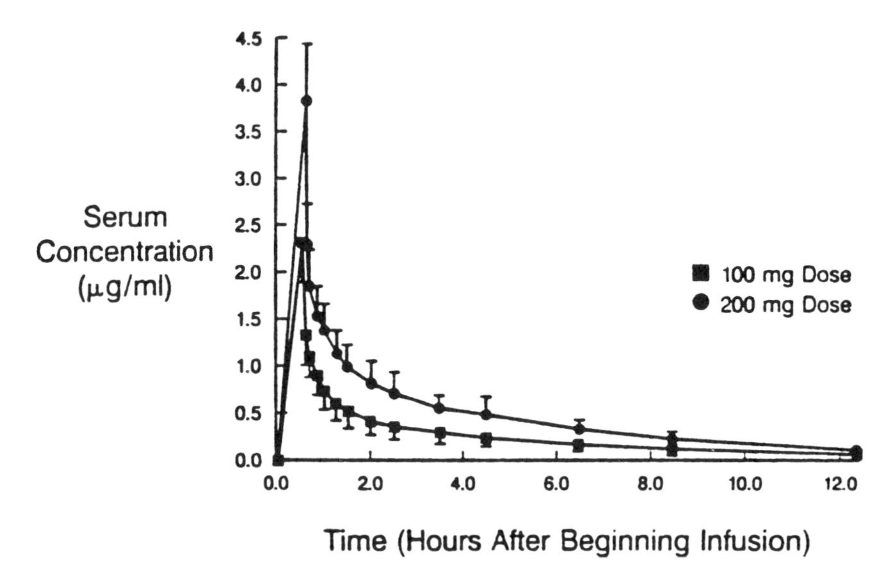

Figure 10. Mean (± standard deviation) concentrations in serum after 100- and 200-mg, 30-min infusions of ciprofloxacin in six normal volunteers. (From reference 16.)

Table 10. Plasma clearance, renal clearance, and half-life of ciprofloxacin[a]

Dose (mg)	Mean ± SD			Reference
	CL_S (liters/h per 1.73 m²)	CL_R (liters/h per 1.73 m²)	$t_{1/2}$ (h)	
100[b,c]	34.0 ± 5.3[d]	28.7 ± 4.1[d]	4.0 ± 0.9	68
25[c,e]	35.4 ± 6.8	20.4[f]	3.6 ± 0.7	28
50[c,e]	31.8 ± 3.0	21.9[f]	4.2 ± 0.7	28
75[c,e]	31.3 ± 5.4	24.8[f]	3.5 ± 0.4	28
100[c,e]	30.1 ± 3.4	18.8[f]	3.7 ± 0.6	29
150[c,e]	29.8 ± 4.0	18.8[f]	3.6 ± 0.3	29
200[c,e]	26.9 ± 4.1	18.4[f]	4.0 ± 0.7	29
50[b,g]	41.2 ± 7.8	25.5 ± 7.0	3.0 ± 0.5	33
100[b,g]	31.8 ± 6.2	20.1 ± 4.0	3.1 ± 1.9	33
100[b,g]	23.0 ± 9.1	15.5 ± 9.1	4.7 ± 0.4	16
200[b,g]	23.7 ± 5.1	15.5 ± 3.0	4.8 ± 1.0	16
200[b,g]	28.5 ± 4.7	16.9 ± 3.0	4.2 ± 0.8	18
100[b,g]	42.0 ± 14.2	NA[h]	2.90 ± 0.75	6
100[b,g,i]	9.60 ± 2.09	4.61 ± 1.14	4.21 ± 0.89	19
150[b,g,i]	9.57 ± 2.02	4.40 ± 1.16	4.62 ± 1.19	19
200[b,g,i]	8.15 ± 1.21	3.80 ± 0.87	4.22 ± 0.63	19

[a] CL_S, Serum clearance. For other abbreviations see Table 1, footnote *a*.
[b] Single-dose study.
[c] Microbiologic assay.
[d] Not normalized to body surface area.
[e] Multiple-dose study.
[f] Mean of 3 days.
[g] High-pressure liquid chromatography assay.
[h] NA, Not available.
[i] Clearance normalized to milliliters per minute per kilogram.

3 to 4 h. Serum clearances ranged from 26.9 ± 4.1 liters/h per 1.73 m^2 (200-mg dose) to 35.4 ± 6.8 liters/h per 1.73 m^2 (25-mg dose), differences unlikely to be of clinical importance. No significant differences were seen across doses for any of the pharmacokinetic parameter values examined, and clinically important drug accumulation was not seen.

One-third to one-half of the serum clearance of ciprofloxacin is accounted for by nonrenal mechanisms. Four metabolites have been characterized. Each of these compounds has limited microbiologic activity, usually one-quarter to one-half the activity of the parent drug. All metabolism occurs by alterations in the piperazine side chain (Fig. 11). Beermann et al. (4) have quantitated the amounts of metabolites formed after intravenous and oral administration of labeled ciprofloxacin (Table 11). Less than 20% of the administered dose is recoverable as metabolites, even when urine and stool are assayed. Levels of the M2 metabolite are slightly higher after oral administration than after the intravenous route, indicating that limited first-pass hepatic metabolism of ciprofloxacin occurs. After intravenous dosing 15% of the parent compound is recovered in the stool, suggesting elimination across the intestinal wall, as was proposed by Rohwedder and Bergan (R. W. Rohwedder and T. Bergan, *Proc. Congr. Bacterial Parasitic Drug Resistance,* 1986).

Protein binding for ciprofloxacin has been determined by Joos et al. (35) to be between 16 and 28% and is independent of concentration and pH.

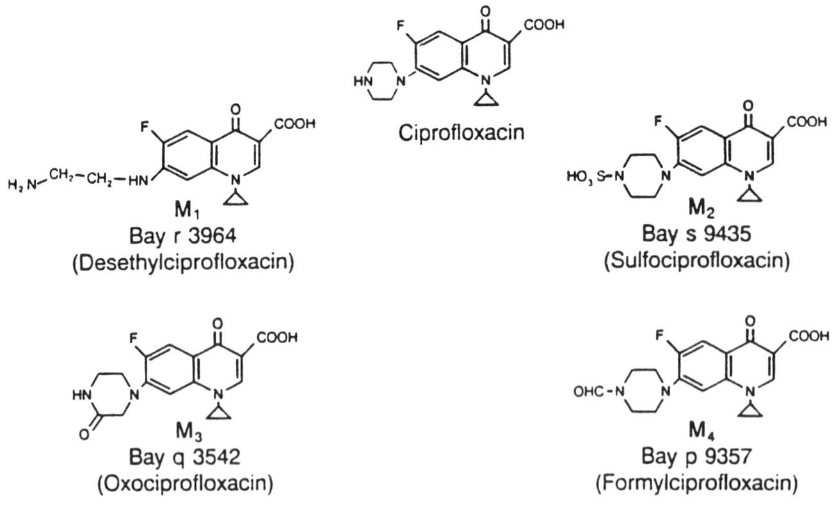

Ciprofloxacin

M_1
Bay r 3964
(Desethylciprofloxacin)

M_2
Bay s 9435
(Sulfociprofloxacin)

M_3
Bay q 3542
(Oxociprofloxacin)

M_4
Bay p 9357
(Formylciprofloxacin)

Figure 11. Structure of ciprofloxacin and its metabolites. (From W. Gau, personal communication.)

Table 11. Excretion of ciprofloxacin and metabolites after oral and intravenous dosing of ^{14}C-labeled ciprofloxacin[a]

Determination	% Dose excreted			
	Oral (259 mg)		Intravenous (107 mg)	
	Urine	Feces	Urine	Feces
M1	1.4	0.5	1.3	0.5
M2	3.7	5.9	2.6	1.3
M3	6.2	1.1	5.6	0.8
M4	ND[b]	ND	ND	ND
Ciprofloxacin[c]	44.7	25.0	61.5	15.2

[a]W. Wingender, personal communication.
[b]ND, Not detected.
[c]High-pressure liquid chromatography

Other investigators, using ultracentrifugation techniques (33), have found values in the range of 40%.

As with other fluoroquinolones, tissue penetration of ciprofloxacin is excellent. LeBel et al. (40) studied penetration into blister fluid in a noninflammatory (suction blister) model, and Wise and Donovan (64) examined penetration into an inflammatory (cantharides blister) system after a 500-mg oral dose. In noninflammatory blister fluid the peak concentration after a single dose was 1.75 ± 0.75 μg/ml, with a concentration after 8 h of 0.45 ± 0.07 μg/ml. The blister fluid half-life was similar to the serum half-life. The penetration, calculated as the AUC ratio for blister to serum, was 88.8 ± 26%. The only clinically significant change seen with multiple dosing was an increase in the steady-state concentration of blister fluid at 8 h to 0.97 ± 0.57 μg/ml. In inflammatory blister fluid, penetration was 117%, due to the longer half-life of the drug in the inflammatory fluid (5.6 h) relative to that in the serum (4.4 h).

Wise et al. (64, 68) have also examined the penetration of ciprofloxacin into inflammatory blister fluid (as above) and noninflammatory peritoneal fluid (obtained at an operation) after intravenous administration. Results for the blister fluid again indicated excellent penetration, with an AUC ratio of 1.21. For the peritoneum, the penetration was estimated at 95%, with no differences between the half-lives in peritoneal fluid and serum.

Alterations in ciprofloxacin pharmacilinetics have been studied in patients with various degrees of renal dysfunction. In 32 of these patients receiving a single oral dose of 750 mg of ciprofloxacin, significant relationships were found between creatinine clearance and both apparent serum clearance and renal clearance (24). Although no relationship was discernible between the terminal elimination rate constant and creatinine clearance, there was a trend toward an increasing terminal elimination half-life with increasing renal dysfunction, with normal volunteers having a mean half-life of 5.2 ± 0.7 h, while patients maintained on hemodialysis had a mean half-life of 6.9 ± 2.9 h between dialyses. Total apparent serum clearance declined from 43.5 ± 9.0 liters/h per 1.73 m^2 in normal volunteers to 25.4 ± 7.8 liters/h per 1.73 m^2 in patients on hemodialysis. A reduction in dose of approximately one-third was recommended for creatinine clearances of 20 to 30 ml/min per 1.73 m^2 or less. These findings were reasonably consistent with those of both Boelaert et al. (7) and Singlas et al. (53) for doses of 250 and 500 mg, respectively.

Drusano et al. (18) examined the influence of various degrees of renal impairment on the pharmacokinetics of ciprofloxacin (200 mg) administered intravenously (Table 12). Significant relationships were found between normalized creatinine clearance and both normalized serum and renal clearance. A hyperbolic relationship between creatinine clearance and terminal elimination half-life was seen. A dose reduction of 50% was recommended for creatinine clearance of 1.2 to 1.8 liters/h per 1.73 m^2 (20 to 30 ml/min per 1.73 m^2) or less. Because of the greater-than-threefold variability in half-life noted in the anephric patients, it was also recommended that the dose be reduced by half while maintaining a 12-h dosing interval in order to avoid in some patients prolonged periods during which concentrations in serum are below the MIC for less susceptible pathogens. Webb et al. (59) found similar outcomes with a dose of 100 mg administered intravenously but recommended that a 50% reduction in dose for severely impaired patients be accomplished by lengthening the dosing interval to 24 h.

The ability of hemodialysis to clear ciprofloxacin has been examined by Boelaert et al. (7) and Singlas et al. (53). Extraction of ciprofloxacin by hemodialysis was 23 to 31%, and dialysis clearance was 40 to 57 ml/min. In one study (7) the half-life fell from a mean of 5.8 h on an interdialysis day to 3.2 h during dialysis.

Table 12. Model-independent pharmacokinetic parameters for 200 mg of ciprofloxacin administered to patients with various degrees of renal dysfunction[a] (from reference 18)

Group[b]	Mean ± SD				
	V_{area} (liters/kg)	CL_S (liters/h per 1.73 m^2)	CL_R (liters/h per 1.73 m^2)	$t_{1/2}$ (h)	% CL_R/CL_S
>6.0	2.49 ± 0.46	26.8 ± 5.7	16.4 ± 3.5	4.27 ± 0.84	61.6 ± 8.7
$\geq3.6, <6.0$	3.19 ± 1.26	26.3 ± 10.3	11.9 ± 5.3	6.12 ± 1.61	44.9 ± 11.9
$\geq0.6, <3.6$	2.38 ± 0.62	15.0 ± 3.8	4.5 ± 2.6	7.70 ± 1.22	29.2 ± 13.2
0	2.73 ± 0.92	15.4 ± 4.3	0.0	8.55 ± 3.27	0.0

[a]For abbreviations see Tables 1, 2, and 10, footnote *a*.
[b]Creatinine clearance in liters per hour per 1.73 m^2.

Golper et al. (27) found little clearance (6.2 ml/min per 1.73 m^2) of ciprofloxacin during chronic ambulatory peritoneal dialysis. Considerable dwell times were necessary before attainment of drug concentrations in the dialysate sufficient to inhibit pathogens such as *Staphylococcus aureus*, which may infect these patients. Shalit et al. (52) reported similar findings. Dialysate concentrations did not exceed 1 μg/ml (the MIC for 90% of the strains of *S. aureus*) until 2.5 h postdose. Consequently, minimum dwell times of 4 to 6 h will be necessary to attain drug concentrations in dialysis fluid adequate for this or other less susceptible pathogens.

Drusano et al. (G. Drusano, A. Forrest, K. Plaisance, P. Garjian, G. Yuen, M. Egorin, M. Didolkar, and H. C. Standiford, *Program Abstr. 27th Intersci. Conf. Antimicrob. Agents Chemother.*, abstr. no. 1269, 1987) and Lettieri et al. (J. Lettieri, R. W. Frost, G. Krol, K. Lasseter, and E. C. Shamblin, *27th ICAAC*, abstr. no. 1268, 1987) examined the effect of altered hepatic function on ciprofloxacin disposition. Drusano et al. administered ciprofloxacin (500 mg) to patients with Childs class A and Childs class B cirrhosis and found no alteration in drug disposition based on liver dysfunction. Indocyanine green dye clearance, an index of liver blood flow, was not predictive of ciprofloxacin clearance. Lettieri et al. examined patients with mild to moderate hepatic cirrhosis documented by biopsy. No evidence was seen for an alteration in drug disposition when compared with normal controls. Both studies concluded that no adjustment in dose was warranted for patients with mild to moderate liver disease. Patients with severe liver disease need to be studied before specific recommendations can be made. Data are also lacking for patients with combined renal and hepatic dysfunction. A major dose adjustment seems likely to be required, because both major clearance routes are impaired in such patients.

The impact of cystic fibrosis on the disposition of ciprofloxacin has been reported by multiple investigators (5, 26, 39). In contrast to many β-lactam drugs and aminoglycosides, total serum clearance and renal clearance are unaltered in cystic fibrosis patients.

Because elderly patients have alterations in lean body mass in comparison with younger patients and because creatinine clearance declines with age, it is important to examine the disposition of drugs in the elderly. LeBel et al. (38) compared the disposition of ciprofloxacin in 12 healthy elderly (mean age, 75 years) and 12 healthy young (mean age, 22 years) volunteers. Mean creatinine clearance among the elderly was 42 ml/min. Significant alterations in many pharmacokinetic parameter values were seen in the elderly. Concentrations in serum were higher throughout the dosing interval, with peak concentrations of 3.24 $\pm$ 0.79 μg/ml in the elderly and 2.26 $\pm$ 0.75 μg/ml in the young. Drug half-lives were also longer (6.83 versus 3.69 h), with lower apparent clearances (394 versus 908 ml/min), both renal and nonrenal. Consequently, the elderly should be observed carefully for side effects when ciprofloxacin is administered. For the treatment of urinary tract infections, 24-h dosing intervals would seem appropriate in this population. If more severe infections are being treated, the drug can be given every 12 h, but care should be taken to monitor carefully for drug toxicity, especially in patients with altered renal function. Differences in ciprofloxacin pharmacokinetics in elderly patients during acute as compared with convalescent phases of illness were not detected (32).

Febrile, neutropenic cancer patients receiving 400 mg of ciprofloxacin intravenously every 12 h appear to have drug half-lives and serum clearance similar to those reported for normal volunteers (54).

Enoxacin

Enoxacin, the first of the new fluoroquinolones which is a naphthyridine, differs from norfloxacin only by the substitution of a nitrogen for a carbon at position 8 of the quinolone ring (see Fig. 1 in chapter 2), a single difference which improves bioavailability.

Wolf et al. (69) examined enoxacin in single and multiple doses in a small number of volunteers. With multiple doses of 400, 600, and 800 mg administered orally twice daily for 2 weeks in four volunteers, maximum concentrations in serum were reached rapidly (1.4 to 1.7 h) but did not increase in a linear fashion with the dose (Tables 13 and 14). AUCs also did not increase in a linear fashion with the dose, being 18.47 ± 4.68 for the 400-mg dose, 27.30 ± 6.48 for the 600-mg dose, and 26.99 ± 12.89 for the 800-mg dose. Half-lives during the first dosing interval ranged from 4.9 to 5.4 h. In all groups by the final study day, maximum concentrations in serum, half-life, and AUC had increased when compared with day 1 (Table 13).

Table 13. Pharmacokinetic parameters of single and multiple doses of enoxacin (from reference 69)

Parameter[a] (unit)	Mean ± SD at dose (mg b.i.d.[b]) of:		
	400	600	800
Single dose (day 1)			
C_{max} (mg/liter)	3.09 ± 0.86	4.18 ± 0.74	3.70 ± 1.30
T_{max} (h)	1.42 ± 0.67	1.67 ± 0.75	1.50 ± 0.89
$t_{1/2}$ (h) (harmonic)	4.90	4.74	5.44
$AUC_{0-\infty}$ (μg·h/ml)	18.47 ± 4.68	27.30 ± 6.48	26.99 ± 12.89
Steady state (day 15)			
C_{max} (mg/liter)	4.53 ± 0.81	5.90 ± 2.07	6.91 ± 2.29
T_{max} (mg/liter)	1.05 ± 0.40	2.02 ± 0.72	2.49 ± 0.54
$t_{1/2}$ (h) (harmonic)	5.66	5.98	6.27
AUC_{0-12} (μg·h/ml)	25.82 ± 7.18	37.30 ± 13.85	47.36 ± 15.81

[a] $AUC_{0-\infty}$, AUC for zero time to infinity; for other abbreviations see Tables 1 and 8, footnote *a*.
[b] b.i.d., Twice a day.

Table 14. Alteration in drug clearance with repeated administration of enoxacin (from reference 69)

Parameter	Mean value (ml/min per kg) at dose (mg b.i.d.[a]) of:		
	400	600	800
Single dose (day 1)			
Total body clearance	5.78	6.00	9.60
Renal clearance	3.60	3.83	5.42
Nonrenal clearance	2.18	2.17	4.18
Steady state (day 15)			
Total body clearance	4.17	4.41	4.85
Renal clearance	2.40	2.80	2.42
Nonrenal clearance	1.77	1.91	2.43
Day 15/day 1 ratio			
Total body clearance	0.72	0.79	0.51
Renal clearance	0.67	0.73	0.45
Nonrenal clearance	0.81	0.88	0.58

[a]b.i.d., Twice a day.

Chang et al. (11) examined enoxacin doses of 200 and 800 mg administered both intravenously and orally in a four-way randomized crossover design study involving eight subjects (Fig. 12). The doses administered intravenously were given at a constant rate of infusion over 1 h. The half-life, AUC (normalized for dose), and both total and renal clearance were dose dependent, confirming the nonlinearity uncovered in the multiple-dose study discussed above. When the doses administered intravenously were examined, mean terminal elimination half-lives were 3.28 h for the 200-mg dose and 4.68 h for the 800-mg dose. Somewhat longer half-lives have been reported by other investigators (67). Total serum clearances were 9.25 and 6.85 ml/min per kg, respectively, with renal clearance accounting for approximately 55% of the total (Table 15), indicating that renal tubular secretion occurs.

Approximately 50% of the administered dose was recovered as enoxacin in the urine, with concentrations in urine remaining above the MIC for clinically relevant pathogens for at least 24 h. A substantial amount of the oxometabolite was also recovered in the urine, averaging 16% of the 200-mg dose and 11% of the 800-mg dose.

The absolute oral bioavailability of enoxacin was high (87 to 91%) and independent of dose (11). As with other quinolones examined, the onset of absorption was delayed by a meal, but other pharmacokinetic parameter values were unaltered (55).

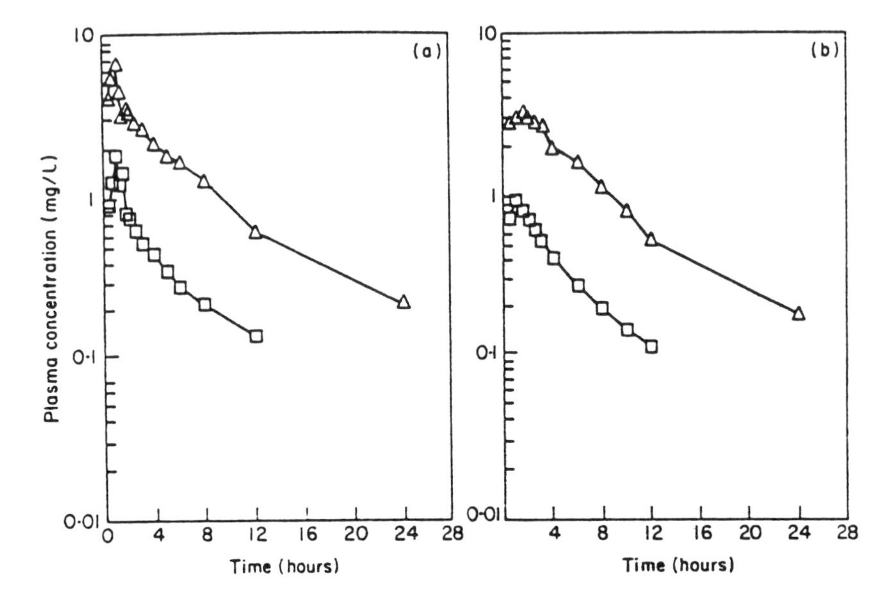

Figure 12. Mean concentrations of enoxacin in plasma of human volunteers after 200 (□)- and 800 (△)-mg intravenous (a) and oral (b) doses. (From reference 11.)

Protein binding of enoxacin and its 4-oxo metabolite was determined by Wijnands et al. (61) in 8 volunteers receiving 400 mg and in 11 volunteers receiving 600 mg as single doses. The parent compound binding averaged 48 ± 11% for the 400-mg dose and 54 ± 12% for the 600-mg dose. The binding for the metabolite was 58 ± 7 and 69 ± 14%, respectively.

Like other fluoroquinolones, enoxacin penetrates well into tissues. Wise et al. (67) examined penetration into inflammatory blister fluid for enoxacin given as a 600-mg oral dose or a 400-mg intravenous infusion over 1 h. Penetration, calculated as a ratio of the AUC in blister fluid to the AUC in serum, was 113 and 133%, respectively.

To evaluate the effect of renal dysfunction on enoxacin disposition, Nix et al. (48) gave single doses of 400 mg to 28 patients with varying renal function. Volumes of distribution and maximum concentrations in serum were independent of renal function. The apparent clearance, renal clearance, and half-life changed with increasing renal dysfunction (Table 16). A maximum decrease in the apparent clearance of 58% was noted. The half-life increased from 4.9 ± 1.2 to 10.5 ± 3.82 h. Renal clearance was shown to be highly correlated with creatinine clearance. These authors concluded that a 50% reduction in the enoxacin dose was appropriate in patients with a creatinine clearance of 30 ml/min or less.

Table 15. Mean pharmacokinetic parameters in human volunteers after single 200- and 800-mg intravenous and oral doses of enoxacin[a] (from reference 11)

Parameter (unit)	Mean RSD[a] (%)			
	Intravenous dose (mg)		Oral dose (mg)	
	200	800	200	800
C_{max} (mg/liter)	1.83 (45)	6.58 (26)	1.02 (25)	3.80 (13)
T_{max} (h)	1.11 (20)	1.00 (13)	1.00 (50)	1.38 (61)
$t_{1/2}$ (h)	3.28 (31)	4.68 (11)	3.16 (18)	4.93 (30)
$AUC_{0-\infty}$ (mg/h per liter)	5.35 (13)	29.08 (23)	4.67 (30)	25.75 (13)
Normalized $AUC_{0-\infty}$ (mg/h per liter)	5.35 (13)	7.27 (23)	4.67 (30)	6.44 (13)
V_B (liters/kg)	2.85 (30)	2.78 (10)		
CL (ml/min per kg)	9.25 (14)	6.85 (18)		
CL_R (ml/min per kg)	4.96 (20)	3.78 (22)	5.30 (21)	3.81 (16)
CL_{NR} (ml/min per kg)	4.92 (19)	3.38 (17)		

[a]RSD, Relative standard deviation; CL_{NR}, nonrenal clearance. For other abbreviations see Tables 1, 4, and 13, footnote *a*.

The decrease in serum protein binding of enoxacin (40.1 to 13.5%) seen in patients with renal failure is not expected to be of clinical importance.

Four hemodialysis patients have been evaluated (47). Dialysis clearance did not contribute extensively to drug elimination, averaging 45.8 ± 18.3 ml/min. Less than five percent of the administered dose (assuming complete bioavailability) was recovered in the dialysate as the parent compound. Oxoenoxacin was not identified in the dialysate, and hence, hemodialysis is not efficient for the removal of this metabolite. Multiple-dose studies are needed to determine the degree of oxoenoxacin accumulation and removal by hemodialysis.

Table 16. Mean pharmacokinetic parameters for nondialysis subjects after oral administration of 400 mg of enoxacin (from reference 48)

Parameter[a] (unit)	Mean for following group[b]:			
	>60	≥30, <60	≥15, <30	<15[c]
C_{max} (mg/liter)	2.25 ± 0.57	2.60 ± 0.54	2.82 ± 0.78	1.93 ± 0.65
T_{max} (h)	1.6 ± 0.96	1.0 ± 0	2.1 ± 1.1	2.4 ± 1.1
AUC (mg·h/liter)[d]	16.2 ± 5.3	26.2 ± 6.2	36.2 ± 11.9	28.2 ± 12.2
Half-life (h)[d]	4.91 ± 1.22	7.45 ± 1.84	10.5 ± 3.82	9.36 ± 3.54
Clearance[d]/fraction absorbed (ml/min)	466 ± 202	264 ± 60	197 ± 50	286 ± 150
Renal[d] clearance (ml/min)	193 ± 47.3	83 ± 27.4	30 ± 8.8	13 ± 5.2
Volume of distribution/fraction absorbed (liters/kg)	2.51 ± 0.62	2.11 ± 0.37	2.30 ± 0.19	3.29 ± 1.68

[a]For abbreviations see Table 1, footnote *a*.
[b]Creatinine clearance in milliliters per minute.
[c]Not on hemodialysis.
[d]Indicates significant difference between groups ($P < 0.05$).

Two studies have addressed the disposition of enoxacin in elderly patients (47, 63). In elderly patients (mean age, 82 years; range, 70 to 100) receiving 200 mg of enoxacin orally twice daily, the mean serum half-life (6.1 ± 2.2 h) was longer than that described for normal volunteers, with a wide observed range (3.2 to 10.1 h) (63). Similarly prolonged and variable half-lives have been reported in other elderly patients receiving enoxacin (47). Peak and trough concentrations in serum increased between days 1 and 3 (peaks, 1.5 μg/ml on day 1 and 2.65 μg/ml on day 3). AUCs for the oxoenoxacin metabolite were 0.35 ± 0.23 of AUCs for enoxacin. Less accumulation was seen for the metabolite relative to the parent compound on days 3 and 5. Although it was concluded that dose alterations for enoxacin were unnecessary in the elderly, the wide range of half-lives (greater than fourfold) and AUCs (greater than ninefold) suggests that caution should be exercised in using conventional doses in this patient population.

Fleroxacin

Fleroxacin (Ro 23-6240, AM 833) is a trifluorinated quinolone with balanced renal and nonrenal clearance. The drug possesses fluorines attached to positions 6 and 8 of the quinolone ring as well as a fluoroethyl substituent at position 1 (see Fig. 1 of chapter 2).

Weidekamm et al. (60) examined the single-dose pharmacokinetics of 200, 400, and 800 mg given orally, as well as 100 mg given intravenously in a 20-min infusion. In addition, the multiple-dose pharmacokinetics of 800 and 1,200 mg given by mouth daily for 10 days was studied (Fig. 13).

The terminal half-life was 8.6 ± 1.3 h (100 mg given intravenously) to 11.8 ± 2.8 h (1,200 mg given orally at the steady state). The volume of distribution was large, with an average value of 110 ± 36 ml/min for the intravenous dose, with renal clearance accounting for slightly more than 60% of the total clearance (Table 17). Protein binding was 23%. Urinary recovery of the parent compound accounted for 50 to 65% of the dose, with *N*-desmethyl and *N*-oxide metabolites accounting for another 6.5 to 11% of the administered dose. Tubular secretion probably has little to do with the renal handling of the drug, because the coadministration of probenecid had no effect on the parameters examined. The ratio of AUC to the dose for the single-dose oral studies revealed that the drug behaved in a dose-linear fashion over the range of 200 to 800 mg. The AUCs for intravenous and oral doses were similar, indicating virtually complete oral absorption.

For multiple dosing of 800 and 1,200 mg, drug accumulation from days 1 to 10 was in the range predicted from the half-life of the drug and the dosing interval. Ratios of AUC for 800 and 1,200 mg on days 1 and 10 also provided good evidence for dose proportionality. The terminal half-life did

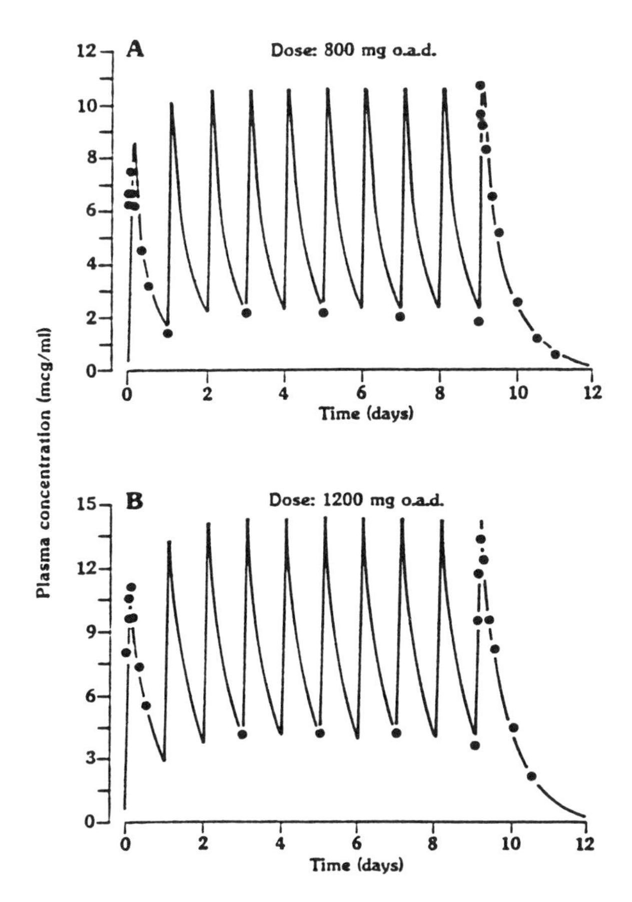

Figure 13. Mean fleroxacin concentrations during multiple dosing of 800 (A) and 1,200 (B) mg once daily (o.a.d.) on 10 consecutive days ($n = 6$). The graph shows the plasma profiles which were generated by iteration over all measured concentration points. (From reference 60.)

show a modest, but statistically insignificant, increase between days 1 and 10. The total serum clearance was lower than that seen in the intravenous dosing study (128 ± 25 versus 168 ± 36 ml/min). Whether this difference represents a trend toward saturability at the higher doses or differences in study populations is unknown. The doses at which potential saturation was observed, however, are above those likely to be used clinically.

Fleroxacin has been evaluated for its ability to penetrate into inflammatory exudate by Wise et al. (65). The AUC for the inflammatory fluid averaged 90 ± 6% of the AUC for serum, indicating excellent penetration.

Table 17. Pharmacokinetic parameters after intravenous (i.v.) and oral (p.o.) administration of fleroxacin (from reference 60)

Parameter[a] (unit)	Mean ± SD for following dose:				
	100 mg i.v. ($n = 6$)	200 mg p.o. ($n = 12$)	400 mg p.o. ($n = 12$)	800 mg p.o. ($n = 12$)	400 mg p.o. + probenecid ($n = 5$)
C_{max} (μg/ml)	2.85 ± 1.17	2.33 ± 0.65	4.36 ± 1.15	7.04 ± 0.84	3.95 ± 0.47
T_{max} (h)	0.33[b]	1.1 ± 0.8	1.3 ± 1.1	1.9 ± 1.0	1.2 ± 0.8
$t_{1/2}$ (h)	8.6 ± 1.3	8.9 ± 1.0	9.2 ± 1.7	10.3 ± 1.4	10.9 ± 1.4
$AUC_{0-\infty}$ (μg·h/ml)	10.2 ± 1.7	20.9 ± 2.4	48.3 ± 9.3	106.1 ± 16.9	61.0 ± 9.1
V_{ss} (liters)	110.1 ± 25.5				
CL_S (ml/min)	168.0 ± 36.0				
CL_R (ml/min)	105.2 ± 27.6				
Urinary excretion (0–60 h) (% of dose)					
Fleroxacin	62.4 ± 7.1	64.6 ± 10.9	49.8 ± 9.2	51.8 ± 13.8	50.1 ± 5.8
N-Desmethyl metabolite	3.6 ± 1.0	4.7 ± 1.6	5.0 ± 1.3	6.5 ± 2.1	6.1 ± 1.7
N-Oxide metabolite	2.9 ± 0.9	3.1 ± 1.7	4.2 ± 1.1	5.0 ± 1.5	4.7 ± 1.7

[a]For abbreviations see Tables 1, 10, and 13, footnote *a*.
[b]Endpoint of infusion.

Summary

A comparison of the pharmacokinetic properties of the quinolone agents is presented in Table 18. To minimize the variability in approaches between laboratories, data are drawn from a single center (Wise et al.).

Finally, recommendations for dose alteration in renal or hepatic dysfunction are presented in Table 19.

Table 18. Comparative pharmacokinetics of the new fluoroquinolones[a] (summarized from references 65 and 66)

Fluoroquinolone (dose [mg])	Mean ± SD				
	C_{max} (μg/ml)	$AUC_{0-\infty}$ (μg·h/ml)	$t_{1/2}$ (h)	Urinary recovery (% [24h])	Blister fluid penetration (% [AUC_{serum}])
Ofloxacin (600, p.o.)	10.7 ± 6.4	57.5 ± 11.3	7.0 ± 1.1	73.0 ± 11.9	125
Pefloxacin (400, i.v.)	NA	56.1 ± 12.1	10.5 ± 1.5	4.9 ± 0.6	69
Norfloxacin (400, p.o.)	1.45 ± 0.1	5.4 ± 1.7	3.2 ± 0.5	27.0 ± 8.6	106
Ciprofloxacin (500, p.o.)	2.3 ± 0.7	9.9 ± 2.4	3.9 ± 0.8	30.6 ± 9.8	117
Enoxacin (600, p.o.)	3.7 ± 0.5	28.8 ± 4.9	6.2 ± 1.1	61.2 ± 7.4	114
Fleroxacin (400, p.o.)	6.1 ± 2.2	78.3 ± 9.4	12.0 ± 1.0	42.5 ± 3.9[b]	90

[a]p.o., Orally; i.v., intravenous; NA, not available. For other abbreviations see Tables 1 and 13, footnote *a*.
[b]Urinary recovery at 72 h was 58.6 ± 5.0%.

Table 19. Dose alteration with organ system dysfunction[a]

Quinolone	Dose alteration[b]	
	Renal	Hepatic[c]
Ofloxacin	50% Reduction at 40–50 ml/min; 75% reduction at 10–20 ml/min	Unnecessary
Pefloxacin	Unnecessary	50% Reduction at PT[d] ↑ of 100%; 75% reduction at PT ↑ of 200%
Norfloxacin	33–50% Reduction at 20–30 ml/min	Unnecessary[e]
Ciprofloxacin	33–50% Reduction at 20–30 ml/min	Unnecessary[e]
Enoxacin	33–50% Reduction at 20–30 ml/min	Insufficient data

[a]Data for the recommendations were drawn from references 13, 18, 20, 22, 23, 46, and 48.
[b]Dose reduction may take place as a dose decrease with the maintenance of the dosing interval or prolongation of the dosing interval with the maintenance of the dose. If interval prolongation is chosen, milligrams per day should be employed in the calculation.
[c]Recommendations for pefloxacin in hepatic failure should be regarded with caution because of the large amount of interindividual variation.
[d]PT, Prothrombin time.
[e]Data are available for mild to moderate hepatic disease only. Investigation is required for severe hepatic disease.

Literature Cited

1. **Adhami, Z. N., R. Wise, and B. Crump.** 1984. The pharmacokinetics and tissue penetration of norfloxacin. *J. Antimicrob. Chemother.* **13:**87–92.
2. **Aronoff, G. E., C. H. Kenner, R. S. Sloan, and S. T. Pottratz.** 1984. Multiple-dose ciprofloxacin kinetics in normal subjects. *Clin. Pharmacol. Ther.* **36:**384–388.
3. **Barre, J., G. Houin, and J. P. Tillement.** 1984. Dose-dependent pharmacokinetic study of pefloxacin, a new antibacterial agent, in humans. *J. Pharm. Sci.* **73:**1379–1382.
4. **Beermann, D., H. Scholl, W. Wingender, D. Forster, E. Beubler, et al.** 1986. Metabolism of ciprofloxacin in man, p. 141–146. *In* H. C. Neu and H. Weuta (ed.), *1st International Ciprofloxacin Workshop, Leverkusen 1985.* Excerpta Medica, Amsterdam.
5. **Bender, S. W., A. Dalhoff, P. M. Shah, R. Strehal, and H. G. Posselt.** 1986. Ciprofloxacin pharmacokinetics in patients with cystic fibrosis. *Infection* **14:**17–21.
6. **Bergan, T., S. B. Thorsteinsson, R. Solberg, L. Bjornskau, I. M. Kostad, and S. Johnsen.** 1987. Pharmacokinetics of ciprofloxacin: intravenous and increasing oral doses. *Am. J. Med.* **82**(Suppl. 4A):97–102.
7. **Boelaert, J., Y. Valcke, M. Schurgers, R. Daneels, M. Rosseneu, M. T. Rosseel, and M. G. Bogaert.** 1985. The pharmacokinetics of ciprofloxacin in patients with impaired renal function. *J. Antimicrob. Chemother.* **16:**87–93.
8. **Brittain, D. C., B. E. Scully, M. J. McElrath, R. Steinman, P. Labthavikul, and H. Neu.** 1985. The pharmacokinetics and serum and urine bactericidal activity of ciprofloxacin. *J. Clin. Pharmacol.* **25:**82–88.

9. **Brumfitt, W., I. Franklin, D. Grady, J. M. T. Hamilton-Miller, and A. Iliffe.** 1984. Changes in the pharmacokinetics of ciprofloxacin and fecal flora during administration of a 7-day course to human volunteers. *Antimicrob. Agents Chemother.* **26:**757–761.
10. **Campoli-Richards, D. M., J. P. Monk, A. Price, P. Benfield, P. A. Todd, and A. Ward.** 1988. Ciprofloxacin: a review of its antibacterial activity, pharmacokinetic properties and therapeutic use. *Drugs* **35:**373–447.
11. **Chang, T., A. Black, A. Dunky, R. Wolf, A. Sedman, J. Latts, and P. G. Welling.** 1988. Pharmacokinetics of intravenous and oral enoxacin in healthy volunteers. *J. Antimicrob. Chemother.* **21**(Suppl. B):49–56.
12. **Cofsky, R. D., L. duBouchet, and S. H. Landesman.** 1984. Recovery of norfloxacin in feces after administration of a single oral dose to human volunteers. *Antimicrob. Agents Chemother.* **26:**110–111.
13. **Danan, G., G. Montay, R. Cunci, and S. Erlinger.** 1985. Pefloxacin kinetics in cirrhosis. *Clin. Pharmacol. Ther.* **38:**439–442.
14. **Davis, R. L., J. R. Koup, J. Williams-Warren, A. Weber, and A. L. Smith.** 1985. Pharmacokinetics of three oral formulations of ciprofloxacin. *Antimicrob. Agents Chemother.* **28:**74–77.
15. **Dow, J., J. Chazal, A. M. Frydman, P. Janny, R. Woehrle, F. Djebbar, and J. Gaillot.** 1986. Transfer kinetics of pefloxacin into cerebro-spinal fluid after one hour iv infusion of 400 mg in man. *J. Antimicrob. Chemother.* **17**(Suppl. B):81–87.
16. **Drusano, G. L., K. I. Plaisance, A. Forrest, and H. C. Standiford.** 1986. Dose ranging study and constant infusion evaluation of ciprofloxacin. *Antimicrob. Agents Chemother.* **30:**440–443.
17. **Drusano, G. L., H. C. Standiford, K. Plaisance, A. Forrest, J. Leslie, and J. Caldwell.** 1986. Absolute oral bioavailability of ciprofloxacin. *Antimicrob. Agents Chemother.* **30:**444–446.
18. **Drusano, G. L., M. Weir, A. Forrest, K. Plaisance, T. Emm, and H. C. Standiford.** 1987. Pharmacokinetics of intravenously administered ciprofloxacin in patients with various degrees of renal function. *Antimicrob. Agents Chemother.* **31:**860–864.
19. **Dudley, M. N., J. Ericson, and S. H. Zinner.** 1987. Effect of dose on serum pharmacokinetics of intravenous ciprofloxacin with identification and characterization of extravascular compartments using noncompartmental and compartmental pharmacokinetic models. *Antimicrob. Agents Chemother.* **31:**1782–1786.
20. **Eandi, M., I. Viano, F. DiNola, L. Leone, and E. Genazzani.** 1983. Pharmacokinetics of norfloxacin in healthy volunteers and patients with renal and hepatic damage. *Eur. J. Clin. Microbiol.* **2:**253–259.
21. **Edlund, C., T. Bergan, K. Josefsson, R. Solberg, and C. E. Nord.** 1987. Effect of norfloxacin on human oropharyngeal and colonic microflora and multiple-dose pharmacokinetics. *Scand. J. Infect. Dis.* **19:113–121.**
22. **Fillastre, J. P., T. Hannedouche, A. Leroy, and G. Humbert.** 1984. Pharmacokinetics of norfloxacin in renal failure. *J. Antimicrob. Chemother.* **14:**439.
23. **Fillastre, J. P., A. Leroy, and G. Humbert.** 1987. Ofloxacin pharmacokinetics in renal failure. *Antimicrob. Agents Chemother.* **31:**156–160.
24. **Forrest, A., M. Weir, K. I. Plaisance, G. L. Drusano, J. Leslie, and H. C. Standiford.** 1988. Relationships between renal function and disposition of oral ciprofloxacin. *Antimicrob. Agents Chemother.* **32:**1537–1540.
25. **Frydman, A. M., Y. Le Roux, M. A. Lefebvre, F. Djebbar, J. B. Fourtillan, and J. Gaillot.** 1986. Pharmacokinetics of pefloxacin after repeated intravenous and oral administration (400 mg bid) in young healthy volunteers. *J. Antimicrob. Chemother.* **17**(Suppl. B):65–79.

26. **Goldfarb, J., G. P. Wormser, M. A. Inchiosa, Jr., G. Guideri, M. Diaz, R. Gandhi, C. Goltzman, and A. V. Macia.** 1986. Single-dose pharmacokinetics of oral ciprofloxacin in patients with cystic fibrosis. *J. Clin. Pharmacol.* **26:**222–226.
27. **Golper, T. A., A. I. Hartstein, V. H. Morthland, and J. M. Christensen.** 1987. Effects of antacids and dialysate dwell times on multiple-dose pharmacokinetics of oral ciprofloxacin in patients on continuous ambulatory peritoneal dialysis. *Antimicrob. Agents Chemother.* **31:**1787–1790.
28. **Gonzalez, M. A., A. H. Moranchel, S. Duran, A. Pichardo, J. L. Magana, B. Painter, and G. L. Drusano.** 1985. Multiple-dose ciprofloxacin dose ranging and kinetics. *Clin. Pharmacol. Ther.* **37:**633–637.
29. **Gonzalez, M. A., A. H. Moranchel, S. Duran, A. Pichardo, J. L. Magana, B. Painter, A. Forrest, and G. L. Drusano.** 1985. Multiple-dose pharmacokinetics of ciprofloxacin administered intravenously to normal volunteers. *Antimicrob. Agents Chemother.* **28:**235–239.
30. **Gonzalez, M. A., F. Uribe, S. D. Moisen, A. P. Fuster, A. Selen, P. G. Welling, and B. Painter.** 1984. Multiple-dose pharmacokinetics and safety of ciprofloxacin in normal volunteers. *Antimicrob. Agents Chemother.* **26:**741–744.
31. **Granneman, G. R., K. M. Snyder, and V. S. Shu.** 1986. Difloxacin metabolism and pharmacokinetics in humans after single oral doses. *Antimicrob. Agents Chemother.* **30:**689–693.
32. **Guay, D. R. P., W. M. Awni, P. K. Peterson, S. Obaid, R. Breitenbucher, and G. R. Matzke.** 1987. Pharmacokinetics of ciprofloxacin in acutely ill and convalescent elderly patients. *Am. J. Med.* **82**(Suppl. 4a):124–129.
33. **Hoffken, G., H. Lode, C. Prinzing, K. Borner, and P. Koeppe.** 1985. Pharmacokinetics of ciprofloxacin after oral and parenteral administration. *Antimicrob. Agents Chemother.* **27:**375–379.
34. **Hughes, P. J., D. B. Webb, and A. W. Asscher.** 1984. Pharmacokinetics of norfloxacin (MK 366) in patients with impaired kidney function—some preliminary results. *J. Antimicrob. Chemother.* **13**(Suppl. B):55–57.
35. **Joos, B., B. Ledergerber, M. Flepp, J.-D. Bettex, R. Luthy, and W. Siegenthaler.** 1985. Comparison of high-pressure liquid chromatography and bioassay for determination of ciprofloxacin in serum and urine. *Antimicrob. Agents Chemother.* **27:**353–356.
36. **Kalager, T., A. Digranes, T. Bergan, and T. Rolstad.** 1986. Ofloxacin: serum and skin blister fluid pharmacokinetics in the fasting and non-fasting state. *J. Antimicrob. Chemother.* **17:**795–800.
37. **Kumasaka, Y., H. Nakahata, K. Imamura, and K. Takebe.** 1981. Fundamental study on AM-715. *Chemotherapy* (Tokyo) **29**(Suppl. 4):56–65.
38. **LeBel, M., G. Barbeau, M. G. Bergeron, D. Roy, and F. Vallee.** 1986. Pharmacokinetics of ciprofloxacin in elderly subjects. *Pharmacotherapy* **6(2):**87–91.
39. **LeBel, M., M. G. Bergeron, F. Vallee, C. Fiset, G. Chasse, P. Bigonesse, and G. Rivard.** 1986. Pharmacokinetics and pharmacodynamics of ciprofloxacin in cystic fibrosis patients. *Antimicrob. Agents Chemother.* **30:**260–266.
40. **LeBel, M., F. Vallee, and M. G. Bergeron.** 1986. Tissue penetration of ciprofloxacin after single and multiple doses. *Antimicrob. Agents Chemother.* **29:**501–505.
41. **Ledergerber, B., J.-D. Bettex, B. Joos, M. Flepp, and R. Luthy.** 1985. Effect of standard breakfast on drug absorption and multiple-dose pharmacokinetics of ciprofloxacin. *Antimicrob. Agents Chemother.* **27:**350–352.
42. **Leroy, A., F. Borsa, G. Humbert, P. Bernadet, and J. P. Fillastre.** 1987. The pharmacokinetics of ofloxacin in healthy adult male volunteers. *Eur. J. Clin. Pharmacol.* **31:**629–630.

43. **Lockley, M. R., R. Wise, and J. Dent.** 1984. The pharmacokinetics and tissue penetration of ofloxacin. *J. Antimicrob. Chemother.* **14:**647–652.
44. **Lode, H., G. Hoffken, P. Olschewski, B. Sievers, A. Kirch, K. Borner, and P. Koeppe.** 1987. Pharmacokinetics of ofloxacin after parenteral and oral administration. *Antimicrob. Agents Chemother.* **31:**1338–1342.
45. **Montay, G., Y. Goueffon, and F. Rogquet.** 1984. Absorption, distribution, metabolic fate, and elimination of pefloxacin mesylate in mice, rats, dogs, monkeys, and humans. *Antimicrob. Agents Chemother.* **25:**463–472.
46. **Montay, G., C. Jacquot, J. Bariety, and R. Cunci.** 1985. Pharmacokinetics of pefloxacin in renal insufficiency. *Eur. J. Clin. Pharmacol.* **29:**345–349.
47. **Naber, K. G., F. Sorgel, F. Gutzler, and B. Bartosik-Wich.** 1986. In vitro activity, pharmacokinetics, clinical safety and therapeutic efficacy of enoxacin in the treatment of patients with complicated urinary tract infections. *Infection* **14**(Suppl. 3)**:**S203–S208.
48. **Nix, D. E., R. W. Schultz, R. W. Frost, A. J. Sedman, D. J. Thomas, A. W. Kinkel, and J. J. Schentag.** 1988. The effect of renal impairment and haemodialysis on single dose pharmacokinetics of oral enoxacin. *J. Antimicrob. Chemother.* **21**(Suppl. B)**:**87–95.
49. **Ozaki, T., H. Uchida, and T. Irikura.** 1981. Studies on metabolism of AM-715 in humans by high performance liquid chromatography. *Chemotherapy* (Tokyo) **29**(Suppl. 4)**:**128–135.
50. **Pecquet, S., A. Andremont, and C. Tancrède.** 1987. Effect of oral ofloxacin on fecal bacteria in human volunteers. *Antimicrob. Agents Chemother.* **31:**124–125.
51. **Plaisance, K. I., G. L. Drusano, A. Forrest, C. I. Bustamante, and H. C. Standiford.** 1987. Effect of dose size on bioavailability of ciprofloxacin. *Antimicrob. Agents Chemother.* **31:**956–958.
52. **Shalit, I., R. B. Greenwood, M. I. Marks, J. A. Pederson, and D. L. Frederick.** 1986. Pharmacokinetics of single-dose oral ciprofloxacin in patients undergoing chronic ambulatory peritoneal dialysis. *Antimicrob. Agents Chemother.* **30:**152–156.
53. **Singlas, E., A. M. Taburet, I. Landru, H. Albin, and J. P. Ryckelinck.** 1987. Pharmacokinetics of ciprofloxacin tablets in renal failure: influence of haemodialysis. *Eur. J. Clin. Pharmacol.* **31:**589–593.
54. **Smith, G. M., M. J. Leyland, I. D. Farrell, and A. M. Geddes.** 1986. Preliminary evaluation of ciprofloxacin, a new 4-quinolone antibiotic, in the treatment of febrile neutropenic patients. *J. Antimicrob. Chemother.* **18**(Suppl. D)**:**165–174.
55. **Somogyi, A. A., F. Bochner, J. A. Keal, P. E. Rolan, and M. Smith.** 1987. Effect of food on enoxacin absorption. *Antimicrob. Agents Chemother.* **31:**638–639.
56. **Swanson, B. N., V. K. Boppana, P. H. Vlaysses, H. H. Rotmensch, and R. K. Ferguson.** 1983. Norfloxacin disposition after sequentially increasing oral doses. *Antimicrob. Agents Chemother.* **23:**284–288.
57. **Tartaglione, T. A., A. C. Raffalovich, W. J. Poynor, A. Espinel-Ingroff, and T. M. Kerkering.** 1986. Pharmacokinetics and tolerance of ciprofloxacin after sequential increasing oral doses. *Antimicrob. Agents Chemother.* **29:**62–66.
58. **Verho, M., V. Malerczyk, E. Dagrosa, and A. Korn.** 1985. Dose linearity and other pharmacokinetics of ofloxacin: a new, broad-spectrum antimicrobial agent. *Pharmatherapeutica* **4:**376–382.
59. **Webb, D. B., D. E. Roberts, J. B. Williams, and A. W. Asscher.** 1986. Pharmacokinetics of ciprofloxacin in healthy volunteers and patients with impaired kidney function. *J. Antimicrob. Chemother.* **18**(Suppl. D)**:**83–87.
60. **Weidekamm, E., R. Portmann, K. Suter, C. Partos, D. Dell, and P. W. Lucker.** 1987. Single- and multiple-dose pharmacokinetics of fleroxacin, a trifluorinated quinolone, in humans. *Antimicrob. Agents Chemother.* **31:**1909–1914.

61. **Wijnands, W. J. A., T. B. Vree, A. M. Baars, and C. L. A. van Herwaarden.** 1988. Pharmacokinetics of enoxacin and its penetration into bronchial secretions and lung tissue. *J. Antimicrob. Chemother.* **21**(Suppl. B):67–77.
62. **Wijnands, W. J. A., T. B. Vree, and C. L. A. van Herwaarden.** 1986. The influence of quinolone derivatives on theophylline clearance. *Br. J. Clin. Pharmacol.* **22**:677–683.
63. **Wise, R., S. L. Baker, M. Misra, and D. Griggs.** 1987. The pharmacokinetics of enoxacin in elderly patients. *J. Antimicrob. Chemother.* **19**:343–350.
64. **Wise, R., and I. A. Donovan.** 1987. Tissue penetration and metabolism of ciprofloxacin. *Am. J. Med.* **82**(Suppl. 4A):103–107.
65. **Wise, R., B. Kirkpatrick, J. Ashby, and D. J. Griggs.** 1987. Pharmacokinetics and tissue penetration of Ro 23-6240, a new trifluoroquinolone. *Antimicrob. Agents Chemother.* **31**:161–163.
66. **Wise, R., D. Lister, C. A. M. McNulty, D. Griggs, and J. M. Andrews.** 1986. The comparative pharmacokinetics of five quinolones. *J. Antimicrob. Chemother.* **18**(Suppl. D):71–81.
67. **Wise, R., D. Lister, C. A. M. McNulty, D. Griggs, and J. M. Andrews.** 1986. The comparative pharmacokinetics and tissue penetration of four quinolones including intravenously admininstered enoxacin. *Infection* **14**(Suppl. 3):S196–S202.
68. **Wise, R., R. M. Lockley, M. Webberly, and J. Dent.** 1984. Pharmacokinetics of intravenously administered ciprofloxacin. *Antimicrob. Agents Chemother.* **26**:208–210.
69. **Wolf, R., R. Eberl, A. Dunky, N. Mertz, T. Chang, J. R. Goulet, and J. Latts.** 1984. The clinical pharmacokinetics and tolerance of enoxacin in healthy volunteers. *J. Antimicrob. Chemother.* **14**(Suppl. C):63–69.
70. **Wolff, M., B. Regnier, C. Daldoss, M. Nkam, and F. Vachon.** 1984. Penetration of pefloxacin into cerebrospinal fluid of patients with meningitis. *Antimicrob. Agents Chemother.* **26**:289–291.
71. **Wolfson, J. S., and D. C. Hooper.** 1985. The fluoroquinolones: structures, mechanisms of action and resistance, and spectra of activity in vitro. *Antimicrob. Agents Chemother.* **28**:581–586.

Chapter 5

Treatment of Urinary Tract Infections with Quinolone Antimicrobial Agents

S. Ragnar Norrby

Department of Infectious Diseases, University of Lund,
Lasarettet S-22185 Lund, Sweden

For many years the quinolones were classified as urinary antiseptics, and it was not unusual that protocols for clinical trials of new antibiotics allowed for concurrent use of these drugs since they were obviously considered to have no systemic effect. We now also know that the older quinolones, e.g., nalidixic acid, cinoxacin, and pipemidic acid, are bactericidal antimicrobial agents with a high degree of activity against many gram-negative and gram-positive aerobic and microaerophilic bacterial species. With the introduction of the fluorinated 4-quinolones, new areas of usage of this group of antimicrobial agents have been opened. However, treatment of urinary tract infections (UTIs) is likely to become one of the main indications for these new agents. This review discusses the new derivatives and their efficacy in treatment of various types of UTI. Treatment of bacterial prostatitis will also be discussed. Emphasis will be put on the five most studied of the newer quinolones: ciprofloxacin, enocaxin, norfloxacin, ofloxacin, and pefloxacin.

In Vitro Activity of Newer Quinolones against Urinary Tract Pathogens

As summarized in Table 1 (for details see chapter 3), all five quinolones reviewed here are highly active against *Escherichia coli, Klebsiella* spp., *Enterobacter* spp., *Citrobacter* spp., *Proteus mirabilis, Proteus vulgaris,* and *Morganella morganii.* For these species, the differences among the com-

Table 1. Summary of in vitro activity of newer quinolones against bacterial species causing UTIs

Species	MIC (μg/ml) for 90% of strains when tested against:				
	Ciprofloxacin (1, 23)[a]	Enoxacin (14, 15)	Norfloxacin (23, 52)	Pefloxacin (15, 38, 61)	Ofloxacin (2)
Escherichia coli	0.016	0.4	0.06	0.12	0.12
Klebsiella pneumoniae	0.06	0.4	0.25	0.5	0.5[b]
Enterobacter cloacae	0.03	0.4	0.13	0.25[c]	0.12
E. aerogenes	0.06	0.8	0.25	ND[d]	0.12
Citrobacter diversus	0.03	0.2	0.13	ND	0.06
C. freundii	0.06	0.2	0.25	0.25	0.5
Serratia marcescens	0.13	0.8	0.5	2	1
Proteus mirabilis	0.03	0.4	0.13	0.25	0.06
P. vulgaris	0.06	1.6	0.13	0.25	0.12
P. stuartii	1	1.6	2	2	1
Morganella morganii	0.016	0.4	0.06	ND	0.12
Pseudomonas aeruginosa	0.5	3.1	2	2	2
P. maltophilia	4	6.3	32	4[e]	16
P. cepacia	8	6.3	32	ND	16
Acinetobacter calcoaceticus subsp. *anitratum*	1	ND	16	1[f]	16[f]
A. calcoaceticus subsp. *lwoffi*	0.25	ND	4	ND	ND
Staphylococcus aureus	0.5	3.1	1	0.5	1
S. epidermidis	0.5	6.3	1	0.5	0.5
S. saprophyticus	0.5	2	4	2	ND
Streptococcus agalactiae	1	25	8	32	4
Enterococcus faecalis	2	25	8	8	8

[a]Reference numbers.
[b]Data for *Klebsiella* spp.
[c]Data for *Enterobacter* spp.
[d]ND, No data.
[e]Data for *Pseudomonas* spp.
[f]Data for *Acinetobacter* spp.

pounds seem to be of little clinical importance. Differences of possible clinical relevance in infections disseminating outside the lower urinary tract exist for *Serratia marcescens* (pefloxacin and ofloxacin are less active than the others, with MICs for 90% of strains of ≥1 μg/ml) and *Proteus vulgaris* (enoxacin is less active, with a MIC for 90% of strains of 1.6 μg/ml). Against *Pseudomonas aeruginosa*, only ciprofloxacin has MICs below 1 μg/ml and the other derivatives inhibit that species at concentrations of 2 μg or more per ml. *Pseudomonas maltophilia, Pseudomonas cepacia*, and *Acinetobacter* spp. are less susceptible and probably resistant to norfloxacin and ofloxacin, while ciprofloxacin, enoxacin, and pefloxacin inhibit these organisms, with MICs for 90% of strains of 0.25 to 6.3 μg/ml (ciprofloxacin is the most active). Against gram-positive organisms causing UTIs, all derivatives are less active than against gram-negative organisms. Ciprofloxacin, norfloxacin, and ofloxacin are the most active compounds, while resistance to enoxacin and pefloxacin is not unusual.

The in vitro activity of quinolones decreases in the presence of Mg^{2+} and Ca^{2+} (7, 57). When MIC determinations have been performed in urine, lower MICs than those found in conventional media have been reported, probably due to a low pH and high concentrations of magnesium in urine (3, 57).

Pharmacokinetics of Quinolones with Special Reference to Treatment of UTIs and Prostatitis

Pharmacokinetic characteristics which may influence the efficacy of a UTI antibiotic are concentrations in plasma, renal tissue, urine, and feces. Unless an antibiotic is intended for use in patients with cystitis alone, concentrations in plasma should be high enough to eliminate pathogens disseminating to blood. Concentrations in renal tissue must exceed the MICs against pathogens causing pyelonephritis. Concentrations in urine should be sufficient to inhibit bacterial regrowth during a majority of the dose interval. Concentrations in feces are of importance for reducing the possibility of early reinfections from the fecal bacterial reservoir. Thus, it is not necessarily true that a high degree of absorption of an antibiotic for treatment of UTIs is better than moderate absorption.

Table 2 summarizes some kinetic data on ciprofloxacin, enoxacin, norfloxacin, and ofloxacin. Pefloxacin has not been included due to its complicated kinetics with a high degree of metabolism and antibacterially active metabolites (25, 45). When the kinetics of the quinolones are compared with their in vitro activity, it is obvious that sufficient concentrations can be achieved in the urine to inhibit all gram-negative urinary tract pathogens, even if the reduced in vitro activity of these compounds in urine is taken into account. This is also the case for gram-positive pathogens when ciprofloxacin, norfloxacin, and ofloxacin are considered. With enoxacin and pefloxacin, the possibility of insufficient concentrations in urine for inhibition of growth of *Streptococcus agalactiae* and enterococci should be considered.

The concentrations of these quinolones in blood are generally low in relation to the doses given. This is more the case with the poorly absorbed derivatives ciprofloxacin and norfloxacin than with enoxacin, ofloxacin, and pefloxacin, which are well absorbed (25). Concentrations above the MICs

Table 2. Summary of pharmacokinetic characteristics of orally administered quinolones (for details see chapter 4)

Quinolone (reference)	Dose (mg)	Pharmacokinetic parameter[a] (mean)			
		C_{max} (μg/ml)	T_{max} (h)	$t_{1/2\beta}$ (h)	UR_{0-24} (% of dose)
Ciprofloxacin (4)	100	0.7	0.9	3.0	12
	250	1.7	1.1	3.2	28
	500	2.3	1.5	3.2	27
	1,000	5.9	1.8	3.4	28
Enoxacin (65)	200	1.2	0.8	4.2	45
	400	2.0	1.5	4.2	47
	800	4.3	1.3	5.4	36
	1,600	8.1	2.0	6.4	42
Norfloxacin (58)	200	0.8	1.1	7.3	55
	400	1.6	1.3	7.4	52
	800	2.4	1.5	6.2	51
	1,600	3.9	1.9	6.8	27
Ofloxacin (44)	100	1.0–1.3	0.5–2.2	3.6–6.7	70–99
	200	2.6	0.8–1.0	5.7–7.0	74–87
	400	4.0–5.6	0.7–1.4	5.0–7.4	74–81
	600	6.8–6.9	1.0–2.8	5.9–6.7	75–94

[a] C_{max}, Maximum concentration of drug in serum; T_{max}, time to C_{max}; $t_{1/2\beta}$, half-life in β-phase; UR_{0-24}, urinary recovery from 0 to 24 h after administration.

for gram-negative pathogens are achieved with moderate doses of all derivatives when *E. coli*, *Klebsiella* spp., *Enterobacter* spp., *Citrobacter* spp., *P. mirabilis*, and *M. morganii* are considered. For *Proteus stuartii*, MICs are higher and peak concentrations in serum above the MICs would be achieved only with ofloxacin at normal doses or by increased doses of the other quinolones. *P. aeruginosa* is less susceptible to all of the quinolones, and only with ciprofloxacin and ofloxacin can blood concentrations above the MIC for most strains be expected. Against staphylococci disseminating to blood from the urinary tract, ciprofloxacin, ofloxacin, and pefloxacin can be expected to be active at normal doses, while norfloxacin and enoxacin are less active against these organisms.

Limited data are available on the penetration of quinolones to renal tissue. Animal experiments have shown norfloxacin concentrations which were three to eight times higher in the kidneys than in serum (46, 49). In humans, Daschner et al. (20) reported ciprofloxacin concentrations which were four to five times higher than the concurrent levels in serum. Thus, therapeutic concentrations for all gram-negative UTI pathogens and most of the gram-positive ones seem to be achievable in renal tissue with all the derivatives discussed here.

In prostatic tissue and prostatic fluid, concentrations of ciprofloxacin and norfloxacin similar to or above concurrent concentrations in serum have been reported by several investigators (5, 9, 12, 18, 19, 30, 35) (Table 3). While there was some variability of concentrations in prostatic tissue in these studies, concentrations in prostatic fluid when studied (9, 18) showed a marked variation. This variability could have been due to contamination with urine, which is difficult to avoid when prostatic fluid is collected. It seems clear, however, that concentrations in prostatic fluid and tissue well above the MICs for most gram-negative pathogens causing prostatitis can be achieved with both ciprofloxacin and norfloxacin.

High concentrations in feces are achieved with the poorly absorbed derivatives, i.e., ciprofloxacin and norfloxacin (10, 13, 22, 51). It should be noted, however, that most probably the quinolones are bound to fecal components, thereby losing much of their antibacterial activity in feces (51, 62). Despite that, the concentrations achieved effectively suppress the aerobic gram-negative fecal flora for several days after the last oral dose (23, 51). With the well-absorbed quinolones, low concentrations in feces are achieved and the suppression of the bowel flora is less pronounced but still considerable (44, 51).

Efficacy in Uncomplicated Cystitis

Despite the fact that more than 90% of all UTIs are uncomplicated cystitis, rather few studies have been performed with the quinolones in patients with such infections. The largest trial was one on the clinical and bacteriological efficacy of norfloxacin at 200 mg twice a day (b.i.d.; 252 evaluable patients) versus norfloxacin at 400 mg b.i.d. (240 evaluable patients) versus trimethoprim-sulfamethoxazole at 160/800 mg b.i.d. (141 evaluable patients) for 7 days in consecutive outpatients with UTIs (52, 60). The short-term bacteriological efficacy (elimination of bacteriuria at 3 to 13 days after treatment) varied between 97.5 and 98.6% in the three groups, and the accumulated bacteriological efficacy (elimination of bacteriuria at 3 to 38 days after treatment) varied between 87.9 and 88.8%; i.e., there were very small differences among the three regimens. However, significantly higher frequencies of adverse reactions were seen in patients randomized to trimethoprim-sulfamethoxazole than in those receiving norfloxacin in this double-blind study. When the two norfloxacin doses were compared, slightly higher frequencies of adverse reactions were seen in the high-dose than in the low-dose group. With norfloxacin, another large study has been performed in female outpatients with uncomplicated cystitis in which 3- versus 7-day treatments with a dose of 400 mg b.i.d. were compared

Table 3. Penetration of ciprofloxacin and norfloxacin into prostatic fluid and prostatic tissue

Reference no.	No. of patients	Antimicrobial agent	Dose (mg)	Concn[a] (μg/ml or μg/g) in:		
				Serum	Prostatic tissue	Prostatic fluid
12	15	Norfloxacin	400	1.2	0.9	ND[b]
5	10	Norfloxacin	400	1.4	1.8	ND
30	12	Ciprofloxacin	500	1.2	3.0	ND
19	15	Ciprofloxacin	750	2.2	2.1	ND
9	3	Ciprofloxacin	500	1.3	2.0	ND
	11	Ciprofloxacin	500	1.7	ND	2.0
35	25	Ciprofloxacin	100[c]	1.2	3.0	ND

[a]All values are means. Serum and tissue or prostatic fluid were obtained concurrently.
[b]ND, No data.
[c]Intravenous administration.

(Swedish/Norwegian Study Group, *Scand. J. Infect. Dis*, in press). In this large randomized, double-blind study, no differences were seen between the groups in terms of elimination of bacteriuria at 3 to 10 days after treatment, while significantly more patients had recurrences at 3 to 4 weeks after treatment in the 3-day group (36 of 193 patients, 18.7%) than in those treated for 7 days (15 of 180 patients, 8.3%).

With the other quinolones the individual clinical trials performed in patients with uncomplicated cystitis are considerably smaller than the norfloxacin studies discussed above. In comparative trials of ofloxacin, higher rates of elimination of bacteriuria were seen in patients receiving ofloxacin than in those receiving nalidixic acid or trimethoprim-sulfamethoxazole, a difference which seems to be due to the better in vitro activity of ofloxacin (8). In U.S. studies of ofloxacin at 100 mg b.i.d. versus 150 mg b.i.d. versus trimethoprim-sulfamethoxazole at 160/800 mg b.i.d. for 7 days, elimination of bacteriuria at 4 to 6 weeks after treatment was seen in 63 of 66 patients (95%), 62 of 67 patients (93%), and 70 of 73 patients (96%), respectively (17). In two small randomized studies better results were obtained with ciprofloxacin than with comparative agents, and again the explanation seems to be markedly better in vitro activity of the quinolone than of the comparative agent (29, 34). In a study by Newsom et al. (50), ciprofloxacin at 100 mg b.i.d. for 5 days was compared with trimethoprim at 200 mg b.i.d. for 5 days. Of 16 patients in each group, only 1 on trimethoprim had persistent bacteriuria at the end of therapy, but 4 patients in the ciprofloxacin group and 3 in the trimethoprim group became reinfected at follow-up 4 to 6 weeks after treatment. In uncontrolled studies of ofloxacin, 443 patients were evaluable for efficacy (44). They received 100 to 600 mg of ofloxacin per day for a mean time of 3.5 days (range, 2 to 17 days). Bacteriological eradication without reinfection at the follow-up was achieved in 94.4% of the patients.

For enoxacin and pefloxacin, limited information on the efficacy of these agents in uncomplicated cystitis is available. In a summary of three studies, presented at the 14th International Congress for Chemotherapy in 1985, Shah and Frech (56) reported 83 to 100% efficacy when 200 to 400 mg of enoxacin was used for 7 days in the treatment of uncomplicated UTIs. Enoxacin at 200 to 400 mg for 1 to 2 weeks eradicated bacteriuria in five of six cases of cystitis (one reinfection) and in five of seven patients with covert bacteriuria (one relapse and one reinfection) (59).

In the above studies treatment was given for several days, normally 5 days or longer. With ofloxacin, single-dose studies showed that with 100 mg of ofloxacin, elimination of bacteriuria was achieved in 30 of 37 patients (81%) in comparison with trimethoprim-sulfamethoxazole and in 89 of 95 patients (94%) in studies versus nalidixic acid (8). In another trial, single-dose treatment of uncomplicated cystitis was evaluated with 100-mg (19 episodes) or 250-mg (19 episodes) doses of ciprofloxacin (26). At 5 days after therapy, three patients in the low-dose group and two in the high-dose group had persistent bacteriuria. Bischoff (6) compared single doses with 400 mg of enoxacin and 3 g of amoxicillin. At 6 weeks after treatment, 35 of 40 patients (82.5%) had sterile urine in the enoxacin group while only 17 of 40 patients (42.4%) had sterile urine in the amoxicillin group. All of these studies included too few patients to allow any conclusions to be drawn as to the efficacy of single-dose treatment of uncomplicated cystitis with the new fluorinated quinolones.

It should be noted that in only very few patients has development of resistance to the quinolones been reported. It is also important to note that superinfections or recurrences with quinolone-resistant bacterial strains were rare in the trials discussed here.

The above studies show that ciprofloxacin, norfloxacin, and ofloxacin are all highly effective agents for the treatment of uncomplicated lower UTIs when low doses are used twice daily for periods of 5 days or longer. The 3-versus 7-day study of norfloxacin demonstrated higher reinfection rates in patients treated for 3 days. Most probably that would also be the case if a single-dose treatment were used. For enoxacin and pefloxacin, the documentation found in this review is still insufficient and does not allow statements on the efficacy of those agents.

Efficacy in Uncomplicated Pyelonephritis

Although pyelonephritis accounts for approximately 8% of all UTIs, clinical trials in patients with this condition normally include few patients.

Considering the in vitro and pharmacokinetic properties of the fluorinated quinolones, they should offer alternatives for oral treatment of upper UTIs. One of the larger sets of data has been compiled on ofloxacin with information from several individual trials (8). In comparative trials versus trimethoprim-sulfamethoxazole, bacterial elimination was achieved in 182 of 223 patients (82%) receiving ofloxacin at 200 mg b.i.d. as compared with 161 of 202 patients (80%) receiving trimethoprim-sulfamethoxazole at 160/800 mg b.i.d. for 7 days. The lack of difference between treatment groups is surprising since 25% of the pathogens were resistant to trimethoprim-sulfamethoxazole and only 3% were resistant to ofloxacin. In an open study Guibert et al. (33) treated 18 patients with pyelonephritis with ciprofloxacin at 500 mg b.i.d. for 10 to 84 days. Similar results have been achieved in another small open study in which ciprofloxacin was used in acute uncomplicated pyelonephritis (13a). Bacteriological elimination without reinfection at 4 to 6 weeks after treatment was reported in 10 patients, 6 had reinfections, and 2 relapsed. In the data file on norfloxacin (Merck Sharp & Dohme, data on file), 22 of 24 patients (92%) with uncomplicated pyelonephritis responded to treatment with norfloxacin at 400 mg b.i.d., with elimination of bacteriuria shortly after the end of treatment. Of seven patients treated with enoxacin at 200 to 400 mg for 1 to 2 weeks, four had no bacteriuria at 5 weeks after treatment, one had persistent infection, and two had relapses (59).

In summary, few data are available on the efficacy of these quinolones in the treatment of acute pyelonephritis. However, the published data, as well as the in vitro activity and pharmacokinetic properties of these agents, make them highly likely to be effective for treatment of acute pyelonephritis.

Efficacy in Complicated UTIs

Many clinical trials of quinolones in UTIs have been performed in patients with complicated lower (and sometimes upper) UTIs. The obvious advantage of the fluorinated quinolones in such patients is that these quinolones have the unique property of being active as oral drugs against *P. aeruginosa* and some other multiresistant gram-negative aerobic and microaerophilic urinary tract pathogens. However, this property makes comparative studies difficult since patients with pseudomonas infections cannot be randomized to other oral antibiotics. The studies will then have to be uncontrolled or comparative versus other fluorinated 4-quinolones or parenteral antibiotics.

In many of the reports reviewed below, there are considerable difficulties in identifying how the authors have defined a complicated UTI. In some trials advanced age in itself has been considered a complicating factor, while in other trials a complicated UTI has been defined as a UTI in a patient with obstructed urine flow, e.g., one with prostatic hyperplasia or tumors, or a patient with factors making the elimination of bacteriuria difficult, e.g., renal stones or residual urine. In the overview below trials in which advanced age only defined a complicated UTI are noted.

Ciprofloxacin

In three comparative studies ciprofloxacin at a dose of 250 mg b.i.d. (68 patients) or 100 mg b.i.d. (33 patients) was compared with trimethoprim-sulfamethoxazole at a dose of 160/800 mg b.i.d. (59 patients), using 7 to 10 days of treatment (40, 55, 64). Eradication of causative pathogens was achieved in 62 patients (91%) receiving ciprofloxacin at 250 mg b.i.d., in 25 (76%) receiving ciprofloxacin at 100 mg b.i.d., and in 44 (75%) receiving trimethoprim-sulfamethoxazole. In another study Naber and Bartosik-Wich (47) compared ciprofloxacin at 500 mg once daily with norfloxacin once daily in groups of 30 patients each. Elimination of bacteriuria was seen in all but one patient in each group. However, 10 patients on ciprofloxacin and 9 on norfloxacin were reinfected or had relapses at late follow-up. Giamarellou et al. (27) treated 19 patients with upper UTIs caused by *P. aeruginosa* with ciprofloxacin at 500 to 750 mg b.i.d. in an open uncontrolled study. Eradication was achieved in 6 patients and 13 relapsed after treatment, 1 with a resistant strain.

Enoxacin

Cox (16) compared enoxacin and sulfamethoxazole-trimethoprim in 56 patients, of whom 46 (23 in each group) were evaluable. The enoxacin dose used was 400 mg b.i.d., and the treatment duration was 14 days. One week after treatment sterilization of the urine was achieved in all patients on sulfamethoxazole-trimethoprim and in 83% of patients on enoxacin. One patient on enoxacin developed a superinfection with a resistant strain of *P. aeruginosa* during treatment. In an open study Huttunen et al. (36) treated 67 elderly patients, of whom 53 had pyelonephritis and 17 had indwelling catheters, with enoxacin at 200 to 400 mg b.i.d. for 4 to 14 days. At 4 to 6 weeks after treatment, 46 patients (69%) had no bacteriuria. Foot et al. (24) used enoxacin at 400 mg b.i.d. for 14 days in 25 evaluable patients with complicated lower UTIs and reported reinfections or relapse in 12 patients at 4 to 6 weeks after the last dose. Elimination of bacteriuria was achieved in 3 of 10 patients with *P. aeruginosa*. Naber et al. (48) treated 25 patients with complicated UTIs with enoxacin at 400 mg b.i.d. for 7 days (1 patient was treated for 14 days). At 2 weeks after treatment, relapses were reported in six patients (two of three patients with *P. aeruginosa*) and one patient had persistent growth of a *Klebsiella pneumoniae* strain.

Norfloxacin

Norfloxacin has been evaluated in several studies in patients with complicated UTIs. Leigh et al. (43) compared norfloxacin at 400 mg b.i.d. with amoxicillin at 250 mg three times a day for 5 to 9 days and found sterile urine in 19 of 20 patients (95%) at 5 to 9 days after treatment. The corresponding elimination rate in the amoxicillin group was 15/20 (75%), a difference explained by a higher frequency of amoxicillin resistance. In two studies norfloxacin at 400 mg b.i.d. for 10 days has been compared with trimethoprim-sulfamethoxazole at 160/800 mg b.i.d. for 10 days (28, 31). Not all of the patients had complicated UTIs, but a majority of the infections were hospital acquired. At 4 to 6 weeks posttreatment 39 of 45 patients (87%) in the norfloxacin group and 38 of 41 patients (93%) in the trimethoprim-sulfamethoxazole group had sterile urine. None of the patients had pseudomonas infection. Sabbaj et al. (54) summarized 16 studies in which norfloxacin at 400 mg b.i.d. was compared with trimethoprim-sulfamethoxazole at 160/800 mg b.i.d., using a 10-day treatment time. At the first follow-up after treatment, 98 of 101 patients (97%) in the norfloxacin group and 113 of 125 patients (90%) in the trimethoprim-sulfamethoxazole group had no bacteriuria. This difference is significant ($P < 0.05$) and, notably, only patients with organisms susceptible to both drugs were included in the analysis. In an open study, Leigh and Emmanuel (42) treated 19 patients with pseudomonas bacteriuria with norfloxacin at 400 to 800 mg for 5 to 10 days. At 5 to 9 days after treatment, 16 (84%) had no bacteriuria.

Ofloxacin

Blomer et al. (8) summarized a large number of comparative studies with ofloxacin in upper and lower complicated UTIs (Table 4). In most trials the doses used were 100 mg b.i.d. in lower UTIs and 200 mg b.i.d. in upper UTIs. It is unclear from the publication when bacteriological efficacy was registered in relation to the end of treatment; neither is it stated whether patients with urine isolates resistant to any of the antibiotics compared were included in the analysis. In comparative Japanese studies 89% of bacteriuria strains were eliminated from 115 patients with complicated UTIs receiving ofloxacin at 200 mg b.i.d. for 5 days while the elimination rate was only 72% in 113 patients randomized to pipemidic acid at 500 mg four times a day for 5 days (39). In a multiclinic study in the United States, Cox et al. (17) com-

Table 4. Summary of bacteriological results in comparative studies of ofloxacin in upper and lower UTIs[a]

Comparative agent	Type of UTI	Elimination of bacteriuria (no. of patients)	
		Ofloxacin	Comparative
Trimethoprim/ sulfamethoxazole	Lower	113/119 (89)[b]	90/101 (89)
	Upper	182/223 (82)	161/202 (80)
Amoxicillin + clavulanic acid	Lower	20/26 (77)	11/22 (50)
	Upper	26/27 (96)	13/23 (57)
Norfloxacin	Lower	71/91 (78)	38/58 (66)
Pipemidic acid	Lower	15/25 (60)	15/29 (52)

[a]A majority of the infections were complicated and/or recurrent (8).
[b]Numbers within parentheses indicate percentages.

pared 10 days of treatment with ofloxacin at 200 mg b.i.d. or trimethoprim-sulfamethoxazole at 160/800 mg b.i.d. in patients with complicated UTIs. At 4 to 6 weeks after treatment, all of 32 patients in the ofloxacin group and 29 of 32 patients in the trimethoprim-sulfamethoxazole group showed elimination of bacteriuria without reinfection. In another comparative study 7-day treatments of complicated UTIs with ofloxacin at 100 mg b.i.d. or ciprofloxacin at 250 mg b.i.d. were compared (41). In the ofloxacin group 12 of 18 isolates from 18 patients were eradicated from the urine at 17 days after treatment. In the ciprofloxacin group only 50% of the isolates from 16 patients were eradicated. Two patients infected with strains of *P. aeruginosa* in the ofloxacin group and one patient in the ciprofloxacin group relapsed. In an open trial 30 patients with UTIs received ofloxacin at 300 mg three times a day for 9 days or longer (21). In 3 of 14 patients infected with *E. coli* strains, 1 of 7 infected with *S. marcescens* strains, 1 of 4 infected with *K. pneumoniae* strains, and 3 of 5 infected with *P. aeruginosa* strains, the bacteria persisted or the patients relapsed after treatment.

Pefloxacin

In an open trial Boerema et al. (11) treated 104 patients, of whom 88 were evaluable for bacteriological efficacy, with complicated UTIs with pefloxacin at 400 mg b.i.d. for 9 to 21 days. At 7 days after therapy, bacteriological eradication was achieved in 84 patients (96%). One of four patients infected with strains of *P. aeruginosa* relapsed.

In all of these trials, failures have only rarely been due to the emergence of resistance to the quinolone used or superinfections with resistant bacterial strains but rather seem to be related to factors with the patients treated facilitating reinfections or making efficient elimination of bacteriuria difficult. It should be noted, however, that in most trials no systematic search for resistance in relapsing infections or reinfections is reported. Guibert et al. (32) reported one relapse with a resistant mutant strain of *S. marcescens* when ciprofloxacin at 500 mg b.i.d. was used for the treatment of cystitis and pyelonephritis in 30 patients in an open uncontrolled study.

In summary, studies of these quinolones, and especially of ciprofloxacin, norfloxacin, and ofloxacin, have demonstrated that the drugs are effective in complicated infections caused by pathogens which are multiresistant. In many such patients the quinolones can replace parenteral antibiotics, which considerably reduces the need for hospitalization and also the costs for antibacterial treatment.

Efficacy in Prostatitis

A few articles evaluating 4-quinolones in patients with prostatitis have been published. Weidner et al. (63) used ciprofloxacin at 500 mg b.i.d. for 2 weeks and reported cure in 5 of 12 patients, probable success in 2, and failure in 3 (1 patient could not be evaluated). In this trial the best results were obtained in patients with *E. coli* as the causative pathogen. In two of three patients with enterococcal infections, the bacteria persisted, and one patient with *P. aeruginosa* prostatitis failed two courses of treatment, however, without emergence of resistance. Guibert et al. (32) treated 26 patients with prostatitis with ciprofloxacin at 500 mg b.i.d. for 28 or 84 days and reported cure without reinfection in 20 patients, cure with reinfection in 4, and relapse in 2. Sabbaj et al. (53) reported eradication in 23 of 25 patients with prostatitis treated with norfloxacin at 400 mg b.i.d. for 4 to 6 weeks, while the same result was reported in 10 of 15 patients randomized to sulfamethoxazole-trimethoprim. One unidentified pathogen in the norfloxacin group and two in the sulfamethoxazole-trimethoprim group persisted and acquired resistance to the trial drug. Bologna et al. (12) used norfloxacin at 400 mg b.i.d. for only 10 days and reported success in 17 of 20 patients. Two of the patients who were considered cured had *P. aeruginosa* infections.

These relatively small studies all indicate that fluorinated quinolones may be effective in the treatment of prostatitis, an assumption supported by the antibacterial spectrum and pharmacokinetic characteristics of these agents.

Summary

The fluorinated 4-quinolones represent a major breakthrough in antibacterial therapy mainly due to their very broad spectrum of activity against gram-negative pathogens. Thus, they offer a possibility for oral treatment of pathogens such as *P. aeruginosa* and *Acinetobacter* spp., species which have previously often been resistant to all oral antibiotics. Comparative studies have not demonstrated that the quinolones are more effective than comparative agents against susceptible pathogens, but they may offer a possibility to reduce the incidence of side effects, as shown in a large comparison of norfloxacin and trimethoprim-sulfamethoxazole (60). It can be concluded that the new quinolone derivatives are drugs of choice for the oral treatment of UTIs caused by multiply resistant bacterial pathogens when the alternative would be a parenteral antibiotic. In other types of UTI, caused by pathogens also susceptible to other oral agents, the quinolones become alternatives to such antibiotics. In prostatitis, the place of the quinolones is not yet clear, and efficacy with only ciprofloxacin and norfloxacin has been documented. However, considering their kinetic properties and antibacterial spectrum, this may become one of the major indications in the future.

A fear when a new group of antimicrobial agents is introduced is that it will be used in too many patients, creating resistance problems. It therefore seems prudent to suggest that the fluorinated 4-quinolones should be used with some restriction in uncomplicated cystitis for which indication they should be looked upon as one of many alternatives.

A comparison of the various derivatives discussed here does not reveal any obvious difficulties in terms of clinical or bacteriological efficacy. From a theoretical point of view, the rather low degree of absorption from the gastrointestinal tract of ciprofloxacin and norfloxacin may increase and prolong the suppression of the fecal aerobic flora, thus reducing the risks of early reinfections. With most other agents such poor absorption would result in an increased risk of diarrhea, but that does not seem to be the case with the 4-quinolones. It should be pointed out that few trials have been performed in which new fluorinated 4-quinolones are compared with each other. It is therefore not possible at present to draw any conclusions as to possible differences among them in terms of efficacy or safety.

Literature Cited

1. **Barry, A. L., and R. N. Jones.** 1987. In vitro activity of ciprofloxacin against Gram-positive cocci. *Am. J. Med.* **82**(Suppl. 4A):27–32.
2. **Barry, A. L., C. Thornsberry, and R. N. Jones.** 1986. In vitro evaluation of A-56619 and A-56620, two new quinolones. *Antimicrob. Agents Chemother.* **29**:40–43.
3. **Bergan, T.** 1987. Quinolones, p. 169–183. *In* P. K. Peterson and J. Verhoef (ed.), *The Antimicrobial Agents Annual* 2. Elsevier/North-Holland Publishing Co., Amsterdam.
4. **Bergan, T., S. B. Thorsteinsson, R. Solberg, L. Bjornskau, I. M. Kolstad, and S. Johnsen.** 1987. Pharmacokinetics of ciprofloxacin: intravenous and increasing oral doses. *Am. J. Med.* **82**(Suppl. 4A):97–102.
5. **Bergeron, M., M. Thabet, R. Roy, C. Lessard, and P. Foucault.** 1985. Norfloxacin penetra-

tion into human renal and prostatic tissue. *Antimicrob. Agents Chemother.* **28:**349–350.

6. **Bischoff, W.** 1986. Vergleichende Untersuchung von Enoxacin mit Amoxicillin bei der akuten unkomplizierten Zystitis der Frau. *Infection* **14**(Suppl. 3):209–210.
7. **Blaser, J., M. N. Dudley, D. Gilbert, and S. H. Zinner.** 1986. Influence of media and method on the in vitro susceptibility of *Pseudomonas aeruginosa* and other bacteria to ciprofloxacin and enoxacin. *Antimicrob. Agents Chemother.* **29:**927–929.
8. **Blomer, R., K. Bruch, and R. N. Zahltne.** 1986. Zusammengefasste Ergebnisse der klinischen Phase II- und III-Studien mit Ofloxacin (HOE 280) in Europa. *Infection* **14**(Suppl. 1):102–107.
9. **Boerema, J., A. Dalhoff, and F. Debruyne.** 1985. Ciprofloxacin distribution in prostatic tissue and fluid following oral administration. *Chemotherapy* (Basel) **31:**13–18.
10. **Boerema, J. B. J., B. J. Olthof, and H. K. F. van Saene.** 1986. Effects of norfloxacin on the faecal flora in patients with complicated urinary tract infections. *Scand. J. Infect. Dis. Suppl.* **48:**27–31.
11. **Boerema, J. B. J., R. Pauwels, J. Scheepers, and W. Cromback.** 1986. Efficacy and safety of pefloxacin in the treatment of patients with complicated urinary tract infections. *J. Antimicrob. Chemother.* **17**(Suppl. B):103–109.
12. **Bologna, M., L. Vaggi, D. Flammini, G. Carlucci, and C. Forchetti.** 1985. Norfloxacin in prostatitis: correlation between HPLC tissue concentrations and clinical results. *Drugs Exp. Clin. Res.* **11:**95–100.
13. **Brumfitt, W., I. Franklin, D. Gradu, J. M. T. Hamilton-Miller, and A. Iliffie.** 1986. Changes in the pharmacokinetics of ciprofloxacin and fecal flora during administration of a 7-day course to human volunteers. *Antimicrob. Agents Chemother.* **26:**757–761.

13a. **Campoli-Richards, D. M., J. P. Monk, A. Price, P. Benfield, P. A. Todd, and A. Ward.** 1988. Ciprofloxacin. A review of its antibacterial effect, pharmacokinetic properties and therapeutic use. *Drugs* **35:**373–447.

14. **Chin, N. X., and H. C. Neu.** 1983. In vitro activity of enoxacin, a quinolone carboxylic acid, compared with those of norfloxacin, new β-lactams, aminoglycosides, and trimethoprim. *Antimicrob. Agents Chemother.* **24:**754–763.
15. **Clarke, A. M., S. J. V. Zemcov, and M. E. Campbell.** 1985. In-vitro activity of pefloxacin compared to enoxacin, norfloxacin, gentamicin and new β-lactams. *J. Antimicrob. Chemother.* **15:**39–44.
16. **Cox, C. E.** 1988. A comparison of enoxacin and co-trimoxazole in the treatment of patients with complicated urinary tract infections. *J. Antimicrob. Chemother.* **21**(Suppl. B):113–118.
17. **Cox, C. E., S. V. Callery, and K. J. Tack.** 1986. Clinical experience with ofloxacin in urinary tract infections. *Infection* **14**(Suppl. 4):303–304.
18. **Dahlhoff, A., and W. Weidner.** 1984. Diffusion of ciprofloxacin into prostatic fluid. *Eur. J. Clin. Microbiol.* **3:**360–362.
19. **Dan, M., J. Golomb, A. Gorea, Z. Braf, and S. Berger.** 1986. Concentration of ciprofloxacin in human prostatic tissue after oral administration. *Antimicrob. Agents Chemother.* **30:**88–89.
20. **Daschner, F. D., M. Westenfelder, and A. Dalhof.** 1986. Penetration of ciprofloxacin into kidney, fat, muscle and skin tissue, p. 81–82. *In* H. C. Neu and D. S. Reeves (ed.), *Current Topics in Infectious Diseases and Clinical Microbiology,* vol. 1. *Ciprofloxacin. Microbiology-Pharmacokinetics-Clinical Experience.* Vieweg & Sohn, Braunschweig/Wiesbaden.
21. **Delia, S., C. De Simone, V. Vullo, and F. Sorice.** 1986. Ofloxacin: clinical evaluation in urinary and respiratory tract infections. *Infection* **14**(Suppl. 4):297-299.
22. **Edlund, C., T. Bergan, K. Josefsson, C. O. Solberg, and C. E. Nord.** 1987. Effect of norfloxacin on human oropharyngeal and colonic microflora and multiple dose pharmacokinetics. *Scand. J. Infect. Dis.* **19:**113–121.

23. **Fass, R. J.** 1983. In vitro activity of ciprofloxacin (Bay o 9867). *Antimicrob. Agents Chemother.* **24:**568–574.
24. **Foot, M., G. Williams, S. Want, M. Roe, H. Quaghebeur, and S. Bates.** 1988. An open study of the safety and efficacy of enoxacin in complicated urinary tract infections. *J. Antimicrob. Chemother.* **21**(Suppl. B):97–104.
25. **Frydman, A. M., Y. Le Roux, M. A. Lefebvre, F. Djebbar, J. B. Fourtillan, and J. Gaillot.** 1986. Pharmacokinetics of pefloxacin after repeated intravenous and oral administration (400 mg bid) in young healthy volunteers. *J. Antimicrob. Chemother.* **17**(Suppl. B):65–79.
26. **Garlando, F., S. Rietiker, M. G. Täuber, M. Flepp, B. Meier, and R. Lüthy.** 1987. Single-dose ciprofloxacin at 100 versus 250 mg for treatment of uncomplicated urinary tract infections in women. *Antimicrob. Agents Chemother.* **31:**354–356.
27. **Giamarellou, H., N. Galankis, C. Dendrinos, J. Stefanou, E. Daphnis, and G. K. Daikos.** 1986. Evaluation of ciprofloxacin in the treatment of *Pseudomonas aeruginosa* infections, p. 103–106. *In* H. C. Neu and D. S. Reeves (ed.), *Current Topics in Infectious Diseases and Clinical Microbiology,* vol. 1. *Ciprofloxacin. Microbiology-Pharmacokinetics-Clinical Experience.* Vieweg & Sohn, Braunschweig/Wiesbaden.
28. **Giamarellou, H., J. Tsagarikis, G. Petrikkos, and G. K. Daikos.** 1983. Norfloxacin versus cotrimoxazole in the treatment of lower urinary tract infections. *Eur. J. Clin. Microbiol.* **2:**266–269.
29. **Goldstein, E. J. C., M. R. Kahn, M. L. Alpert, B. P. Ginsberg, F. L. Greenway, and D. M. Citron.** 1987. Ciprofloxacin versus cinoxacin in therapy of urinary tract infections. *Am. J. Med.* **82**(Suppl. 4A):284–287.
30. **Grabe, M., A. Forsgren, and T. Björk.** 1986. Concentrations of ciprofloxacin in serum and prostatic tissue in patients undergoing transurethral resection. *Eur. J. Clin. Microbiol.* **5:**211–212.
31. **Guerra, J. G., E. Falconi, J. C. Palomino, L. Benavente, and E. A. de Mayolo.** 1983. Clinical evaluation of norfloxacin versus cotrimoxazole in urinary tract infections. *Eur. J. Clin. Microbiol.* **2:**260–265.
32. **Guibert, J., D. Destrée, C. Konopka, and J. Acar.** 1986. Ciprofloxacin in the treatment of urinary tract infections due to enterobacteria. *Eur. J. Clin. Microbiol.* **5:**247–248.
33. **Guibert, J., D. Destrée, C. Konopka, and J. Acar.** 1986. Ciprofloxacin in the treatment of urinary tract infection due to enterobacteria, p. 118–119. *In* H. C. Neu and D. S. Reeves (ed.), *Current Topics in Infectious Diseases and Clinical Microbiology,* vol. 1. *Ciprofloxacin. Microbiology-Pharmacokinetics-Clinical Experience.* Vieweg & Sohn, Braunschweig/Wiesbaden.
34. **Henry, N. K., H. J. Schultz, N. C. Grubbs, S. M. Muller, D. M. Ilstrup, and W. R. Wilson.** 1986. Comparison of ciprofloxacin and co-trimoxazole in the treatment of uncomplicated urinary tract infection in women. *J. Antimicrob. Chemother.* **18**(Suppl. D):103–106.
35. **Hoogkamp-Korstanje, J., H. Oort, J. Schipper, and T. van der Wal.** 1984. Intraprostatic concentration of ciprofloxacin and its activity against urinary pathogens. *J. Antimicrob. Chemother.* **14:**641–645.
36. **Huttunen, M., K. Kunnas, and P. Saloranta.** 1988. Enoxacin treatment of urinary tract infections in elderly patients. *J. Antimicrob. Chemother.* **21**(Suppl. B):105–112.
37. **Ichihara, N., H. Tachizawa, M. Tsumura, T. Ume, and K. Sato.** 1984. Phase I study on DL-8280. *Chemotherapy* (Tokyo) **32**(Suppl. 1):118–149.
38. **King, A., and I. Phillips.** 1986. The comparative in-vitro activity of pefloxacin. *J. Antimicrob. Chemother.* **17**(Suppl. B):1–10.
39. **Kishi, H., H. Nito, I. Saito, et al.** 1984. Comparative studies of DL-8280 and pipemidic acid in complicated urinary tract infections by double-blind method. *Acta Urol. Jpn.* **30:**1307–1355.
40. **Kosmidis, J., E. Macrygiannis, and B. Aboudabash.** 1985. Ciprofloxacin in urinary tract

infections: its efficacy as compared with that of co-trimixazole, p. 310–313. *In* H. C. Neu and H. Weuta (ed.), *1st International Ciprofloxacin Workshop, Leverkusen 1985.* Excerpta Medica, Amsterdam.

41. **Kromann-Andersen, B., P. Sommer, C. Pers, V. Larsen, and F. Rasmussen.** 1986. Clinical evaluation of ofloxacin versus ciprofloxacin in complicated urinary tract infections. *Infection* **14**(Suppl. 4):305–306.
42. **Leigh, D. A., and F. X. S. Emmanuel.** 1984. The treatment of *Pseudomonas aeruginosa* urinary tract infections with norfloxacin. *J. Antimicrob. Chemother.* **13**(Suppl. B):85–88.
43. **Leigh, D. A., E. C. Smith, and J. Mariner.** 1984. Comparative study using norfloxacin and amoxycillin in the treatment of complicated urinary tract infections in geriatric patients. *J. Antimicrob. Chemother.* **13**(Suppl. B):79–83.
44. **Monk, J. P., and D. M. Campoli-Richards.** 1987. Ofloxacin. A review of its antibacterial activity, pharmacokinetic properties and therapeutic use. *Drugs* **33**:346–391.
45. **Montay, G., Y. Goueffon, and F. Roquet.** 1984. Absorption, distribution, metabolic fate, and elimination of pefloxacin in mice, rats, dogs, monkeys, and humans. *Antimicrob. Agents Chemother.* **25**:463–472.
46. **Murayama, S., K. Hirai, A. Ito, Y. Abe, and T. Irikura.** 1981. Studies on absorption, distribution and excretion of AM-715 in animals by bioassay method. *Chemotherapy* (Toyko) **29**(Suppl. 4):98–104.
47. **Naber, K. G., and B. Bartosik-Wich.** 1986. Ciprofloxacin versus norfloxacin in the treatment of complicated urinary tract infections: in vitro activity, serum and urine concentrations, safety and therapeutic efficacy, p. 314–317. *In* H. C. Neu and H. Weuta (ed.), *1st International Ciprofloxacin Workshop, Leverkusen 1985.* Excerpta Medica, Amsterdam.
48. **Naber, K. G., F. Sörgel, F. Gutzler, and B. Bartosik-Wich.** 1986. In vitro activity, pharmacokinetics, clinical safety and therapeutic efficacy of enoxacin in the treatment of patients with complicated urinary tract infections. *Infection* **14**(Suppl. 3):203–208.
49. **Nagatsu, Y., K. Endo, and T. Irikura.** 1981. Studies on the fate of ^{14}C-labelled AM-715. *Chemotherapy* (Toyko) **29**(Suppl. 4):105–118.
50. **Newsom, S. W. B., P. Murphy, and J. Matthew.** 1986. A comparative study of ciprofloxacin and trimethoprim in the treatment of urinary tract infections in geriatric patients. *J. Antimicrob. Chemother.* **18**(Suppl. D):111–115.
51. **Reeves, D. S.** 1986. The effect of quinolone antibacterials on the gastrointestinal flora compared with that of other antibacterials. *J. Antimicrob. Chemother.* **18**(Suppl. D):89–102.
52. **Rylander, M., S. R. Norrby, and R. Svärd.** 1987. Norfloxacin versus co-trimoxazole for treatment of urinary tract infections in adults: microbiological results of a coordinated multicentre study. *Scand. J. Infect. Dis.* **19**:551–557.
53. **Sabbaj, J., V. L. Hoagland, and T. Cook.** 1986. Norfloxacin versus co-trimoxazole in the treatment of recurring urinary tract infections in men. *Scand. J. Infect. Dis. Suppl.* **48**:48–53.
54. **Sabbaj, J., V. L. Hoagland, and W. J. Shih.** 1985. Multiclinic comparative study of norfloxacin and trimethoprim-sulfamethoxazole for treatment of urinary tract infections. *Antimicrob. Agents Chemother.* **27**:297–301.
55. **Schmicker, R., and G. Nauman.** 1985. Efficacy of ciprofloxacin in urinary tract infections in geriatric patients, p. 305–309. *In* H. C. Neu and H. Weuta (ed.), *1st International Ciprofloxacin Workshop, Leverkusen 1985.* Excerpta Medica, Amsterdam.
56. **Shah, P. M., and K. Frech.** 1986. Overview of the clinical experience with the quinolones, p. 29–43. *In* A. Percival (ed.), *Quinolones—Their Future in Clinical Practice.* Royal Society of Medicine, London.
57. **Smith, J. T., and N. T. Ratcliffe.** 1986. Einfluss von pH-Wert und Magnesium auf die antibakterielle Aktivität von Chinolonpräparaten. *Infection* **14**(Suppl. 1):31–35.
58. **Swanson, B. N., V. K. Boppana, P. H. Vlasse, H. H. Rotmensch, and R. K. Ferguson.** 1983.

Norfloxacin disposition after sequentially increasing oral doses. *Antimicrob. Agents Chemother.* **23**:284–288.

59. **Thomas, M. G., and R. B. Ellis-Pegler.** 1985. Enoxacin treatment of urinary tract infections. *J. Antimicrob. Chemother.* **15**:759–763.
60. **Urinary Tract Infection Study Group.** 1987. Coordinated multicenter study of norfloxacin versus trimethoprim-sulfamethoxazole treatment of symptomatic urinary tract infections. *J. Infect. Dis.* **155**:170–177.
61. **Van Caekenberghe, D. L., and S. R. Pattyn.** 1984. In vitro activity of ciprofloxacin compared with those of other new fluorinated piperazinyl-substituted quinolone derivatives. *Antimicrob. Agents Chemother.* **25**:518–521.
62. **van Saene, J. J. M., H. K. E. van Saene, and C. F. Lerk.** 1986. Inactivation of quinolones by faeces. *J. Infect. Dis.* **153**:998–1000.
63. **Weidner, W., H. Schiefer, and A. Dalhoff.** 1987. Treatment of chronic bacterial prostatitis with ciprofloxacin. Results of a one-year follow-up study. *Am. J. Med.* **82**(Suppl. 4A):280–283.
64. **Williams, A. H., and R. N. Grüneberg.** 1986. Ciprofloxacin and co-trimoxazole in urinary tract infections. *J. Antimicrob. Chemother.* **18**(Suppl. D):107–110.
65. **Wolf, R., R. Eberl, A. Dunky, N. Merrtz, T. Chang, J. R. Goulet, and J. Latts.** 1984. The clinical pharmacokinetics and tolerance of enoxacin in healthy volunteers. *J. Antimicrob. Chemother.* **14**(Suppl. C):63–69.

Chapter 6

Treatment of Sexually Transmitted Diseases with Quinolone Antimicrobial Agents

Gina A. Dallabetta and Edward W. Hook III

Division of Infectious Diseases, Johns Hopkins Hospital, Baltimore, Maryland 21205

Introduction

Therapy of bacterial sexually transmitted diseases (STDs) poses a variety of challenges to clinicians. These include the need to treat syndromes caused by a number of different pathogens, each with differing antimicrobial susceptibilities; the need for treatment of patients with simultaneous infection by coexistent pathogens; and the continuing emergence of antimicrobial resistance among some STD pathogens. A further consideration which affects the choice of therapy for STD patients is that these infections are most often treated in outpatient settings, thereby introducing the need to use medications which are well tolerated. Although beta-lactam and tetracycline antibiotics have long been the mainstays of STD therapy, the newer quinolone antibiotics have considerable promise for therapy of some common STDs.

The newer quinolones are related to nalidixic acid and contain a 4-quinolone nucleus with 6-fluoro and 7-piperazino substitutions (78). Their mechanism of action is similar to that of nalidixic acid, inhibiting DNA gyrase during bacterial replication (61, 66). These new antibiotics have broad-spectrum activity against gram-negative and gram-positive aerobic organisms and, unlike early quinolones, the newer drugs are well absorbed from the gastrointestinal tract. Half-lives are variable and range from about 3 to 4 h for norfloxacin, ciprofloxacin, and rosoxacin, 6 to 7 h for enoxacin and ofloxacin, 10 to 11 h for pefloxacin, and 24 h for difloxacin (Table 1) (29, 44). The adverse effects of the quinolones appear to be low, especially when given for a short duration. Potential toxicities of this class of drugs include

Table 1. Pharmacokinetics and MICs of selected quinolones

Drug	Half-life[a] (h)	Dose (mg)	Peak level in serum[b] (μg/ml)	MIC range (μg/ml) against:	
				N. gonorrhoeae[c]	*C. trachomatis*[d]
Norfloxacin	3–4	400	1.6	0.007–0.25	8–50
		800	2.4		
Ciprofloxacin	3–4	250	1.4	0.002–0.16	0.5–5.0
		500	2.3		
Enoxacin	6–7	600	3.5	0.013–0.06	8–40
Ofloxacin	6–7	600	11	0.002–0.06	0.2–8

[a]References 29 and Neu, *Antimicrob. Newsl.* **4**:9–16, 1987.
[b]References 16, 20, 21, 36, 67, and 76.
[c]References 3, 8, 17, 34, 35, 48, 60, 75, 78, and Neu, *Antimicrob. Newsl.* **4**:9–16, 1987.
[d]References 3, 4, 7, 26–28, 30, 31, 38, 49, 56, 58, 60, 64, and 78.

gastrointestinal and central nervous system toxicity, crystalluria with high doses, and cartilage toxicity (seen in some experimental animals) (29). Because of cartilage toxicity, the quinolones are not recommended for use in children or in pregnant women. The broad range of activity against STD pathogens and potential for activity when administered orally have contributed to interest in the newer quinolones for STD treatment.

Neisseria gonorrhoeae

Since 1976 when *N. gonorrhoeae* producing beta-lactamase was first described (47, 51), the prevalence of clinically significant antimicrobial resistance has increased worldwide and has compromised the utility of penicillins and tetracyclines for gonorrhea therapy. Alternative antimicrobial therapies for resistant gonococcal infections currently include spectinomycin and cephalosporins, both of which are administered by injection. The quinolones are active in vitro against gonococci with all currently recognized forms of antimicrobial resistance, including those with plasmid-mediated penicillin or tetracycline resistance as well as chromosomally mediated antibiotic resistance. Reported MIC ranges for these agents are usually far below peak levels achieved in serum (3, 8, 17, 25, 34, 35, 48, 60, 75, 78; H. C. Neu, *Antimicrob. Newsl.* **4**:9–16, 1987).

Of the newer quinolones only norfloxacin and ciprofloxacin are currently marketed in the United States, the others being under continued investigation. Although data regarding the utility of quinolones for STD therapy are continuing to become available, current data are generally consistent and suggest efficacy for gonorrhea therapy. A number of the studies reviewed below were comparative clinical trials; however, because the efficacy rates of the newer quinolones are high and generally comparable to standard therapy, only the quinolone data will be discussed.

Norfloxacin

The reported MIC range of norfloxacin for *N. gonorrhoeae* is 0.007 to 0.25 μg/ml (3, 34, 35, 48, 76, 78). Peak norfloxacin levels in serum of 1.6 and 2.4 μg/ml are achieved following single oral doses of 400 and 800 mg, respectively (65), well above the susceptibilities reported for *N. gonorrhoeae.*

In one of the first publications describing the utility of newer quinolones for gonorrhea therapy, Crider et al. reported a 100% cure rate in 59 naval personnel with gonococcal urethritis treated with two 600-mg doses of norfloxacin separated by a 4-h interval (15). In this study 46% of the isolates were penicillinase-producing *N. gonorrhoeae* (PPNG) and 26% were chromosomally resistant to penicillin. No adverse effects were noted. The utility of the same regimen was subsequently confirmed by Kaplowitz et al., who cured 100% of 24 men with urethritis and 7 women with cervicitis (one of whom had a simultaneous rectal infection) (33). The same investigators then treated 46 additional patients with single 800-mg doses of norfloxacin; 26 of 27 (96%) men with urethritis and 18 of 19 (95%) women with cervicitis (8 of whom had simultaneous rectal infections) were cured. The authors stated that the two failures in their study appeared to be due to reinfection.

Two subsequent studies in which single 800-mg doses were used have been published: in Canada 100% cure rates were observed in 31 men with urethritis and 25 women with cervicitis (6 of whom had simultaneous rectal infections) (53), while in Rwanda 119 (97%) of 122 patients with uncomplicated gonorrhea (urethritis or cervicitis) were cured (5). Forty-three percent of the Rwandan isolates produced beta-lactamase.

Thus, either two 600-mg doses, given 4 h apart, or single 800-mg doses of norfloxacin appear to be highly effective for gonococcal urethritis, cervicitis, and possibly coexistent rectal infection in women. The studies cited were performed at locations around the world and include substantial numbers of patients infected with antibiotic-resistant *N. gonorrhoeae*. There are insufficient data to ascertain the effectiveness of norfloxacin for treating pharyngeal gonorrhea or rectal gonorrhea in men; in the studies cited only five patients had proven pharyngeal gonorrhea, four of whom were culture negative at the time of follow-up exam.

Toxicity was relatively uncommon in the studies listed and did not appear to have been limiting. Adverse reactions reported for norfloxacin-treated patients included dizziness, nausea, and headache.

Ciprofloxacin

Studies evaluating the utility of ciprofloxacin for gonorrhea therapy have utilized a variety of dosing schedules. Reported ciprofloxacin MICs for *N. gonorrhoeae* range from 0.002 to 0.16 μg/ml (3, 34, 35, 38, 60, 76, 78), and levels in serum of 1.4 and 2.4 μg/ml are achieved after single oral doses of 250 and 500 mg, respectively (16, 20).

In a dose ranging study Szarmach et al. treated 100 men with gonococcal urethritis with doses of ciprofloxacin from 2,000 to 100 mg, evaluating not only drug efficacy but also the rate of microbiologic cure (68). Culturing infected volunteers hourly for 9 h after therapy, these investigators found that a single 2-g dose had eradicated viable *N. gonorrhoeae* from each of 10 men at 3 h after therapy, while in 30 men treated with single 100-mg doses, 100% eradication was not achieved until 7 h. At 3 h 53% of volunteers treated with the 100-mg dose were still culture positive. Despite apparent differences in eradication rates, all patients enrolled in this study were cured of their gonorrhea. In another dose ranging study, there were 100% cure rates for 10 men treated with 200 mg twice daily for 3 days, for 10 men receiving 400 mg twice daily for 1 day, and for 10 men receiving 400 mg as a single dose (44). Other studies have also supported the utility of multiple-dose schedules. Ciprofloxacin dosed at 500 mg twice daily for 7 days in the therapy of men with urethritis irrespective of etiology cured 100% of gonococcal infections in 29 participants (2).

Multiple-dose therapy is inconvenient, however, when single-dose therapy is equally effective. Tegelberg-Stassen et al. successfully treated 79 men with gonococcal urethritis with 500 mg of ciprofloxacin (70), and Loo et al. eradicated gonococcal infections by using the same dose in 5 men with urethritis and in 5 men with proctitis (37). Several additional studies suggest that a single 250-mg dose of ciprofloxacin is as effective as 500 mg for uncomplicated gonorrhea (37, 52, 59, 71). Using 250-mg doses, Tegelberg-Stassen eradicated 100% of gonococcal urethritis in 85 men (70); 100% of urethral infections were also cured in 49 men with gonococcal urethritis, as were 4 of 5 rectal infections in men studied by Roddy et al. (52); and 100% cure rates were similarly reported for 47 gonococcal urethral and 7 rectal infections in men studied by Loo et al. (37).

Toxocity in these studies appeared to be mild and included headaches, dizziness, nausea, and a mild elevation in serum glutamic oxalacetic transaminase noted in a few patients.

In summary, ciprofloxacin as single doses of either 500 or 250 mg appears to be highly effective for gonococcal urethritis and possibly for rectal infections in men. At present there are no published studies reporting the efficacy of ciprofloxacin for gonococcal cervicitis, although persumably cure rates would be similar to those reported for urethritis. As with norfloxacin, there are insufficient data to ascertain the effectiveness of ciprofloxacin for pharyngeal gonorrhea. In the above studies there were a total of four documented pharyngeal infections; all were cured after treatment with 250 mg of ciprofloxacin.

Ofloxacin

Ofloxacin for treatment of gonorrhea has been evaluated in a limited number of studies published to date. Reported ofloxacin MICs for *N. gonorrhoeae* range from 0.002 to 0.06 μg/ml (3, 35, 48, 78) and are far below the peak concentration in serum of 11 μg/ml, reported after a single oral dose of 600 mg (36).

Two studies, in which 600-mg single-dose therapy was used, reported 100% efficacy for the treatment of gonococcal urethritis in 28 men from Hong Kong, where the PPNG rate was 25%, and in 21 men in the United States (13, 32). Cure rates were also 100% with 400-mg single-dose therapy for gonococcal urethritis in 31 men in Hong Kong (PPNG rate, 25%), 110 men in Thailand (PPNG rate, 45%), and 32 men in France (1, 13, 41). Similar results were found in Thailand with single 200-mg doses for 110 men with gonococcal urethritis (1).

Adverse effects were reported from only one study and consisted of blurred vision and dizziness in two patients (13).

Thus, ofloxacin also appears to be effective for therapy of gonococcal urethritis when single doses of 600, 400, and 200 mg are used. The studies evaluated volunteers from diverse areas of the world, including some with significant PPNG prevalence. There are insufficient published data to ascertain the effectiveness of ofloxacin for gonococcal cervicitis or rectal or pharyngeal infections.

Enoxacin

Three clinical studies have been published which evaluated enoxacin for gonorrhea therapy. MICs of enoxacin reported for gonococci range from 0.013 to 0.06 μg/ml (3, 48, 60, 76, 78), far below the peak level in serum of about 3.5 μg/ml after a single 600-mg oral dose (77).

In a small pilot study, Notowicz et al. successfully treated all of 22 patients (19 men and 3 women) with enoxacin; 11 patients received 600 mg as a single dose, and 11 received two 400-mg doses separated by 4 h (43).

In a larger study in which 200- and 400-mg doses for therapy of uncomplicated gonococcal urethritis were evaluated, van der Willigen et al. reported cures in only 69 (90%) of 77 men treated with 400 mg of enoxacin and 72 (92%) of 78 men receiving 200-mg doses (72). Two rectal infections were detected among volunteers for the 200-mg group and were cured. These investigators also reported that the MICs for isolates from one patient who failed therapy increased from 0.03 to 2.0 μg/ml after therapy with 400 mg of enoxacin. In another study with the same doses, Tegelberg-Stassen et al. reported higher success rates after treatment of 86 women with uncomplicated gonococcal cervicitis, 46 treated with 400 mg and 40 treated with 200 mg (71); cure rates were 100 and 98.7%, respectively.

Side effects were reported in two of the three studies, occurring in 3 to 4% of patients. The adverse effects included nausea, headaches, vomiting, and rash.

Although the studies are few, enoxacin in a dose of 600 mg may be reasonably effective for therapy of uncomplicated gonococcal urethritis and cervicitis. The data for lower doses are less clear and further study is needed. There are no data to ascertain the effectiveness of enoxacin for pharyngeal or rectal gonorrhea. It also appears that enoxacin may have a somewhat higher frequency of side effects than does norfloxacin, ciprofloxacin, or ofloxacin. In addition, the possible induction of resistance after therapy in even a single case is noteworthy and raises concerns regarding the possibility of development of resistance among patients receiving enoxacin therapy. These issues should be addressed in future studies.

Rosoxacin

Rosoxacin was one of the earliest of the newer quinolones evaluated for gonorrhea therapy. MICs of 0.03 to 0.25 μg/ml have been reported (17, 75), substantially below the peak concentrations in serum of 3.9 and 6.4 μg/ml achieved after a single oral dose of 150 and 250 mg, respectively (21). Several clinical trials were performed which established the efficacy of rosoxacin for uncomplicated gonococcal infections (14, 24, 46). Despite high efficacy similar to that of its successors, the utility of rosoxacin was limited by unacceptable toxicity, mainly vestibular dysfunction.

Summary

Each of the quinolones discussed above appears to be effective in treating uncomplicated gonococcal urethritis and, possibly, cervicitis and rectal infections, although the drugs differ widely with respect to the numbers of patients treated in published clinical trials to date. There are no clinical trials reporting the use of quinolones in the treatment of disseminated gonococcal infection. However, given the high activity of these drugs against nearly all *N. gonorrhoeae* strains, they are likely to prove effective for treatment of this syndrome. The use of rosoxacin and possibly enoxacin may be limited by drug toxicity. Nonetheless, these drugs are effective in single oral doses even against PPNG and chromosomally resistant gonorrhea, thereby making them attractive alternatives in appropriate clinical situations.

Single doses of norfloxacin, 800 mg, can be recommended for uncomplicated cervical and urethral gonorrhea. There is not enough evidence yet to recommend its use for patients with rectal and pharyngeal infections. Ciprofloxacin as a single dose of 250 mg is also effective in treating uncomplicated urethritis and possibly cervicitis. The other quinolones also look promising but are not now approved in the United States. The specter of rapidly developing quinolone resistance has been described only once but is worrisome and should be monitored with increasing use of these agents.

Chlamydia trachomatis and NGU

In vitro studies

Nongonococcal urethritis (NGU) and its counterpart in women, mucopurulent cervicitis, are STD syndromes associated with *C. trachomatis* infection. In addition, NGU is also associated with the isolation of *Ureaplasma urealyticum*, and in 20 to 30% of patients with NGU no pathogen can be isolated (6). The potential utility of the newer quinolone antibiotics for treatment of *C. trachomatis*-associated syndromes such as NGU should be considered both in terms of activity against identified pathogens and in terms of resolution of clinical findings. The response rates of men with NGU treated by using currently recommended regimens (12) such as tetracycline (or doxycycline) and erythromycin vary somewhat according to which organisms are present. In patients from whom *C. trachomatis* is isolated, clinical response rates are 60 to 85% (64), while for patients from whom *U. urealyticum* or no pathogen is isolated, response rates are somewhat lower, in the range of 50% (6, 23). Less can be said about mucopurulent cervicitis, the recently described counterpart of NGU for women (9), with respect to response to therapy.

In vitro susceptibility testing for *C. trachomatis* is not completely standardized, limiting the ability to compare MICs reported for different antimicrobial agents tested in different laboratories. Nonetheless, as a group, the quinolones show moderate activity against *C. trachomatis*. The MICs of norfloxacin, ciprofloxacin, ofloxacin, and enoxacin for *C. trachomatis* have been reported in the range of 8 to 50, 0.5 to 5.0, 0.2 to 8, and 8.0 to 40 μg/ml, respectively (Table 1) (3, 4, 7, 26–28, 30, 31, 38, 49, 50, 56, 58, 60, 64, 78). From these data and knowledge of achievable concentrations in serum, one would predict that norfloxacin and enoxacin would have little antichlamydial effect, that ciprofloxacin would be intermediate, and that ofloxacin would be relatively active.

The reported MICs of the newer quinolones for *U. urealyticum* tend to be slightly higher than those for *C. trachomatis* (10). MICs for *U. urealyticum* are reported in the range of 8 to 64 μg/ml for norfloxacin, 0.5 to > 64 μg/ml for ciprofloxacin, 3 to 16 μg/ml for ofloxacin, and 4 to 64 μg/ml for enoxacin (3, 10, 45, 62, 78).

C. trachomatis is often present as a coinfecting pathogen in patients with uncomplicated genital gonorrhea and is often isolated from patients with postgonococcal urethritis (PGU) (urethritis persisting after gonorrhea therapy, seen most often after single-dose therapy with beta-lactam antibiotics). Chlamydia isolation rates are 15 to 23% in men with gonococcal urethritis and 30 to 47% in women with gonococcal cervicitis (63). As the newer quinolones were being developed, there was conjecture that some of these agents might eradicate *C. trachomatis* as well when used for single-dose treatment of gonorrhea. As has been documented for other antichlamydial agents, however, single-dose therapy is not reliably effective. In two studies of ciprofloxacin in which either 250- or 500-mg single-dose therapy for gonorrhea was used, *C. trachomatis* persisted in 17 of 18 men with documented coinfections (37, 52). Similarly, in a study of single-dose enoxacin therapy with either 200 or 400 mg for uncomplicated gonorrhea in women, *C. trachomatis* was isolated from 16 of 73 (22%) patients before therapy and 13 of 72 (18%) patients after therapy. In this study cultures from eight women with initially positive cultures were negative after treatment, while cultures from six additional patients were positive only after treatment (71).

The frequency of PGU in patients receiving single-dose quinolone therapy for gonorrhea is similar to the 25 to 40% reported after other single-dose gonorrhea therapy (22). For example, Tegelberg-Stassen et al. reported PGU rates of 36 and 27% after treatment for uncomplicated gonorrhea with either 250 or 500 mg, respectively, of ciprofloxacin for single-dose therapy of gonorrhea (70). Similarly, Ariyarit et al. treated 181 men with single-dose ofloxacin, 200 or 400 mg, and noted an overall PGU rate of 24% (1). Patients receiving enoxacin in single doses of either 200 or 400 mg for uncomplicated gonococcal urethritis had PGU rates of 42 and 26%, respectively (72).

Since single-dose therapy does not reliably prevent PGU, it is similarly unlikely to affect NGU in patients without gonorrhea. A limited number of trials, using more prolonged dosing schedules, have been conducted.

Ciprofloxacin

To date, ciprofloxacin has been the most extensively studied of the quinolones for NGU therapy. Ciprofloxacin (500 mg orally twice daily for 7 days) was used to treat 56 patients with acute urethritis (2). Among the participants, 22 were infected with gonorrhea alone, 13 had only *C. trachomatis* isolated, 7 were coinfected with both pathogens, and 14 had neither pathogen isolated. *C. trachomatis* was reisolated from 14 (70%) of the 20 men initially infected, and 17 (85%) had persistent urogenital symptoms. Of 14 men with nonchlamydial NGU, 10 (70%) also failed therapy. The authors report that several of the participants had sexual exposure during the study period.

A large double-blind study in which 750 mg of ciprofloxacin was compared with 100 mg of doxycycline, both adminstered twice daily for 7 days as therapy for NGU, was reported by Fong et al. (18). Of the 145 patients, *C. trachomatis* was isolated from 42 (29%), *U. urealyticum* alone was isolated from 32 (22%), 20 (14%) were coinfected with both organisms, and 51 (35%) had negative cultures. Overall, both clinical and bacteriologic cure rates were 61% for the tetracycline-treated group and 52% for the ciprofloxacin-treated group, a statistically nonsignificant difference. Significant differences were noted, however, for patients from whom *C. trachomatis* alone was isolated; only 46% of ciprofloxacin-treated patients were cured, as compared with 75% of those receiving doxycycline. *C. trachomatis* was reisolated in all 12 patients with persistence or recurrence in the ciprofloxacin-treated group but in only 3 of the 5 in the doxycycline-treated group.

Side effects were frequent but not limiting. Fong et al. reported gastrointestinal complaints, nausea, epigastric discomfort, diarrhea, and cramps in 71% of the ciprofloxacin-treated patients, similar to the 74% rate reported for the doxycycline-treated group (18). Other adverse effects included dizziness and insomnia.

Even with extended therapy (7 days), ciprofloxacin does not appear to be effective in eradicating chlamydial urethritis. It may be comparable to other tetracycline antibiotics for therapy of nonchlamydial NGU. However, as chlamydial NGU is not routinely differentiated from nonchlamydial NGU, this distinction is unlikely to have an impact on the clinical use of the drug.

Norfloxacin

Bowie et al. studied norfloxacin (400 mg) administered orally twice daily for 10 days in 96 men with NGU (8). Of 79 evaluable men, 17 (22%) were infected with *C. trachomatis* alone, 24 (30%) were infected with *U. urealyticum* alone, 10 (13%) were coinfected with both organisms, and in 28 (35%) neither pathogen could be isolated. Ninety-five percent of *C. trachomatis*-infected volunteers followed for an average of 6 weeks were clinical failures, and *C. trachomatis* was reisolated from 81% of those. The outcome for patients with either *U. urealyticum*-associated NGU or no pathogen isolated was somewhat better; at follow-up, *U. urealyticum* had been eradicated in 63% of initially infected volunteers. Fifty-six percent of *U. urealyticum*-

infected patients and 69% of patients from whom no pathogen could be isolated were clinically improved. Side effects, which were not limiting, occurred in 18% overall and included abdominal complaints (nausea, loose bowels, flatulence, indigestion, and hyperacidity), mood alteration, drowsiness, lightheadedness, and headache.

Ofloxacin

In vitro susceptibility data suggest that ofloxacin is among the more active of the quinolones against *C. trachomatis*. However, only a few published clinical studies involving small numbers of patients have examined its efficacy for NGU to date. Morel et al. treated 6 men with chlamydial urethritis with ofloxacin (200 mg) orally twice daily for 14 days and an additional 12 men with nonchlamydial urethritis for 7 days (40). In this group of 18 men, 11 men had gonorrhea, 4 had *C. trachomatis*, 1 was coinfected, and 2 had neither pathogen isolated. At follow-up, clinical improvement and negative cultures were noted in all patients (except one with *Trichomonas vaginalis* infection).

In a similar comparative trial of 40 patients with documented chlamydial infections (23 males and 17 females), in which therapy consisted of either 100 mg of doxycycline or 200 mg of ofloxacin, both administered twice daily for 10 days, Fransen et al. found 100% cure rates in the doxycycline-treated group and 2 clinical failures and 1 microbiologic failure (9% overall) in the 21 patients in the ofloxacin-treated group (19). No side effects were noted.

Although these two studies are small, it appears that extended ofloxacin therapy may be effective for treatment of chlamydial infections. Clearly, more studies with larger numbers need to be conducted before further recommendations can be made.

Summary

The quinolone antibiotics evaluated to date are not effective as single-dose therapy against *C. trachomatis* or for prevention of PGU. In studies evaluating multiple-dose therapy both norfloxacin and ciprofloxacin were less effective than doxycycline. Preliminary data from a very limited number of patients suggest that ofloxacin, a quinolone with relatively low MICs for *C. trachomatis* and high levels in serum (36), may be effective for NGU and chlamydial infections. However, larger studies are needed to confirm preliminary observations. At present, no quinolone can be recommended for the therapy of chlamydial NGU or mucopurulent cervicitis (for which there are virtually no published studies).

PID and Epididymitis

There are no current studies assessing the efficacy of the quinolones for the treatment of pelvic inflammatory disease (PID). Therapy for PID should be directed for all potential causative organisms, *N. gonorrhoeae*, *C. trachomatis*, facultative gram-negative organisms, and anaerobes. The quinolones may prove to be an effective component of PID therapy but are as yet untested. The quinolones have little activity against anaerobes and variable activity against *C. trachomatis*. Consequently, it is unlikely that they will constitute monotherapy for PID.

Similarly, there are no studies assessing the efficacy of the quinolones in the therapy of epididymitis. Therapy for this syndrome in men under 40 years of age should be directed against *C. trachomatis* and *N. gonorrhoeae*. In homosexual or bisexual men enteric flora may also be involved.

Syphilis

There are few data evaluating the efficacy of quinolone antibiotics against *Treponema pallidum*. Veller-Fornasa et al. reported an experiment in which 26 rabbits were inoculated intratesticularly with virulent *T. pallidum* and then treated with ofloxacin or penicillin (73). Treatment was initiated 3 days after infection; at 7 days all ofloxacin-treated rabbits, as well as the untreated control rabbits, developed typical syphilitic orchitis and serological reactivity. This experiment suggests that ofloxacin and presumably the quinolones as a class have no effect on incubating, experimental syphilis. The quinolones should not be used to treat syphilis infections at present.

Chancroid

Chancroid, a sexually transmitted genital ulcer disease caused by *Haemophilus ducreyi*, is a major STD outside the United States and has been increasing in incidence in the United States during the last 5 years (57). Recently, in certain areas of the United States, it has become a relatively common cause of genital ulceration. As with gonorrhea, the therapy of chancroid has been complicated by reports of resistance to traditional antimicrobial agents (54). The quinolone antibiotics in general appear to be very active in vitro against *H. ducreyi*, including multiresistant strains isolated in The Gambia, The Netherlands, Thailand, and France. The following MICs (in micrograms per milliliter) have been reported (55, 65, 69, 74): norfloxacin, 0.03 to 0.16; ciprofloxacin, 0.001 to 0.16; ofloxacin, 0.03 to 0.06; enoxacin, 0.12; rosoxacin, 0.004 to 0.12; and pefloxacin, 0.06 to 0.12.

Naamara et al. reported on a double-blind, randomized clinical trial in Kenya in which single-dose ciprofloxacin (500 mg), ciprofloxacin (500 mg) twice daily for 3 days, and trimethoprim-sulfamethoxazole (160/800 mg) twice daily for 3 days in 139 men were compared (42). Bacteriologic and clinical cure was achieved in 41 (93%) of 44 men receiving single doses of ciprofloxacin, in 40 (93%) of 43 men treated with ciprofloxacin for 3 days, and in 41 (96%) of 45 men of the trimethoprim-sulfamethoxazole-treated group. Side effects were reported in two patients in the 3-day ciprofloxacin-treated group and consisted of diarrhea and nausea.

Rosoxacin was evaluated in 107 men from Kenya, using either 300-mg single doses or 150 mg twice daily for 3 days for therapy of chancroid (21). At 1 month after treatment, 38 (95%) of 40 evaluable men treated for 3 days were cured, with eight of nine buboes resolving, but only 14 (61%) of 23 evaluable men in the single-dose group were cured, with three of six buboes resolving. Side effects were common, occurring in 25% of the patients on the 3-day regimen, and consisted of dizziness, headaches, generalized pruritus, nausea, and diarrhea.

In a smaller German study Mensing reported cure of seven men with chancroid, using enoxacin (400 mg) twice daily until resolution of the lesion, a period of 7 to 12 days (39). No side effects were noted.

Summary

The new quinolones which have been studied appear to be effective for treatment of chancroid. Few clinical trials, however, have been reported. As for gonorrhea, the utility of rosoxacin is somewhat compromised because of high rates of side effects. Other quinolones with in vitro activity against *H. ducreyi* may also prove to be effective in treating this disease.

Bacterial Vaginosis

Current data suggest that bacterial vaginosis (previously described as haemophilus, gardnerella, or nonspecific vaginitis) is an overgrowth syndrome in which the normal, lactobacillus-predominant bacterial flora changes to contain smaller concentrations of lactobacilli and increased concentrations of *Gardnerella vaginalis*, *Mobiluncus* species, and vaginal anaerobes. The MICs of cufloxacin, ofloxacin, and ciprofloxacin against *G. vaginalis* have ranged from 2 to >16 µg/ml and were 2 to 4 µg/ml against 90% of *Mobiluncus* species (see chapter 3).

A single clinical trial of ciprofloxacin for treatment of bacterial vaginosis has been reported by Carmona et al. (11). Of 22 women receiving ciprofloxacin (500 mg) twice daily for 7 days, 16 (72.2%) had clinical and "bacteriologic" cures (eradication of *Bacteroides* species, gardnerella, and *Corynebacterium* sp.), 4 (18.2%) had clinical improvement without microbiologic improvement, and 2 (9%) failed to show a response. These cure rates are comparable to those reported after therapy with metronidazole (500 mg) twice daily for 7 days, the currently recommended treatment for bacterial vaginosis (12). Side effects reported included gastric pain, dizziness, somnolence, and pruritus.

Conclusion

As a new class of antibiotics, the quinolones offer promising alternatives for oral therapy of some of the STDs caused by bacteria. They are very active against *N. gonorrhoeae*, including antibiotic-resistant strains, and appear to be very effective for therapy of uncomplicated urethritis and cervicitis. Further studies are needed to define their efficacy at other mucosal sites. Similarly, *H. ducreyi*, the causative agent of chancroid and another genital pathogen with increasing antimicrobial resistance, appears to be susceptible to short courses of quinolone therapy. Unfortunately, *C. trachomatis*, the most common STD in the United States, does not appear to be susceptible to quinolones in single doses, nor does extended therapy with norfloxacin or ciprofloxacin eradicate infection. The quinolones are also not efficacious in incubating syphilis.Thus, on the basis of published studies, the newer quinolones are a useful addition to the antimicrobial armamentarium necessary for the control of STDs. Their ultimate utility, however, will be determined not only by their clinical and antimicrobial efficacy but also by their cost and the rate at which STD pathogens develop quinolone resistance in the future.

Acknowledgment. G.A.D. was supported in part by the Andrew W. Mellon Foundation.

Literature Cited

1. **Ariyarit, C., K. Panikabutra, A. Chirwarakorn, C. Wongba, and A. Buatiang.** 1986. Efficacy of ofloxacin in uncomplicated gonorrhoea. *Infection* **14**(Suppl. 4):S311–S313.
2. **Arya, O. P., D. Hobson, and C. A. Hart.** 1986. Evaluation of ciprofloxacin 500 mg twice daily for one week in treating uncomplicated gonococcal, chlamydial, and non-specific urethritis in men. *Genitourin. Med.* **62:**170–174.
3. **Aznar, J., M. C. Caballero, M. C. Lozano, C. de Miguel, J. C. Palomares, and E. J. Perea.** 1985. Activities of new quinolone derivatives against genital pathogens. *Antimicrob. Agents Chemother.* **27:**76–78.
4. **Bailey, J. M., C. Heppleston, and S. J. Richmond.** 1984. Comparison of the in vitro activities of ofloxacin and tetracycline against *Chlamydia trachomatis* as assessed by indirect immunofluorescence. *Antimicrob. Agents Chemother.* **26:**13–16.

5. **Bogaerts, J., W. M. Tello, L. Verbist, P. Piot, and J. Vandepitte.** 1987. Norfloxacin versus thiamphenicol for treatment of uncomplicated gonorrhea in Rwanda. *Antimicrob. Agents Chemother.* **31**:434–437.
6. **Bowie, W. R.** 1984. Urethritis in males, p. 638–650. *In* K. K. Holmes, P. A. Mardh, P. F. Sparling, et al. (ed.), *Sexually Transmitted Diseases.* McGraw-Hill Book Co., New York.
7. **Bowie, W. R., C. E. Shaw, D. G. W. Chan, J. Boyd, and W. A. Black.** 1986. In vitro activity of difloxacin hydrochloride (A-56619), A-56620, and cefixime (CL 284,635; FK 027) against selected genital pathogens. *Antimicrob. Agents Chemother.* **30**:590–593.
8. **Bowie, W. R., V. Willetts, and L. Sibau.** 1986. Failure of norfloxacin to eradicate *Chlamydia trachomatis* in nongonococcal urethritis. *Antimicrob. Agents Chemother.* **30**:594–597.
9. **Brunham, R. C., J. Paavonen, C. E. Stevens, N. Kiviat, C. Kuo, C. W. Critchlow, and K. K. Holmes.** 1984. Mucopurulent cervicitis: the ignored counterpart in women of urethritis in men. *N. Engl. J. Med.* **311**:1–6.
10. **Cantet, P., H. Renaudin, and C. Quentin.** 1983. Activite comparee in vitro de sept quinolones sur Ureaplasma urealyticum *Pathol. Biol.* **31**:501–503.
11. **Carmona, O., S. Hernandex-Gonzales, and R. Kobelt.** 1987. Ciprofloxacin in the treatment of nonspecific vaginitis. *Am. J. Med.* **82**(Suppl. 4A):321–323.
12. **Centers for Disease Control.** 1985. Sexually transmitted diseases: treatment guidelines, 1985. *Morbid. Mortal. Weekly Rep.* **34**:755–1085.
13. **Chan, A. S., K. C. Tang, K. K. Fung, and K. Ng.** 1986. Single dose ofloxacin in treatment of uncomplicated gonorrhea. *Infection* **14**(Suppl. 4):S314–S315.
14. **Cohen, A. I., M. F. Rein, and R. C. Noble.** 1984. A comparison of rosoxacin with ampicillin and probenacid in the treatment of uncomplicated gonorrhea. *Sex. Transm. Dis.* **11**:24–27.
15. **Crider, S. R., S. D. Colby, L. K. Miller, W. O. Harrison, S. B. Krebs, and S. W. Berg.** 1984. Treatment of penicillin-resistant Neisseria gonorrhoeae with oral norfloxacin. *N. Engl. J. Med.* **311**:137–140.
16. **Crump, B., R. Wise, and J. Dent.** 1983. Pharmacokinetics and tissue penetration of ciprofloxacin. *Antimicrob. Agents Chemother.* **24**:784–786.
17. **Dobson, R. A., J. R. O'Connor, S. A. Poulin, R. B. Kundsin, T. F. Smith, and P. E. Came.** 1980. In vitro antimicrobial activity of rosoxacin against *Neisseria gonorrhoeae, Chlamydia trachomatis,* and *Ureaplasma urealyticum. Antimicrob Agents Chemother.* **18**:738–740.
18. **Fong, A., W. Linton, B. Simbul, R. Thorup, B. McLaughlin, T. Rahm, and P. Quinn.** 1987. Treatment of nongonococcal urethritis with ciprofloxacin. *Am. J. Med.* **82**(Suppl. 4A):311–316.
19. **Fransen, L., D. Avonts, and P. Piot.** 1986. Treatment of genital chlamydial infection with ofloxacin. *Infection* **14**(Suppl. 4):S318–S323.
20. **Gonzalez, M. A., F. Uribe, S. D. Moisen, A. P. Fuster, A. Selen, P. G. Welling, and B. Painter.** 1984. Multiple-dose pharmacokinetics and safety of ciprofloxacin in normal volunteers. *Antimicrob. Agents Chemother.* **26**:741–744.
21. **Haase, D. A., J. O. Ndinya-Achola, R. A. Nash, L. J. D'Costa, D. Hazlett, S. Lubwama, H. Nsanze, and A. R. Ronald.** 1986. Clinical evaluation of rosoxacin for the treatment of chancroid. *Antimicrob. Agents Chemother.* **30**:39–41.
22. **Handsfield, H. H.** 1984. Gonorrhea and uncomplicated gonococcal infections, p. 205–220. *In* K. K. Holmes, P. A. Mardh, P. F. Sparling, et al. (ed.), *Sexually Transmitted Diseases.* McGraw-Hill Book Co., New York.
23. **Handsfield, H. H., R. E. Alexander, S. P. Wang, A. H. B. Pedersen, and K. K. Holmes.** 1976. Differences in the therapeutic response of chlamydia-positive and chlamydia-negative forms of nongonococcal urethritis. *J. Am. Vener. Dis. Assoc.* **2**(3):5–9.

24. **Handsfield, H. H., F. N. Judson, and K. K. Holmes.** 1981. Treatment of uncomplicated gonorrhea with rosoxacin. *Antimicrob. Agents Chemother.* **20:**625–629.
25. **Hart, C. A., S. J. How, and D. Hobson.** 1984. Activity of ciprofloxacin against genital tract pathogens. *Br. J. Vener. Dis.* **60:**316–318.
26. **Heessen, F. W., and H. L. Muytjens.** 1984. In vitro activities of ciprofloxacin, norfloxacin, pipemidic acid, cinoxacin, and nalidixic acid against *Chlamydia trachomatis. Antimicrob. Agents Chemother.* **25:**123–124.
27. **Heppleston, C., S. J. Richmond, and J. Bailey.** 1985. Antichlamydial activity of quinolone carboxylic acids. *J. Antimicrob. Chemother.* **15:**645–647.
28. **Hirai, K., and T. Une.** 1986. Antichlamydial activity of ofloxacin. *Microbiol. Immunol.* **30:**445–450.
29. **Hooper, D. C., and J. S. Wolfson.** 1985. Minireview: the fluoroquinolones: pharmacology, clinical uses, and toxicities in humans. *Antimicrob. Agents Chemother.* **28:**716–721.
30. **How, S. J., D. Hobson, C. A. Hart, and E. Quayle.** 1985. A comparison of the in-vitro activity of antimicrobials against *Chlamydia trachomatis* examined by Giemsa and a fluorescent antibody stain. *J. Antimicrob. Chemother.* **15:**399–404.
31. **How, S. J., D. Hobson, C. A. Hart, and R. E. Webster.** 1985. An in-vitro investigation of synergy and antagonism between antimicrobials against *Chlamydia trachomatis. J. Antimicrob. Chemother.* **15:**533–538.
32. **Judson, F. N., B. S. Beals, and K. J. Tack.** 1986. Clinical experience with ofloxacin in sexually transmitted disease. *Infection* **14**(Suppl. 4):S309–S310.
33. **Kaplowitz, L. G., N. Vishniavsky, T. Evans, S. Vartirarian, H. Dalton, M. Simpson, and R. Gruminger.** 1987. Norfloxacin in the treatment of uncomplicated gonococcal infections. *Am. J. Med.* **82**(Suppl. 6B):35–39.
34. **King, A., and I. Phillips.** 1986. The comparative in-vitro activity of pefloxacin. *J. Antimicrob. Chemother.* **17**(Suppl. B):1–10.
35. **Liebowitz, L. D., J. Saunders, G. Fehler, R. C. Ballard, and H. J. Koornhof.** 1986. In vitro activity of A-56619 (difloxacin), A-56620, and other new quinolone antimicrobial agents against genital pathogens. *Antimicrob. Agents Chemother.* **30:**948–950.
36. **Lockley, M. R., R. Wise, and J. Dent.** 1984. The pharmacokinetics and tissue penetration of ofloxacin. *J. Antimicrob. Chemother.* **14:**647–652.
37. **Loo, P. S., G. L. Ridgeway, and J. D. Oriel.** 1985. Single dose ciprofloxacin for treating gonococcal infections in men. *Genitourin. Med.* **61:**302–305.
38. **Lyon, M. D., K. R. Smith, M. S. Saag, and C. G. Cobbs.** 1987. Brief report: in vitro activity of ciprofloxacin against Neisseria gonorrhoeae. *Am. J. Med.* **82**(Suppl. 4A):40–41.
39. **Mensing, H.** 1985. Treatment of chancroid with enoxacin. *Acta Dermato Venereol.* **65:** 455–457.
40. **Morel, P., I. Casin, A. Bianchi, M. Dolivo, and Y. Perol.** 1986. Traitement par ofloxacine (RU 43280) des urethrites masculines bacteriennes non compliquees. *Pathol. Biol.* **34:** 502–504.
41. **Morel, P., F. Lassau, I. Casin, A. Baury, and Y. Perol.** 1987. Traitement minute des ur^e^trites masculines gonococciques par l'ofloxacine. *Pathol. Biol.* **35:**642–643.
42. **Naamara, W., F. A. Plummer, R. M. Greenblat, L. J. D'Costa, J. O. Ndinya-Achola, and A. Ronald.** 1987. Treatment of chancroid with ciprofloxacin: a prospective, randomized clinical trial. *Am. J. Med.* **82**(Suppl. 4A):317–320.
43. **Notowicz, A., E. Stolz, and B. van Klingeren.** 1984. A double blind study comparing two dosages of enoxacin for the treatment of uncomplicated urogenital gonorrhoea. *J. Antimicrob. Chemother.* **14**(Suppl. C):91–94.
44. **Okazaki, T., H. Kiyota, H. Gotoh, and S. Onodera.** 1986. Bacteriological and clinical study of ciprofloxacin in gonococcal urethritis. *Jpn. J. Antibiot.* **39:**2685–2689.

45. **Osada, Y., and H. Ogawa.** 1983. Antimycoplasmal activity of ofloxacin (DL-8280). *Antimicrob. Agents Chemother.* **23:**509–511.
46. **Panikabutra, K., C. Ariyarit, A. Chitwarakorn, and C. Saensanoh.** 1984. Rosoxacin in the treatment of uncomplicated gonorrhea in men. *Br. J. Vener. Dis.* **60:**231–234.
47. **Perine, P. L., R. S. Morton, P. Piot, M. S. Siegel, and G. M. Antal.** 1979. Epidemiology and treatment of penicillinase-producing *Neisseria gonorrhoeae*. *Sex. Transm. Dis.* **6:**152–158.
48. **Phillips, I.** 1987. Quinolones in the treatment of gonorrhoea. *Quinolones Bull.* **3:**1–4.
49. **Richmond, S., C. Heppleston, and J. Bailey.** 1986. The action of quinolone carboxylic acids on *Chlamydia trachomatis* in vitro. *Biochem. Soc. Trans.* **14:**503–504.
50. **Ridgeway, G. L., G. Mumtaz, F. G. Gabriel, and J. D. Oriel.** 1984. The activity of ciprofloxacin and other 4-quinolones against *Chlamydia trachomatis* and *Mycoplasmas* in vitro. *Eur. J. Clin. Microbiol.* **3:**344–346.
51. **Roberts, M., L. P. Elwell, and S. Falkow.** 1977. Molecular characterization of two beta-lactamase-specifying plasmids isolated from *Neisseria gonorrhoeae*. *J. Bacteriol.* **131:**557–563.
52. **Roddy, R. E., H. H. Handsfield, and E. W. Hook III.** 1986. Comparative trial of single-dose ciprofloxacin and ampicillin plus probenecid for treatment of gonococcal urethritis in men. *Antimicrob. Agents Chemother.* **30:**267–269.
53. **Romanowski, B.** 1986. Norfloxacin in the therapy of gonococcal infections. *Scand. J. Infect. Dis.* **48**(Suppl.)**:**40–45.
54. **Ronald, A. R., and W. L. Albritton.** 1984. Chancroid and Haemophilus ducreyi, p. 385–393. *In* K. K. Holmes, P. A. Mardh, P. F. Sparling, et al. (ed.), *Sexually Transmitted Diseases*. McGraw-Hill Book Co., New York.
55. **Sanson-Le Pors, M. J., I. M. Casin, M.-C. Thebault, G. Arlet, and Y. Perol.** 1986. In vitro activities of U-6336, a spectinomycin analog; roxithromycin (RU 28965), a new macrolide antibiotic; and five quinolone derivatives against *Haemophilus ducreyi*. *Antimicrob. Agents Chemother.* **30:**512–513.
56. **Schachter, J., and J. Moncada.** 1987. In vitro activity of ciprofloxacin against *Chlamydia trachomatis*. *Am. J. Med.* **82**(Suppl. 4A)**:**42–43.
57. **Schmid, G. P., L. L. Sanders, J. H. Blount, and R. Alexander.** 1987. Chancroid in the United States: reestablishment of an old disease. *J. Am. Med. Assoc.* **258:**3265–3286.
58. **Segreti, J., H. A. Kessler, K. S. Kapell, and G. M. Trenholme.** 1987. In vitro activity of A-56268 (TE-031) and four other antimicrobial agents against *Chlamydia trachomatis*. *Antimicrob. Agents Chemother.* **31:**100–101.
59. **Shahmanesh, M., S. R. Shulka, I. Phillips, A. Westwood, and R. N. Thin.** 1986. Ciprofloxacin for treating urethral gonorrhoea in men. *Genitourin. Med.* **62:**86–87.
60. **Shapiro, M. A., C. L. Heifetz, and J. C. Sesnie.** 1987. Comparative in-vitro activity of enoxacin against penicillinase- and non-penicillinase-producing Neisseria gonorrhoeae. *Sex. Transm. Dis.* **14:**111–112.
61. **Shen, L. L., and A. G. Pernet.** 1985. Mechanism of inhibition of DNA gyrase by analogues of nalidixic acid: the target of the drug is DNA. *Proc. Natl. Acad. Sci. USA* **82:**307–311.
62. **Simon, C., and U. Linder.** 1983. In vitro activity of norfloxacin against *Mycoplasma hominis* and *Ureaplasma urealyticum*. *Eur. J. Clin. Microbiol.* **2:**479–480.
63. **Stamm, W. E., and K. K. Holmes.** 1984. *Chlamydia trachomatis* infections in the adult, p. 258–270. *In* K. K. Holmes, P. A. Mardh, P. F. Sparling, et al. (ed.), *Sexually Transmitted Diseases*. McGraw-Hill Book Co., New York.
64. **Stamm, W. E., and R. Suchland.** 1986. Antimicrobial activity of U-70138F (paldimycin), roxithromycin (RU 965), and ofloxacin (ORF 18489) against *Chlamydia trachomatis* in cell culture. *Antimicrob. Agents Chemother.* **30:**806–807.

65. **Strum, A. W.** 1987. Comparison of antimicrobial susceptibility patterns of fifty-seven strains of *Haemophilus ducreyi* isolated in Amsterdam from 1978 to 1985. *J. Antimicrob. Chemother.* **19:**187–191.
66. **Sugina, A., C. L. Peebles, K. N. Kreuzer, and N. R. Cozzarelli.** 1977. Mechanism of action of nalidixic acid: purification of Escherichia coli nal A gene product and its relationship to DNA gyrase and a novel nicking enzyme. *Proc. Natl. Acad. Sci. USA* **74:**4767–4771.
67. **Swanson, B. N., V. K. Boppana, P. H. Vlasses, H. H. Rotmensch, and R. K. Ferguson.** 1983. Norfloxacin disposition after sequentially increasing oral doses. *Antimicrob. Agents Chemother.* **23:**284–288.
68. **Szarmach, H., H. Weuta, J. Podziewski, J. Skarzynski, and A. Wronski.** 1986. Ciprofloxacin in acute male gonorrhea. *Arzneimittelforschung* **36:**1840–1842.
69. **Taylor, D. N., P. Echeverria, S. Hanchalay, C. Pitarangsi, L. Slootmans, and P. Piot.** 1985. Antimicrobial susceptibility and characterization of outer membrane proteins of *Haemophilus ducreyi* isolated in Thailand. *J. Clin. Microbiol.* **21:**442–444.
70. **Tegelberg-Stassen, M. J., J. C. van der Hoek, L. Mooi, J. H. T. Wagenvoort, T. van Joost, M. F. Michel, and E. Stolz.** 1986. Treatment of uncomplicated gonococcal urethritis in men with two dosages of ciprofloxacin. *Eur. J. Clin. Microbiol.* **5:**244–246.
71. **Tegelberg-Stassen, M. J., A. H. van der Willigen, J. C. van der Hoek, J. H. Wagenvoort, H. J. van Vliet, B. van Klingeren, T. van Joost, M. F. Michel, and E. Stolz.** 1986. Treatment of uncomplicated urogenital gonorrhea in women with a single oral dose of enoxacin. *Eur. J. Clin. Microbiol.* **5:**395–398.
72. **Van der Willigen, A. H., J. C. S. van der Hoek, J. H. T. Wagenvoort, H. J. A. van Vliet, B. van Klingeren, W. O. Schalla, J. S. Knapp, T. van Joost, M. F. Michel, and E. Stolz.** 1987. Comparative double-blind study of 200- and 400-mg enoxacin given orally in the treatment of acute uncomplicated urethral gonorrhea in males. *Antimicrob. Agents Chemother.* **31:**535–538.
73. **Veller-Fornasa, C., M. Tarantello, R. Cipriani, L. Guerra, and A. Peserico.** 1987. Effect of ofloxacin on *Treponema pallidum* in incubating experimental syphilis. *Genitourin. Med.* **86:**214.
74. **Wall, R. A., D. C. Mabey, C. S. Bello, and D. Felmingham.** 1985. The comparative in-vitro activity of twelve 4-quinolone antimicrobials against *Haemophilus ducreyi*. *J. Antimicrob. Chemother.* **16:**165–168.
75. **Warren, C. A., K. P. Shannon, and I. Phillips.** 1981. In-vitro anti-gonococcal activity of rosoxacin (WIN 35213). *Br. J. Vener. Dis.* **57:**33–35.
76. **Wise, R., J. M. Andrews, and G. Danks.** 1984. In-vitro activity of enoxacin (CI-919), a new quinolone derivative, compared with that of other antimicrobial agents. *J. Antimicrob. Chemother.* **13:**237–244.
77. **Wise, R., R. Lockley, J. Dent, and M. Webberly.** 1984. Pharmacokinetics and tissue penetration of enoxacin. *Antimicrob. Agents Chemother.* **26:**17–19.
78. **Wolfson, J. S., and D. C. Hooper.** 1985. Minireview. The fluoroquinolones: structures, mechanisms of action and resistance, and spectra of activity in vitro. *Antimicrob. Agents Chemother.* **28:**581–586.

Chapter 7

Therapy of Respiratory Tract Infections with Quinolone Antimicrobial Agents

Brian E. Scully

Division of Infectious Diseases, Department of Medicine, College of Physicians and Surgeons, Columbia University, New York, New York 10032

Introduction

This chapter will begin with comments on microbiologic activity in vitro and pharmacokinetics of the new quinolones and then will review in detail the efficacy of these drugs for the therapy of infections of the lower and upper respiratory tracts. The discussion will be confined to those quinolones for which therapeutic data exist. At the time of this writing, oral ciprofloxacin has been approved in the United States and Europe, ofloxacin has been approved in Europe and Japan, and pefloxacin has been approved in France. Enoxacin is still under investigation. Norfloxacin has been approved in Europe, Japan, and the United States, but only for the therapy of urinary tract infections.

Microbiologic Activity and Pharmacokinetics

Table 1 summarizes the activities in vitro of ciprofloxacin, ofloxacin, enoxacin, and pefloxacin against classic respiratory pathogens and gram-negative bacilli commonly encountered in nosocomial pneumonia (see also chapter 3). In general, MICs of greater than 2 μg/ml are considered resistant, but when differences in MICs among the various drugs are interpreted, consideration should be given to differences in their pharmacokinetics and penetration in tissue as indicated in Table 2 (see also chapter 4). It should also be noted that the levels of ciprofloxacin, ofloxacin, and enoxacin in lung or bronchial tissue are often higher than the peak concentrations in serum.

Table 1. Comparative in vitro activity of quinolones against respiratory pathogens[a]

Organism	Drug MIC range (μg/ml)			
	Ciprofloxacin	Ofloxacin	Enoxacin	Pefloxacin
Streptococcus pneumoniae	0.2–3.2 (2.0)[b]	1.0–2.0 (2.0)	1.6–16 (16)	4.0–16 (8.0)
S. pyogenes	0.25–4.0 (2.0)	0.4–4.0 (2.0)	0.8–25 (25)	4.0–128 (16)
S. viridans	0.5–4.0 (2.0)	1.0–4.0 (2.0)	3.1–25 (25)	4.0–64 (32)
Staphylococcus aureus (including MRSA[c])	0.1–2.0 (0.8)	0.12–1.0 (0.5)	0.5–6.3 (3.1)	0.25–1.0 (0.5)
Haemophilus influenzae	0.004–0.016 (0.01)	0.016–0.3 (0.3)	0.01–0.12 (0.05)	0.03–0.06 (0.6)
Branhamella catarrhalis	0.01–0.5 (0.05)	0.02–0.1 (0.05)	0.02	0.125–0.25 (0.25)
Neisseria meningitidis	<0.01	0.03	<0.1	0.03
Mycoplasma pneumoniae	0.5–2.5	0.4–1.6 (1.6)	5.0	
Legionella spp.	0.01–0.5 (0.12)	0.03–0.5 (0.5)	0.08–1.0 (0.2)	0.25–1.0 (1.0)
Klebsiella spp.	0.08–0.5 (0.25)	0.3–1.0 (1.0)	0.06–4.0 (0.8)	0.06–4.0 (2.0)
Pseudomonas aeruginosa	0.06–1.6 (0.8)	0.8–4.0 (4.0)	0.4–50 (3.1)	1.0–8.0 (8.0)
Acinetobacter spp.	0.04–6.3 (1.0)	0.02–50 (1.0)	0.03–2.0 (1.0)	0.4–6.3 (6.3)
Escherichia coli	0.3–0.5 (0.25)	0.03–2.0 (1.0)	0.025–8.0 (0.4)	0.03–8.0 (2.0)
Enterobacter spp.	0.01–1.0 (0.13)	0.03–4.0 (1.0)	0.06–8.0 (1.0)	0.06–16 (2.0)
Bacteroides spp.	0.5–32 (16)	0.5–8.0 (8.0)	1.0–50 (32)	2.0–32 (32)
Fusobacteria	0.03–4.0 (2.0)	0.5–4.0 (4.0)		1.0–64 (32)
Peptococci and peptostreptococci	0.03–16 (2.0)	0.25–2.0 (2.0)	4.0–8.0 (8.0)	1.0–32 (8.0)
Mycobacterium tuberculosis	0.25–0.5 (0.5)	0.5–1.0 (1.0)		

[a]Data were compiled from references 4, 8, 9, 12, 17, 34, 35, 37, 46, 47, 54, 55, and 61.
[b]MICs for 90% of strains are given within parentheses.
[c]MRSA, Methicillin-resistant *S. aureus*.

Table 2. Concentrations of quinolones in serum, sputum, and lung tissue[a]

Drug	Dosage	Half-life in serum (h)	Peak concn in serum (μg/ml)	Concn in sputum (μg/ml)	Concn in lung tissue (μg/g)
Ciprofloxacin	500 mg p.o. BID	3–4	3.9	1.4	1.2–17
	1,000 mg p.o. BID		7.8	3.1	
	750–1,000 mg p.o. BID		2.5–3.5	1.0–2.0	
	100 mg i.v.[b]		0–85		2.6
Ofloxacin	400 mg p.o.	5–7	8.0	5.0	
	200 mg p.o.		4.0	3.0	7.0
Enoxacin	600 mg p.o. QD	6	4.1	3.7	7.0
Pefloxacin	400 mg p.o. BID	10	10–15	5–13	

[a]Data were compiled from references 6, 27, 57, 59, 65, and Morel et al. (C. Morel, M. Vergnaud, M. M. Langeard, and N. Monrocq, *Proc. 13th Int. Congr. Chemother.*, p. 32–38, 1983).

[b]i.v., Intravenously.

Excellent activity is shared by all of these drugs against *Haemophilus influenzae, Branhamella catarrhalis, Legionella* species, *Escherichia coli,* and most strains of *Klebsiella* spp. and *Enterobacter* spp. Ciprofloxacin is the most active agent in vitro, but the clinical significance of this enhanced potency in vitro is doubtful in view of the superior concentrations in serum and sputum of some of the other agents, in particular, ofloxacin. Ciprofloxacin and ofloxacin inhibit most strains of *Streptococcus pneumoniae* and *Streptococcus pyogenes* at 1 to 2 μg/ml, a range which is considered of only intermediate susceptibility. None of the other agents possesses useful activity against these organisms. Ciprofloxacin is by far the most active agent against *Pseudomonas aeruginosa,* with MICs for 90% of strains of 0.5 to 0.8 μg/ml. The other quinolones are less active, though many strains are of intermediate susceptibility. The excellent penetration into sputum of ofloxacin and perhaps pefloxacin might allow for the effective therapy of such strains. Limited data are available concerning *Mycoplasma pneumoniae,* but both ciprofloxacin and ofloxacin appear to possess intermediate activity. *Bacteroides fragilis, Bacteroides melaninogenicus,* and other anaerobic organisms are resistant to all of the quinolones.

Both ciprofloxacin and ofloxacin have moderate activity against *Mycobacterium tuberculosis,* inhibiting 90% of strains at 0.5 and 1 μg/ml, respectively (17). Ciprofloxacin, but not ofloxacin, may also have useful activity against *Mycobacterium avium-Mycobacterium intratracellulare* (MIC for 90% of strains, 2 μg/ml).

When quinolones are combined with beta-lactams or aminoglycosides in vitro, additive activity is most often demonstrated and synergy is not uncommon. This interaction has been best shown for ciprofloxacin against *P. aeruginosa* (44). Antagonism appears to be a rare phenomenon.

Clinical Studies

Lower respiratory tract

Ciprofloxacin. Oral ciprofloxacin has been studied extensively in the therapy of lower respiratory tract infections in the United States, Europe, and Japan. Comparative studies (Table 3) in patients with mainly bronchitis suggest that ciprofloxacin is at least equivalent to trimethoprim-sulfamethoxazole, doxycycline, ampicillin, or amoxicillin and superior to cephalexin in achieving clinical improvement and sterilization of sputum. Dosages of 500 or 750 mg orally (p.o.) twice a day (BID) were used in all of these studies. Wollschlager et al. (67) reported a comparative, double-blind study of 82 patients with bacterial bronchitis who received either ciprofloxacin (750 mg p.o. BID) or ampicillin (500 mg p.o. four times a day).

Table 3. Oral ciprofloxacin for treatment of lower respiratory tract infections—comparative studies

Reference	Diagnosis	Treatment[a]	Response (%)	
			Clinical	Bacteriologic
Wollschlager et al. (67)	Bronchitis	CIP, 750 mg BID	41/42 (98)	40/42 (95)
		AMP, 500 mg QID[b]	40/45 (89)	30/40 (75)
Gleadhill et al. (22)	Pneumonia or bronchitis	CIP, 500 mg	21/26 (81)	15/18 (83)
		AMX, 250 mg TID	18/22 (82)	8/13 (62)
Magnani et al. (42)	Pneumonia or bronchitis	CIP, 500 mg BID	15/15 (100)	11/15 (73)
		TMP-SMX, 160 mg/800 mg	12/15 (80)	5/12 (42)
Bantz et al. (2)	Bronchitis	CIP, 250 mg TID	53/55 (96)	
		DOX, 100 mg BID	55/55 (100)	
Feist et al. (16)	Lower respiratory tract infection	CIP, 500 mg BID	27/28 (96)	23/28 (82)
		CEP, 1,000 mg BID	22/23 (96)	14/23 (61)

[a]Abbreviations: CIP, ciprofloxacin; AMP, ampicillin; AMX, amoxicillin; TMP-SMX, trimethoprim-sulfamethoxazole; DOX, doxycycline; CEP, cephalexin.
[b]QID, Four times a day.

Both drugs were effective clinically, but ciprofloxacin was more effective in sterilizing the sputum (40 of 42 versus 30 of 47; $P = 0.002$). In this study all five strains of *S. pneumoniae* isolated were eradicated by ciprofloxacin. Gleadhill et al. (22) also found similar results when comparing ciprofloxacin (500 mg p.o. BID) with amoxicillin (250 mg p.o. three times a day [TID]). The clinical responses were 82% for either regimen, with bacteriologic responses of 87% for ciprofloxacin versus 66% for amoxicillin ($P = 0.4$). In nearly one-half of the patients (20 of 48), a pathogen was not isolated before therapy. There has been one small study by Magnani et al. (42) in which ciprofloxacin was compared with trimethoprim-sulfamethoxazole (15 patients each in the therapy of pneumonia and bronchitis). Similar clinical responses were seen with both agents. Bantz et al. (2) compared low-dose ciprofloxacin (250 mg p.o. BID) with doxycycline (100 mg p.o. BID) in a study of predominantly bronchitis in a general practice setting. Almost all patients responded to either therapy, but no cultures were performed. In the study by Feist et al. (16), ciprofloxacin (750 mg p.o. BID) gave clinical and bacteriologic results that were superior to those with cephalexin (1 g p.o. BID), but cephalexin should not be considered a drug of first choice for the treatment of bronchitis. In this study seven of nine isolates of *S. pneumoniae* were eradicated by ciprofloxacin.

In open studies (Table 4) a wider variety of infections was treated, including several hundred patients with pneumonia. Kobayashi (36) has reviewed the results of the Japanese trials of ciprofloxacin. The dosages of ciprofloxacin varied between 200 and 1,200 mg p.o. daily (QD), with most patients receiving 600 mg QD (300 mg p.o. BID). Of 432 patients with bronchitis, bronchiectiasis, or panbronchiolitis, a satisfactory clinical response was obtained in 318 (74%). Of 110 patients with pneumonia, 89 responded. *Haemophilus* spp. and enteric gram-negative bacilli were almost invariably eradicated, but *S. pneumoniae* and *Staphylococcus aureus* persisted in about 30% of cases. *P. aeruginosa* persisted in 75%. The relationship of response to dosage was not discussed. Davies et al. (10) examined the effect of different doses of ciprofloxacin in patients with exacerbations of chronic bronchitis. While about 80% of patients were well at the end of a 10-day course of therapy, this figure fell to 56% 7 days later. Relapses were associated with the persistence or reappearance of *S. pneumoniae* and *P. aeruginosa* in the sputum. Dosages of 1,000 mg p.o. BID provided results that were only marginally superior to those obtained with 500 mg p.o. BID. Fass (13) obtained a 100% clinical response rate in 42 patients with either bronchitis or pneumonia. *P. aeruginosa* persisted in 5 of 12 patients, with the development of resistance in 4. All seven strains of *S. pneumoniae* were eradicated in this study. The study by Hoogkamp-Korstanje and Klein (28) of ciprofloxacin in exac-

Table 4. Oral ciprofloxacin for treatment of lower respiratory tract infections—open trials

Reference	Diagnosis	Ciprofloxacin treatment	Response (%)	
			Clinical	Bacteriologic
Kobayashi (36)	Lower RTI[a] (total)	200–1,200 mg QD (most 300 mg BID)	397/542 (73)	254/391 (65): *S. aureus*, 26/42 (62); *S. pneumoniae*, 29/42 (69); *E. coli*, 7/9 (78); *Klebsiella* spp., 13/17 (76); *H. influenzae*, 109/125 (87); *P. aeruginosa*, 21/94 (22)
	Pneumonia		89/110 (81)	
	Bronchitis		149/214 (70)	
	Bronchiectasis		94/121 (78)	
	Panbroncholitis		36/66 (55)	
	Other		29/31 (94)	
Davies et al. (10)	Bronchitis	Total	42/75 (56)	43/75 (57)
		500 mg p.o. BID	10/19 (53)	
		750 mg p.o. BID	21/38 (55)	
		1,000 mg p.o. BID	11/18 (61)	
Fass (13)	Lower RTI (total)	500 or 750 mg BID	42/42 (100)	35/42 (83): *S. pneumoniae*, 7/7 (100); *H. influenzae*, 21/21 (100); *P. aeruginosa*, 7/12 (58); beta streptococci, 0/2 (0); *Enterobacteriaceae*, 6/7 (86)
	Bronchitis		22/22 (100)	
	Pneumonia		20/20 (100)	
Hoogkamp-Korstanje and Klein (28)	Bronchitis	500 mg BID	26/27 (96)	22/27 (81): *S. pneumoniae*, 5/9 (56); *H. influenzae*, 14/14 (100); *Enterobacteriaceae*, 7/8 (88)

[a]RTI, Respiratory tract infection.

erbations of chronic bronchitis found an excellent clinical response, but *S. pneumoniae* persisted in four of nine patients.

The reported experience with intravenous ciprofloxacin has been small but encouraging. Bassaris et al. (3) treated 78 patients with intravenous (200 mg BID) followed by oral (500 mg BID) ciprofloxacin. The clinical responses were satisfactory in 36 patients, but etiologic pathogens were identified in only 11. Giamarellou and Galanakis (20) reported 15 seriously ill patients with gram-negative nosocomial respiratory infections. Eleven were intubated and in an intensive care unit. Satisfactory clinical responses were noted in all, but *Pseudomonas* spp. persisted in the sputum in 6 of 10 patients.

No patients with bacteremic pneumococcal pneunonia treated with ciprofloxacin have been reported at this time.

Ciprofloxacin has excellent activity in vitro against *Legionella pneumophila,* and good results have been obtained in animal models of infection (54, 55), but as yet there have been no data on the treatment of *Legionella* infection in humans reported.

Ofloxacin. There have been many comparative and open studies in which ofloxacin has been evaluated for the therapy of lower respiratory tract infections. The results of the comparative studies are summarized in Table 5. Fujimori et al. (19) compared ofloxacin (200 mg p.o. TID) with cefaclor (250 mg p.o. TID) in the therapy of acute and chronic bronchitis. The overall results were superior for ofloxacin, mainly due to its efficacy in *H. influenzae* and *P. aeruginosa* infections. Monk and Campoli-Richards (43) reviewed unpublished European data in which ofloxacin was compared with trimethoprim-sulfamethoxazole. Clinical results were similar for both groups. Bacteriologic data were not given. Harazin et al. (25) compared ofloxacin with doxycycline in bronchitis and pneumonia. Responses were similar with both programs, but in only 55 of 230 patients was a pathogen isolated before therapy.

The results of open trials are summarized in Table 6. Grassi et al. (24) reported the results of ofloxacin therapy of 667 patients in a multicenter Italian trial. Ofloxacin dosages of 200, 300, or 400 mg p.o. BID were used. The clinical response rate was 92% and was identical for each of the three dosages. The bacteriologic eradication rate was 79%. *P. aeruginosa* frequently persisted (44%). The Japanese experience with ofloxacin has been reviewed by Saito et al. (53). A total of 553 patients with a variety of lower respiratory tract infections treated with 300 to 600 mg of ofloxacin p.o. QD were evaluated. Satisfactory clinical responses were found in 418 (76%). *P. aeruginosa* and sometimes *S. pneumoniae* persisted in the sputum, whereas *Haemophilus* spp. and enteric gram-negative bacilli were nearly always eradicated. Maesens et al. (41) treated 113 patients with exacerbations of chronic bronchitis with several different dosage programs of ofloxacin (400 to 1,200 mg

Table 5. Oral ofloxacin for treatment of lower respiratory tract infections—comparative trials

Reference	Diagnosis	Treatment[a]	Response (%)	
			Clinical	Bacteriologic
Fujimori et al. (19)	Bronchitis (acute, 37; chronic, 172)	OFL, 200 mg BID for 14 days	82/103 (80)	90/103 (87)
		CEF, 250 mg TID for 14 days	52/105 (50)	50/105 (48)
Monk and Campoli-Richards (43)	Bronchitis	OFL, 100–400 mg BID for 7–20 days	26/32 (81)	
		TMP-SMX	24/28 (86)	
	Pneumonia	OFL 100–400 mg BID	18/19 (95)	
		TMP-SMX	22/24 (92)	
Harazin et al. (25)	Bronchitis	OFL, 200–400 mg BID	47/52 (90)	
		DOX, 100 mg BID	33/36 (92)	
	Pneumonia	OFL, 200–400 mg BID	60/62 (97)	
		DOX, 100 mg BID	62/69 (90)	

[a]Abbreviations: OFL, ofloxacin; CEF, cefaclor, TMP-SMX, trimethoprim-sulfamethoxazole; DOX, doxycycline.

Table 6. Oral ofloxacin for treatment of lower respiratory tract infections—open trials

Reference	Diagnosis (no. of patients)	Ofloxacin therapy	Response (%)	
			Clinical	Bacteriologic
Grassi et al. (24)	Lower RTI[a]: pneumonia (234), bronchitis (381), pleurisy (18), lung abscess (4), other (30)	200, 300, or 400 mg BID	612/667 (92)	287/370 (78): *S. aureus*, 40/47 (85); *H. influenzae*, 11/11 (100); *Klebsiella* spp., 18/20 (90); *P. aeurginosa*, 36/69 (52)
Saito et al. (53)	Lower RTI	300-600 mg QD	418/553 (76)	279/377 (74): *S. aureus*, 16/20 (80); *S. pneumoniae*, 26/34 (76); *E. coli*, 11/11 (100); *Klebsiella* spp., 25/27 (93); *P. aeruginosa*, 32/99 (32); *H. influenzae*, 99/102 (97)
	Chronic bronchitis		150/182 (82)	
	Panbronchiolitis		56/80 (70)	
	Secondary infection, chronic respiratory disease		28/38 (74)	
	Pneumonia		102/127 (80)	
	Lung abscess		4/7 (57)	
	Pyothorax		1/2 (50)	
Maesens et al. (41)	Bronchitits	800–1,200 mg QD	82/113 (73)	See text
Giamarellou and Tsagarakis (21)	Lower RTI		22/25 (88)	
	Bronchitis	200 mg BID for 10 days	14/16 (88)	13/16 (81)
	Pneumonia	400 mg BID for 21 days	7/7 (100)	7/7 (100)
	Lung abscess	400 mg BID for 21 days	1/2 (50)	1/2 (50)
Scully and Neu (*15th ICC*)	Lower RTI	400 mg BID	14/14 (100)	7/14 (50)

[a]RTI, Respiratory tract infection.

p.o. QD). They reported a satisfactory clinical response rate of 72% 1 week after completion of therapy. Failures were due to the persistence in sputum of *S. pneumoniae* and less frequently *Pseudomonas* spp. The best results were obtained when 800 mg was given p.o. as a single dose—perhaps because of higher concentrations of ofloxacin in the sputum. Minor rises in ofloxacin MICs against *S. pneumoniae* and *P. aeruginosa* were noted. The studies of Giamarellou and Tsagarakis (21) and Scully and Neu (B. E. Scully and H. C. Neu, *Proc. 15th Int. Congr. Chemother.*, abstr. no. 1137, 1987) show good clinical efficacy in bronchitis and pneumonia, but the persistence of *P. aeruginosa* with the development of resistance to olfloxacin was noted in the latter study.

Enoxacin. Ishigami (30) has summarized the results of the studies of enoxacin in Japan (Table 7). Most patients were treated with 600 mg of enoxacin p.o. QD. Clinical response rates were 80% in pneumonia but only 67% in exacerbations of chronic lung disease. The microbiologic results did not correlate with the site of infection, and as might be expected, eradication rates for *S. pneumoniae* were low (6%). Small studies by Joet et al. (32), Wijnands et al. (63), and Davies et al. (11) have also been published (Table 7). Overall, the results show only a modest efficacy for enoxacin. Resistance developed frequently in *S. pneumoniae* and *P. aeruginosa.* Adverse interactions with theophylline were common.

Pefloxacin. Only very limited data have been published for pefloxacin. Lauwers et al. (39) reported their experience in which intravenous pefloxacin was used to treat 14 patients with gram-negative bacillary and staphylococcal pneumonia acquired in an intensive care unit. Satisfactory clinical responses were observed in 13 patients, with microbiologic cure in 12. Superinfection with pefloxacin-resistant *Serratia marcescens* occurred in two patients. Giamarellou et al. (H. Giamarellou, N. Galankis, G. Perdikaris, C. H. Tsagaraki, and P. Sfikakis, *Program Abstr. 27th Intersci. Conf. Antimicrob. Agents Chemother.*, abstr. no. 113, 1987) recently compared pefloxacin (intravenous and p.o.) with ceftazidime and found similar results in the treatment of a small number of patients with gram-negative pneumonia. Maesens et al. (41) reported 47 patients with exacerbations of chronic bronchitis treated with pefloxacin (400 mg p.o. BID). Only 28 of 47 patients had satisfactory clinical responses 1 week after completion of therapy. Resistance to *S. pneumoniae* and *P. aeruginosa* occurred frequently.

Cystic fibrosis

Chronic infection of the lower respiratory tract with *P. aeruginosa* is a major complication of cystic fibrosis. By a complex interplay of direct toxic and indirect immune system-mediated damage to the lung, respiratory failure eventually develops. Recent studies have attested to the benefits of

Table 7. Oral enoxacin for treatment of lower respiratory tract infections

Study	Infection	Dose	Response (%)	
			Clinical	Bacteriologic
Ishigami (30)	Pneumonia	600 mg QD	44/51 (86)	Not given
	Bronchitis		63/82 (77)	
	Chronic respiratory infection		194/291 (67)	
Joet et al. (32)	Pneumonia, bronchitis, or bronchiectasis	400 mg BID	17/20 (85)	12/20 (60)
Wijnands et al. (63)	Bronchitis or bronchiectasis	400 or 600 mg BID	39/43 (91)	24/43 (56); *Pseudomonas* spp., 6/23 (26)
Davies et al. (11)	Bronchitis	400 mg BID	7/11 (64)	Not given
		600 mg BID	3/6 (50)	

antipseudomonal therapy, which produces a temporary improvement in the symptoms of pulmonary infection and also seems to slow the progression of the lung disease. The antipseudomonal quinolones, by virtue of their good oral absorption, have the potential to reduce the frequency and length of hospitalizations for these patients. To date, ciprofloxacin and ofloxacin have been studied.

Ciprofloxacin. Table 8 summarizes the results of seven studies in which oral ciprofloxacin was evaluated in the treatment of adults with cystic fibrosis. The majority of patients received either 500 mg p.o. TID or 750 mg p.o. BID for 2 to 3 weeks. Several patients received longer courses of therapy. These studies show that if susceptible organisms are present, ciprofloxacin is effective in treating pulmonary exacerbations, as judged by changes in clinical symptoms and pulmonary function tests. In the comparative studies with azlocillin-aminoglycoside, improvement was similar in both groups, except for the study by Hodson et al. (26), in which improvement appeared to be more sustained in patients treated with ciprofloxacin. The development of resistance in *P. aeruginosa* during therapy was usual but not uncommonly resolved after treatment was stopped. Resistance is more pronounced in patients receiving longer courses of therapy (>3 weeks). Patients retreated with ciprofloxacin did not always respond as well as during the first course. The reasons for some of these failures were unclear, as resistance was not always detected. No serious toxicity was noted in any of these studies.

Experience in children with cystic fibrosis has been limited because of the fears of quinolone-induced cartilage damage and arthropathy. Scully et al. (58) have used ciprofloxacin to treat several children as young as 6 years of age with good efficacy and no toxicity. Raeburn et al. (49) report efficacy, but one case of transient arthropathy occurred among the 30 children treated.

Ciprofloxacin is thus an effective, well-tolerated therapy of pulmonary exacerbations of cystic fibrosis associated with *P. aeruginosa.* There are concerns about the development of resistance with prolonged and repeated use. Further studies are required to determine the precise role of ciprofloxacin for the treatment of patients with cystic fibrosis. For the moment, therapeutic courses should be limited to 10 to 20 days, and ciprofloxacin probably should not be used in consecutive therapies.

Ofloxacin. Because of its inferior antipseudomonal activity in vitro, ofloxacin has been studied less extensively in cystic fibrosis than has ciprofloxacin. Kurz et al. (38) compared ofloxacin (400 mg p.o. BID) with ciprofloxacin (500 mg p.o. BID) for 10 days. Twenty patients each were treated. Results were similar, with two-thirds responding to either therapy. More recently, Jensen et al. (31) reported a double-blind crossover study in which ofloxacin (400 mg p.o. BID) was compared with ciprofloxacin (750

Table 8. Oral ciprofloxacin for treatment of lower respiratory tract infections in patients with cystic fibrosis

Reference	Design[a]	No. of patients	Outcome	Resistance
Bosso et al. (7)	Randomized CIP (750 mg BID) vs AZL + TOB	20 (10/10)	Equivalent clinical and bacterial response, 8/9 (CIP) vs 7/10 (AZL + TOB)	None
Rubio (52)	Open comparative, CIP (750 mg BID) for 14 days vs AZL + TOB	23 (37 courses)	Equivalent, 24/26 vs 11/11	1 *Pseudomonas cepacia* strain
Hodson et al. (26)	Controlled CIP (500 mg) vs AZL + TOB	40	Equivalent, 19/20 vs 18/20; improvement sustained longer in CIP-treated patients	20%, not sustained, similar in both groups
Shalit et al. (60)	Randomized comparison, CIP (750 mg BID for 14 days) vs CIP (1,000 mg BID for 14 days)	29: 14 vs 15	Equivalent, 62% substantial improvement	45%
Goldfarb et al. (23)	Open, 750 mg every 8 h for 21 days	30	All responded	Elevated MICs in all patients for *P. aeruginosa*
Scully et al. (58)	Open, 750 mg every 8–12 h	18 (39 courses)	Overall, 82%; 96% to first course	50%
Bender et al. (5)	Open, 500–1,000 mg for 3–8 wk	19; 13 repeated course	Initial course, 84%	Modest, most with prolonged therapy

[a]Abbreviations: CIP, ciprofloxacin; AZL, azlocillin; TOB, tobramycin.

mg p.o. BID). Again, similar response rates of 91 and 96%, respectively, were observed. Resistance to both agents occurred during therapy but resolved within 3 months. As part of a large open study, Scully and Neu (*15th ICC*) treated nine patients. Seven responded clinically. Resistance in *P. aeruginosa* was noted frequently. On the basis of limited data, it appears that ofloxacin will prove to be effective in the therapy of exacerbations of cystic fibrosis lung disease due to susceptible *P. aeruginosa*. The development of resistance occurs, as with ciprofloxacin.

Tuberculosis

In view of its antituberculosis activity, ofloxacin has been used to treat a number of patients with recalcitrant and drug-resistant tuberculosis. Tsukamura et al. (62) treated 17 patients with ofloxacin (300 mg p.o. QD) for 6 to 8 months. Patients also received additional drugs, but ofloxacin was often the only active agent, because of multiple drug resistance. In five patients (29%), sputum converted to negative, but resistance was noted in the patients with persistently positive sputa. Suzuyama et al. (Y. Suzuyama, K. Hara, A. Saito, K. Yamaguchi, S. Kohno, and Y. Shigeno, *Program Abstr. 26th Intersci. Conf. Antimicrob. Agents Chemother.*, abstr. no. 45, 1986) treated 38 patients with 800 mg p.o. QD. Sputum converted to negative in 10 patients. Thus, ofloxacin may play a role in the management of patients with drug-resistant tuberculosis. Other quinolones have not been studied.

Ichiyama and Tsukamura (29) reported a patient with cavitary lung disease due to *Mycobacterium fortuitum* who was successfully treated with ofloxacin (300 mg p.o. QD) as the sole therapy.

Upper respiratory tract infections

The bacterial spectrum of the quinolones suggests that these drugs should not be agents of first choice in the therapy of pharyngitis, otitis media, or acute sinusitis. Nevertheless, a number of studies have been performed. Ofloxacin is the best studied agent. There are no data concerning the efficacy of pefloxacin in these infections.

Ofloxacin. (i) Otitis media. Lenarz (40) compared ofloxacin (200 mg p.o. BID) with penicillin for the therapy of acute otitis media and, with local therapy, for the treatment of chronic otitis. Response rates of 86% for ofloxacin and 79% for penicillin were obtained in acute otitis due mainly to *S. pneumoniae* and *S. aureus*. For chronic otitis, 91% of patients responded to ofloxacin and 86% responded to local therapy. The most common pathogens were *P. aeruginosa*, *Proteus* spp., and *S. aureus*. Concentrations of ofloxacin in ear fluid and tissue were 5 to 6 μg/ml, twice those in plasma. Data from Japan on about 100 patients treated with ofloxacin (200 mg p.o.

TID) for otitis, reviewed by Kawamura et al. (33), indicate clinical response rates of 60 to 70% and bacterial eradication rates of 50 to 60%.

(ii) Pharyngitis or tonsillitis. There has been one comparative study by Sasaki et al. (56) in which ofloxacin (200 mg p.o. TID) was compared with amoxicillin (250 mg p.o. TID); 110 and 111 patients were treated, respectively. Clinical and bacteriologic response rates were similar. Interestingly, clinical response rates of 71 and 75%, respectively, were less than the bacteriologic eradication rates of 98 and 99%, respectively, suggesting that many patients may have had viral, rather than bacterial, infections. Saito et al. (53) reviewed the total Japanese experience. Of 161 patients with pharyngitis or tonsillitis, 143 responded clinically to ofloxacin. Of 29 strains of *S. pyogenes,* 28 were eradicated. Again, it seems likely that many patients in this study had viral, rather than bacterial, infections.

(iii) Sinusitis. Too few patients with sinusitis have been treated with ofloxacin to make a meaningful assessment of its role. The data of Yamamoto et al. (68), who found a 56% clinical response rate in 16 patients, are not favorable.

Enoxacin. Baba et al. (1) used enoxacin (200 mg p.o. TID) to treat otitis media in 141 patients. The clinical response rate was 51%, and the bacteriologic response rate was 54%. The total Japanese experience with enoxacin (300 to 600 mg p.o. QD) in uncontrolled studies of ear, nose, and throat infections was a 70% clinical response rate. Federspil et al. (15), using 400 mg p.o. BID, obtained an 82% response rate in 59 patients with various infections of the ear, nose, and throat.

Ciprofloxacin. Federspil (14) treated 26 patients with ear, nose, and throat infections with ciprofloxacin (750 mg p.o. TID). Many patients had chronic otitis, with *P. aeruginosa* and *S. aureus* isolated in 11 and 5 patients, respectively. A total of 23 patients responded clinically, and 20 appeared to be bacteriologic cures.

Eight seriously ill patients with malignant otitis externa due to *P. aeruginosa* have been treated by Yu et al. (V. L. Yu, G. Stoehn, J. Rubin, A. Matador, and D. Kamarie, *Program Abstr. 27th Intersci. Conf. Antimicrob. Agents Chemother.,* abstr. no. 188, 1987), using ciprofloxacin (750 mg p.o. BID) and rifampin (600 mg p.o. BID) for 6 to 8 weeks. At limited follow-up, seven patients appear to have been cured.

Nasopharyngeal carriage of *Neisseria meningitidis* and *S. aureus*

N. meningitidis. Ciprofloxacin and ofloxacin are highly active in vitro against *N. meningitidis,* inhibiting the majority of strains at concentrations in the range of 0.004 to 0.02 μg/ml. Both drugs achieve excellent penetration into saliva, resulting in drug concentrations manyfold above the MIC for *N.*

meningitidis. Two studies have shown that ciprofloxacin is highly effective in curing the meningococcal carrier state. Renkonen et al. (51), using 250 mg p.o. BID for 2 days, eradicated the meningococcus in 54 of 56 patients, and Pugsley et al. (48), using 500 mg p.o. BID for 5 days, showed 100% efficacy in 21 patients. Ofloxacin has not been studied but would be expected to have similar efficacy.

S. aureus. *S. aureus*, including methicillin-resistant strains, is typically inhibited by 0.5 to 1.0 μg of ciprofloxacin or ofloxacin per ml (see also chapter 10). Mulligan et al. (45) reported their experience treating methicillin-resistant *S. aureus* colonization with ciprofloxacin (750 mg p.o. BID). All patients were seriously ill. The duration of therapy was 7 to 28 days. Colonization was eradicated in 11 of 14 evaluable patients. Relapses occurred, and the development of resistance in persisting strains and also in relapsing strains was noted. A number of patients tolerated the drug poorly, and only 8 of 20 patients begun on therapy had successful long-term eradication. Further studies are needed.

Toxicity

While overall the new quinolones have been well tolerated in patients with respiratory tract infections, there is concern about toxicity as a result of interactions with theophylline (see also chapter 14). Wijnands et al. (64) have shown that enoxacin reduces the clearance of theophylline, leading to increased theophylline concentrations in serum and the potential for toxicity. Headaches, hallucinations, and even convulsions have occurred in some patients receiving both enoxacin and theophylline. A similar, but milder, interaction has been noted by some investigators with ciprofloxacin. Raoof et al. (50) noted elevations of serum theophylline into the toxic range for 6 of 33 patients receiving theophylline and ciprofloxacin. Milder elevations were noted in 14 patients. Theophylline toxicity has not been noted in the other clinical studies of ciprofloxacin, however. Wijnands et al. (66) studied healthy volunteers and concluded that ciprofloxacin and pefloxacin, as well as enoxacin, reduce theophylline clearance and cause elevations of theophylline concentrations in serum.

Notably, ofloxacin has been studied and does not appear to clinically alter theophylline pharmacokinetics significantly (18, 66).

Conclusions

Ciprofloxacin and ofloxacin have been studied extensively as therapies for respiratory tract infections, particularly infections of the lower respiratory tract. The reported experience with enoxacin and pefloxacin is much less extensive. In general, when these agents have been compared with standard therapies, equivalent results have been obtained. It should be noted that bacteriologic data were often incomplete, and it is likely that some patients evaluated had viral infections or trivial bacterial infections which would have cleared without antimicrobial therapy. There are as yet no data showing efficacy of the quinolones in human legionella or mycoplasma infections. At this time, in view of their potential for selecting resistant bacteria, the quinolones should be reserved for patients with respiratory tract infections shown to be due to bacteria resistant to the standard antibiotics and for patients with multiple drug allergies.

Ciprofloxacin

Ciprofloxacin is effective in bacterial bronchitis and bronchiectasis due to *Haemophilus* spp., *B. catarrhalis,* and enteric gram-negative bacilli. Results in infections due to *S. pneumoniae* have varied; bacterial and clinical relapses appear to be common after short courses (7 days) of therapy, but overall results appear to be comparable to those obtained with standard therapy. The treatment of patients with bacteremic pneumococcal pneumonia has not been reported. Exacerbations of chronic obstructive lung disease, in particular, in patients with cystic fibrosis with infections due to *P. aeruginosa,* respond well to ciprofloxacin. Resistance develops in a substantial minority of patients, and therefore therapy should not be prolonged or repeated too frequently. The exact role and best strategy for the use of ciprofloxacin in these patients have yet to be determined. Limited experience with intravenous ciprofloxacin in gram-negative nosocomial pneumonia has been favorable. Ciprofloxacin is effective in clearing *N. meningitidis* carriers.

Ofloxacin

Ofloxacin has been effective in a variety of regimens for the therapy of infective exacerbations of chronic lung disease. Like ciprofloxacin, ofloxacin is especially effective in infection due to *Haemophilus* spp., *B. catarrhalis,* and enteric gram-negative bacilli. Relapses occurred commonly in infections due to *S. pneumoniae,* though programs utilizing 800 mg p.o. as a single QD dose may provide superior results. Infections due to ofloxacin-susceptible *P. aeruginosa* respond clinically, but the emergence of resistance as with ciprofloxacin is not uncommon and may persist for many months. A small number of studies show fair to good efficacy of ofloxacin for the treatment of otitis, pharyngitis, sinusitis, and recalcitrant drug-resistant tuberculosis. There are insufficient comparative data for a good assessment of the role of ofloxacin in these infections, however. Ofloxacin, unlike other quinolones, does not cause elevations in theophylline concentrations in serum.

Enoxacin

The intrinsic resistance of most *S. pneumoniae* and *P. aeruginosa* strains to enoxacin, coupled with the theophylline interaction, makes enoxacin an unsuitable drug for the therapy of exacerbations of chronic lung disease. The results of limited studies in upper respiratory tract infections are also not encouraging.

Pefloxacin

There are limited data for pefloxacin. Oral pefloxacin, as might be expected from its activity in vitro, has had poor efficacy for the treatment of exacerbations of chronic bronchitis. Better results, however, have been seen in a small number of patients treated for nosocomial gram-negative bacillary pneumonia. Most often, patients were treated with intravenous pefloxacin initially. Pefloxacin has not been studied in infections of the upper respiratory tract.

Literature Cited

1. **Baba, S., H. Kinoshita, Y. Mori, T. Iwasawa, H. Sugimori, et al.** 1984. A comparative double-blind study of AT-2266 with pipemidic acid in the treatment of suppurative otitis media. *Chemotherapy* (Tokyo) **32**(Suppl. 3):1061–1083.
2. **Bantz, P. M., J. Goote, W. Peters-Hartel, J. Stahmann, J. Tunin, R. Kasten, and H. Bruck.** 1987. Low dose ciprofloxacin in respiratory tract infections. A randomized comparison with doxycycline in general practice. *Am. J. Med.* **82**(Suppl. 4A):208–210.
3. **Bassaris, H. P., C. Chrysanthopoulos, A. Skoutelis, and V. Politi-Makrypoulia.** 1985. Treatment of pneumonias with ciprofloxacin, p. 241–244. *In* H. C. Neu and H. Weuta (ed.), *Proceedings of the 1st International Ciprofloxacin Workshop.* Excerpta Medica, Amsterdam.
4. **Bauernfeind, A., and U. Ullmann.** 1984. In vitro activity of enoxacin, norfloxacin and nalidixic acid. *J. Antimicrob. Chemother.* **14**(Suppl. C):33–37.
5. **Bender, S. W., H. G. Posselt, R. Wonne, B. Stover, R. Strehl, P. M. Shah, and A. Bauernfeind.** 1986. Ciprofloxacin treatment of patients with cystic fibrosis and *Pseudomonas* bronchopneumonia, p. 272–278. *In* H. C. Neu and H. Wauta (ed.), *Proceedings of the 1st International Ciprofloxacin Workshop.* Excerpta Medica, Amsterdam.
6. **Bergogne-Berezin, E.** 1985. Pharmacokinetic parameters of quinolones in respiratory tract infections. *Quinolones Bull.* **1**:17–19.
7. **Bosso, J. A., P. G. Black, and J. M. Matsen.** 1987. Ciprofloxacin versus tobramycin plus azlocillin in pulmonary exacerbations in adult patients with cystic fibrosis. *Am. J. Med.* **82**(Suppl. 4A):180–184.
8. **Chin, N.-X., and H. C. Neu.** 1983. In vitro activity of enoxacin, a quinolone carboxylic acid, compared with those of norfloxacin, new β-lactams, aminoglycosides, and trimethoprim. *Antimicrob. Agents Chemother.* **24**:754–763.
9. **Chin, N.-X., and H. C. Neu.** 1984. Ciprofloxacin, a quinolone carboxylic acid compound

active against aerobic and anaerobic bacteria. *Antimicrob. Agents Chemother.* **25:** 319–326.

10. **Davies, B. I., F. P. V. Maesen, C. Bair, and J. P. Teengs.** 1986. Clinical efficacy of various dosage regimens of ciprofloxacin in chronic bronchitis, p. 248–251. *In* H. C. Neu and H. Weuta (ed.), *Proceedings of the 1st International Ciprofloxacin Workshop.* Excerpta Medica, Amsterdam.
11. **Davies, B. I., F. B. V. Maesen, J. P. Teengs, and C. Baur.** 1986. The quinolones in chronic bronchitis. *Pharm. Weekbl.* **8:**53–59.
12. **Fallon, R., and W. Brown.** 1985. In vitro sensitivity of legionellas, meningococci and mycoplasmas to ciprofloxacin and enoxacin. *J. Antimicrob. Chemother.* **15:**787–789.
13. **Fass, R. J.** 1987. Efficacy and safety of oral ciprofloxacin in the treatment of serious respiratory infections. *Am. J. Med.* **82**(Suppl. 4A):202–207.
14. **Federspil, P.** 1986. Efficacy and safety of ciprofloxacin in ear, nose and throat infections, p. 399–401. *In* H. C. Neu and H. Wueta (ed.), *Proceedings of the 1st Ciprofloxacin Workshop.* Excerpta Medica, Amsterdam.
15. **Federspil, P. J., A. Lind, E. Tiesler, and B. Schmidt.** 1986. Clinical and pharmacokinetic studies on enoxacin in ear, nose and throat infections. *Infection* **14**(Suppl. 3):211–212.
16. **Feist, H., N. Vetter, M. Drlicek, I. Otupal, and H. Weuta.** 1986. Comparative study of ciprofloxacin and cephalexin in the treatment of patients with lower respiratory tract infections, p. 265–267. *In* H. C. Neu and H. Weuta (ed.), *Proceedings of the 1st International Ciprofloxacin Workshop.* Excerpta Medica, Amsterdam.
17. **Fenton, C. H., and M. H. Cynamon.** 1986. Comparative in vitro activity of ciprofloxacin and other 4-quinolones against *Mycobacterium tuberculosis* and *Mycobacterium intracellulare. Antimicrob. Agents Chemother.* **29:**386–388.
18. **Fourtillan, J. B., J. Branier, B. Saint-Salvi, J. Salmon, A. Surjus, et al.** 1986. Pharmacokinetics of ofloxacin and theophylline alone and in combination. *Infection* **14**(Suppl. 1):67–69.
19. **Fujimori, I., Y. Kobayashi, M. Obana, A. Saito, and M. Tomizawa.** 1984. Comparative clinical study of ofloxacin and cefaclor in bacterial bronchitis. *Kansenshogaku Zasshi* **58:**832–861.
20. **Giamarellou, H., and N. Galanakis.** 1987. Use of intravenous ciprofloxacin in difficult-to-treat infections. *Am. J. Med.* **82**(Suppl. 4A):346–351.
21. **Giamarellou, H., and J. Tsagarakis.** 1987. Efficacy and tolerance of oral ofloxacin in treating various infections. *Drugs* **34**(Suppl. 1):119–123.
22. **Gleadhill, I. C., W. Ferguson, and R. C. Lowry.** 1986. Efficacy and safety of ciprofloxacin in patients with respiratory infections in comparison with amoxycillin. *J. Antimicrob. Chemother.* **18**(Suppl. D):133–138.
23. **Goldfarb, J., R. C. Stern, M. D. Reed, T. S. Yamashita, C. M. Myers, and J. L. Blumer.** 1987. Ciprofloxacin monotherapy for acute pulmonary exacerbations of cystic fibrosis. *Am. J. Med.* **82**(Suppl. 4A):174–179.
24. **Grassi, C., G. C. Grassi, and P. Mangiarotti.** 1987. A multicenter study on clinical efficacy of ofloxacin in lower respiratory tract infections. *Drugs* **34**(Suppl. 1):80–82.
25. **Harazin, H., J. Winner, and H. P. Mittermayer.** 1987. An open randomized comparison of ofloxacin and doxycycline in lower respiratory tract infections. *Drugs* **34**(Suppl. 1): 71–73.
26. **Hodson, M. E., C. M. Roberts, R. J. A. Batland, M. J. Smith, and J. Batten.** 1987. Oral ciprofloxacin compared with conventional intravenous treatment for *Pseudomonas aeruginosa* infection in adults with cystic fibrosis. *Lancet* **i:**235–237.
27. **Honeybourne, D., R. Wise, and J. M. Andrews.** 1987. Ciprofloxacin penetration into lungs. *Lancet* **1:**1040.

28. **Hoogkamp-Korstanje, J. A. A., and S. Klein.** 1986. Efficacy of ciprofloxacin in ambulatory patients with respiratory tract infections, p. 257–259. *In* H. C. Neu and H. Weuta (ed.), *Proceedings of the 1st International Ciprofloxacin Workshop.* Excerpta Medica, Amsterdam.
29. **Ichiyama, S., and M. Tsukamura.** 1987. Ofloxacin and the therapy of pulmonary disease due to *Mycobacterium fortuitum. Chest* **92:**1110–1112.
30. **Ishigami, J.** 1985. Clinical efficacy of enoxacin by disease. *Res. Clin. Forums* **7:**107–108.
31. **Jensen, T., S. S. Redersen, C. H. Nielsen, H. Hoiby, and C. Koch.** 1987. The efficacy and safety of ciprofloxacin and ofloxacin in chronic *Pseudomonas aeruginosa* infection in cystic fibrosis. *J. Antimicrob. Chemother.* **20:**585–594.
32. **Joet, F. P., J. Nourrit, Y. Frances, and A. Arnaud.** 1985. Exoxacin in the treatment of bronchopulmonary infections. *Eur. J. Clin. Microbiol.* **4:**512–513.
33. **Kawamura, S., Y. Fujimaki, T. Iwasawa, T. Saski, O. Yanai, et al.** 1984. A comparative double-blind study of DL-8280 and pipemidic acid in suppurative otitis media. *Otol. Fukuoka* **30:**642–670.
34. **King, A., K. Shannon, and I. Phillips.** 1984. The in vitro activity of ciprofloxacin compared with that of norfloxacin and nalidixic acid. *J. Antimicrob. Chemother.* **13:**325–331.
35. **King, A., K. Shannon, and I. Phillips.** 1985. The in vitro activities of enoxacin and ofloxacin compared with that of ciprofloxacin. *J. Antimicrob. Chemother.* **15:**551–558.
36. **Kobayashi, H.** 1987. Clinical efficacy of ciprofloxacin in the treatment of patients with respiratory infection in Japan. *Am. J. Med.* **82**(Suppl. 4A):169–173.
37. **Kumada, T., and H. C. Neu.** 1985. In vitro activity of ofloxacin, a quinolone carboxylic acid, compared to other quinolones and other antimicrobial agents. *J. Artimicrob. Chemother.* **16:**563–574.
38. **Kurz, C. C., W. Marget, K. Harms, and R. M. Bertele.** 1986. Kreuzstudie über die Wirksamkeit von Ofloxacin and Ciprofloxacin bei oraler Anwendung. *Infection* **14**(Suppl. 1):82–86.
39. **Lauwers, S., A. Vincken, A. Naessens, and D. Pierard.** 1986. Efficacy and safety of pefloxacin in the treatment of severe infections in patients hospitalized in intensive care units. *J. Antimicrob. Chemother.* **17**(Suppl. B):111–116.
40. **Lenarz, T.** 1987. Chemotherapy of otitis media with ofloxacin. *Drugs* **34**(Suppl. 1): 139–143.
41. **Maesens, F. P. V., B. I. Davies, W. H. Geraedts, and C. Baur.** 1987. The use of quinolones in respiratory tract infections. *Drugs* **34**(Suppl. 1):74–79.
42. **Magnani, C., S. Fregni, G. Valli, R. Consentini, and A. Bisseti.** 1985. Comparative efficacy of ciprofloxacin and co-trimoxazole in respiratory tract infections, p. 260–264. *In* H. C. Neu and H. Wueta (ed.), *Proceedings of the 1st International Ciprofloxacin Workshop.* Excerpta Medica, Amsterdam.
43. **Monk, J. P., and D. M. Campoli-Richards.** 1987. Ofloxacin: a review of its antibacterial activity, pharmacokinetic properties and clinical use. *Drugs* **33:**346–391.
44. **Moody, J. A., D. M. Gerding, and L. R. Peterson.** 1987. Evaluation of ciprofloxacin's synergism with other agents by multiple in vitro methods. *Am. J. Med.* **82**(Suppl. 4A):44–54.
45. **Mulligan, M. E., P. J. Ruane, L. Johnstone, P. Wong, J. P. Wheelock, et al.** 1987. Ciprofloxacin for eradication of methicillin-resistant *Staphylococcus aureus* colonization. *Am. J. Med.* **82**(Suppl. 4A):215–219.
46. **Osade, Y., and H. Ogawa.** 1983. Antimycoplasmal activity of ofloxacin (DL-8280). *Antimicrob. Agents Chemother.* **23:**509–511.
47. **Phillips, I., and A. King.** 1986. The comparative in vitro activity of pefloxacin. *J. Antimicrob. Chemother.* **17**(Suppl. B):1–10.
48. **Pugsley, M. P., D. L. Dworzack, E. A. Horowitz, T. A. Cuevas, W. E. Sanders, and C. C.**

Sanders. 1987. Efficacy of ciprofloxacin in the treatment of nasopharyngeal carriers of *Neisseria meningitidis. J. Infect. Dis.***156:**211–213.

49. **Raeburn, J. A., J. R. Gavan, W. M. McCrae, A. P. Greening, P. S. Collier, M. D. Hodson, and M. C. Goodchild.** 1987. Ciprofloxacin therapy in cystic fibrosis. *J. Antimicrob. Chemother.* **20:**295–296.
50. **Raoof, S., C. Wollschlager, and F. Khanv.** 1987. Ciprofloxacin increases serum levels of theophylline. *Am. J. Med.* **82**(Suppl. 4A)**:**115–118.
51. **Renkonen, O., A. Swanen, and R. Visakorpi.** 1987. Effect of ciprofloxacin on the carrier rate of *Neisseria meningitidis* in army recruits in Finland. *Antimicrob. Agents Chemother.* **31:**962–963.
52. **Rubio, T. T.** 1987. Ciprofloxacin comparative data in cystic fibrosis. *Am. J. Med.* **82**(Suppl. 4A)**:**185–188.
53. **Saito, A., M. Katsu, A. Saito, and R. Soejima.** 1987. Ofloxacin in respiratory tract infection: a review of results of clinical trials in Japan. *Drugs* **34**(Suppl. 1)**:**83–89.
54. **Saito, A., H. Koga, H. Shigeno, K. Watanabe, K. Mori, S. Kohno, Y. Shigeno, Y. Suzuyama, K. Yamaguchi, M. Hirota, and K. Hara.** 1986. The antimicrobial activity of ciprofloxacin against *Legionella* species and the treatment of experimental legionella pneumonia in guinea pigs. *J. Antimicrob. Chemother.* **18:**251–260.
55. **Saito, A., K. Sawatari, Y. Fukuda, M. Negasawa, H. Koga, A. Tomonaga, H. Nakazato, K. Fujita, Y. Shigeno, Y. Suzuyama, K. Yamaguchi, K. Izumikawa, and K. Hara.** 1985. Susceptibility of *Legionella pneumophila* to ofloxacin in vitro and in experimental *Legionella* pneumonia in guinea pigs. *Antimicrob. Agents Chemother.* **28:**15–20.
56. **Sasaki, T., T. Unno, T. Tomiyama, O. Yamai, T. Iwasawa, et al.** 1984. Evaluation of clinical effectiveness and safety of DL-8280 in acute lacunar tonsillitis—in comparison with amoxicillin by double-blind method. *Otol. Fukuoka* **30:**484–513.
57. **Schlenkoff, D., A. Dalhoft, J. Knopf, and W. Opferkuch.** 1986. Penetration of ciprofloxacin into human lung tissue following intravenous injection. *Infection* **14:**299–300.
58. **Scully, B. E., M. Nakatomi, C. Ores, S. Davidson, and H. C. Neu.** 1987. Ciprofloxacin therapy of cystic fibrosis. *Am. J. Med.* **82**(Suppl. 4A)**:**196–201.
59. **Shah, P. M., R. Strehl, H. G. Posselt, and S. W. Bender.** 1984. Ciprofloxacin bei Mukoviszidose-zystische Fibrose (CF) cine pharmakokinetische Untersuchung. *Fortschr. Antimikrob. Antineoplastischen Chemoterapie* **3:**685–690.
60. **Shalit, I., H. R. Stutman, M. I. Marks, S. A. Chartrand, and B. C. Hilman.** 1987. Randomized study of two dosage regimens of ciprofloxacin for treating chronic bronchopulmonary infection in patients with cystic fibrosis. *Am. J. Med.* **82**(Suppl. 4A)**:**189–195.
61. **Sutter, U. L., Y. Y. Kwok, and J. Bulkacz.** 1985. Comparative activity of ciprofloxacin against aerobic bacteria. *Antimicrob. Agents Chemother.* **27:**427–428.
62. **Tsukamura, M., E. Nakamura, S. Yoshii, and H. Amano.** 1985. Therapeutic effect of a new antibacterial substance ofloxacin (DL 8280) on pulmonary tuberculosis. *Am. Rev. Respir. Dis.* **131:**352–356.
63. **Wijnands, W. J. A., A. J. A. van Griethuysen, T. B. Vree, B. van Klingeren, and C. L. A. van Herwaarden.** 1986. Enoxacin in lower respiratory tract infections. *J. Antimicrob. Chemother.* **18:**719–727.
64. **Wijnands, W. J. A., C. L. A. van Herwaarden, and T. B. Vree.** 1986. Enoxacin raises plasma theophylline concentrations. *Lancet* **ii:**108–109.
65. **Wijnands, W. J. A., T. B. Vree, A. M. Baars, and C. L. H. van Heerwaarden.** 1988. Pharmacokinetics of enoxacin and its penetration into bronchial secretions and lung tissue. *J. Antimicrob. Chemother.* **21**(Suppl. B)**:**67–77.
66. **Wijnands, W. J. A., T. B. Vree, and C. L. A. van Herwaarden.** 1986. The influence of quinolone derivatives on theophylline clearance. *Br. J. Clin. Pharmacol.* **22:**677–683.

67. **Wollschlager, C. M., S. Raoof, F. Kanh, J. J. Guarneri, V. Labombardi, and Q. Afzal.** 1987. Controlled comparative study of ciprofloxacin versus ampicillin in treatment of bacterial respiratory tract infections. *Am. J. Med.* **82**(Suppl. 4A):164–168.
68. **Yamamoto, M., M. Ohyama, K. Katsuda, T. Nobori, M. Hashimoto, et al.** 1984. The tissue concentrations of DL-8280 in the paranasal sinus mucosa and its clinical efficiency on sinusitis. *Otol. Fukuoka* **30**:477–483.

Chapter 8

Quinolone Antimicrobial Agents in the Management of Bacterial Enteric Infections

Herbert L. DuPont

Medical School, University of Texas Health Science Center, Houston, Texas 77225

While fluid and electrolyte therapy has received appropriate emphasis as lifesaving in the management of acute diarrhea in areas where severe diarrhea and malnutrition coexist, the value of antimicrobial treatment is less established except in certain forms of organism-specific illness. It is generally agreed upon that antimicrobial therapy is absolutely indicated for typhoid fever and appears to be useful in shigellosis and enterotoxigenic *Escherichia coli* (ETEC) diarrhea. The drugs commonly employed in enteric infection are of questionable value in the management of campylobacteriosis and intestinal salmonellosis. These antimicrobial agents have limitations in treating bacterial diarrhea, including the occurrence of a high frequency of resistance among enteric pathogens (ampicillin, tetracyclines) or because of ineffectiveness against both *Campylobacter* and *Shigella* strains (trimethoprim-sulfamethoxazole [TMP-SMX]).

In view of their activity against aerobic enteric gram-negative bacilli and low frequency of resistance, the quinolones are logical agents to evaluate in intestinal and biliary tract infectious diseases. This review will focus on in vitro susceptibility of enteric bacterial pathogens to the quinolones, and then efficacy trials looking at this group of drugs in the therapy of enteric infections will be reviewed.

In Vitro Susceptibility of Enteric Pathogens to Quinolones

Ampicillin and tetracycline resistance is currently found to be widespread among enteric bacterial pathogens in most regions of the world (42).

TMP-SMX resistance has been documented among enteropathogens in selected areas (7, 8, 42, 48). In vitro studies of antimicrobial susceptibility, performed more than 10 years ago with the first-generation quinolones (nalidixic acid, cinoxacin, and oxolinic acid), demonstrated a high degree of activity against bacterial enteropathogens (11, 22). Several studies have been performed more recently to examine the in vitro activity of the newer, more active quinolones against the broad-range bacterial enteropathogens isolated from diverse areas of the world (12, 17, 20, 26, 28, 36, 41, 44, 49, 51, 53, 54). In Table 1 the results of in vitro susceptibility testing of a variety of bacterial enteropathogens to selected antimicrobial agents, including several of the newer quinolone preparations, are presented (12, 20). Clearly, the quinolone antimicrobial agents studied (norfloxacin, enoxacin, and ciprofloxacin) showed the lowest MICs (highest susceptibility) when *Campylobacter jejuni*, as well as the other pathogens, was tested. Not shown in the table, the three quinolones employed in these in vitro studies were found to have a MIC for 90% of strains of <0.5 μg/ml for 29 strains of miscellaneous bacterial pathogens, including *Aeromonas hydrophila* (6 strains), *Vibrio parahaemolyticus* (9 strains), and *Yersinia enterocolitica* (14 strains). The only other drug tested in our in vitro studies with such a high degree of in vitro activity against the wide range of strains tested was furazolidone. The quinolones have been shown in other studies to have excellent in vitro activity against *Aeromonas* (49), *Plesiomonas* (44), and *Vibrio* (41) species and *Campylobacter pylori* (4, 5, 25, 39, 50, 52). These drugs show variable to low activity against strains of *Clostridium difficile* (13, 17, 51).

Clinical Evaluations of Quinolones in Enteric Infections

Shigellosis

Soon after the first quinolone, nalidixic acid, was introduced, it was successfully employed in the therapy of common forms of shigellosis (40). When ampicillin- and TMP-SMX-resistant Shiga dysentery (illness due to *Shigella dysenteriae* type 1) emerged as a problem in selected areas, nalidixic acid was found to be a useful therapy (7, 8, 48). In a limited trial oxolinic acid appeared to have some value in the therapy of experimentally induced human shigellosis (21). One study of shigellosis in a small group of hospitalized children dampened enthusiasm for this class of drugs by suggesting that nalidixic acid was of only marginal effectiveness in severe disease (29). A total of 17 infants received nalidixic acid and 19 were treated with ampicillin. Of these, 4 of 17 (24%) receiving nalidixic acid and none of the ampicillin-treated subjects were still experiencing diarrhea after 5 days of therapy.

Table 1. MICs of various antimicrobial agents against 90% of strains tested

Bacterial enteropathogen	No. of strains tested	MIC (μg/ml) against 90% of strains tested with[a]:							
		AMP	DOX	ERY	TMP-SMX	FUR	NOR	ENO	CIP
Salmonella spp.	50	128	64	64	1	1	≤0.5	≤0.5	≤0.5
Shigella spp.	80	>128	16	64	4	2	≤0.5	≤0.5	≤0.5
ETEC	50	>128	32	128	8	1	≤0.5	≤0.5	≤0.5
C. jejuni	30	8	8	≤0.5	>128	≤0.5	1	2	≤0.5

[a]Abbreviations: AMP, ampicillin; DOX, doxycycline; ERY, erythromycin; TMP-SMX, trimethoprim-sulfamethoxazole; FUR, furazolidone; NOR, norfloxacin; ENO, enoxacin; CIP, ciprofloxacin.

Several studies carried out more recently with the newer, more active quinolones have established their value as clinically efficacious in patients with shigellosis. Norfloxacin was shown to be as effective as nalidixic acid in the treatment of Shiga dysentery (48). Norfloxacin eliminated *Shigella* strains from stools more rapidly than nalidixic acid did in this study. In a clinical trial (23) in which ciprofloxacin was compared with TMP-SMX versus a placebo in the therapy of traveler's diarrhea, 15 adults with shigellosis were included. The average hours of diarrhea after initiation of therapy when the causal agent was a *Shigella* strain was 28 h for the ciprofloxacin group (five patients) compared with 16 h for those receiving TMP-SMX (four persons) and 84 h for those receiving a placebo (six persons).

Salmonellosis

Conventional antimicrobial therapy appears to be of limited, if any, benefit in treating intestinal salmonellosis. In fact, antimicrobial agents may encourage the development of resistance and increase the duration of postconvalescent shedding of the infecting strain (3). Recently, Pichler et al. (46) demonstrated that ciprofloxacin effectively reduced the duration of diarrheal illness and period of time the infecting strain could be isolated from stool specimens in adults with intestinal salmonellosis. A total of 16 patients with salmonellosis received ciprofloxacin, and 21 were given a placebo. The mean duration of diarrhea was 1.9 days for the active drug group and 3.4 days for the group receiving a placebo treatment ($P < 0.01$). Other preliminary reports have provided evidence that the newer quinolones could reduce the period of intestinal shedding of *Salmonella* strains in those excreting the organism asymptomatically (18, 37).

In view of a low MIC of the quinolones for *Salmonella typhi*, representatives from this class of drug have been evaluated in typhoid fever and typhoid carriers. In an open trial ciprofloxacin was effective in the therapy of 38 patients with blood culture-proved typhoid fever (47). Ciprofloxacin was given in a dose of 500 mg every 12 h for an average of 14 days. There was one treatment failure in a patient developing acute cholecystitis. Defervescence occurred on the average after 4.2 days of therapy. In a separate clinical study, norfloxacin successfully eradicated *S. typhi* carriage when compared with a placebo in a group of chronic typhoid carriers (28a). Chronic typhoid carriers were treated with either 400 mg of norfloxacin or a matching placebo every 12 h for 28 days. Of 12 treated with norfloxacin, 11 had negative stool and bile cultures for *S. typhi* after therapy. All 12 placebo-treated subjects had positive cultures at the completion of therapy. These investigators showed a high rate of organism eradication equally well in those with or without cholelithiasis. Several cases of successful eradication of *S. typhi* intestinal carriage have been reported after administration of ciprofloxacin (18, 33) or ofloxacin (37).

Campylobacteriosis

An important advantage of the quinolones when compared with other available antimicrobial agents used in diarrhea is the high degree of in vitro activity against *C. jejuni*. Few data are available to establish the values of antimicrobial therapy of any sort in intestinal *Campylobacter* infection. Chronic infection of marmosets was successfully treated with ciprofloxacin (27). A preliminary report in humans suggested that ciprofloxacin could reduce symptoms and the duration of excretion of the causative agent in patients with *Campylobacter* infection (46). Nineteen adults with intestinal campylobacteriosis who received ciprofloxacin experienced on the average 1.1 days of diarrhea after therapy was initiated compared with 2.2 days for 11 placebo-treated subjects ($P < 0.01$). It remains to be determined whether resistance is likely to occur among infecting strains of *C. jejuni* during therapy with the newer quinolones. Emergence of resistance of the infecting *Campylobacter* strain during therapy has been described for nalidixic acid (1).

Gastric infection secondary to *C. pylori* has been treated with the combination of ofloxacin with either ranitidine or bismuth subsalicylate (4, 5, 25). In one of the studies, therapy with the quinolone led to more rapid healing and eradication of the organism from posttreatment biopsy (4). In the other two studies, resistance to the quinolone occurred, limiting the value of the treatment (5, 25).

Traveler's diarrhea and ETEC diarrhea

Ciprofloxacin was shown to be as effective as TMP-SMX, and both were significantly more effective than a placebo in treating traveler's diarrhea (20, 23). The average duration of diarrhea was 29 or 20 h after the initiation of therapy with ciprofloxacin or TMP-SMX, respectively, compared with 81 h for those receiving a placebo. The duration of diarrhea for ETEC diarrhea, the most common form of traveler's diarrhea, for the three groups was 33, 26, and 84 h, respectively.

In a separate study norfloxacin was shown to prevent 88% of diarrhea cases that would be expected to occur without prophylaxis in a placebo-controlled trial among U.S. travelers during a 2-week stay in Mexico (34). The drug led to a significant decline in normal aerobic flora during prophylaxis. After norfloxacin was discontinued, fecal flora returned to pretreatment levels. No resistant gram-negative aerobic flora was found during weekly quantitative cultures before, during, or after therapy.

Endemic diarrhea of presumed bacterial origin

In studies conducted in seven countries, two doses of norfloxacin, 400 mg twice daily or 400 mg three times daily for 5 days, were shown to be equivalent to conventional TMP-SMX treatment for acute diarrhea in adults (19). Clinical cure occurred in 89% (lower dose) and 91% (higher dose) of those treated with norfloxacin, compared with 78% for those receiving TMP-SMX. Cure rates were greater in all treatment groups when the patient's stool specimen contained numerous leukocytes. This is the second study we have carried out showing that antimicrobial agents show the greatest clinical effect in patients with fecal leukocytes (43). This undoubtedly relates to the fact that patients in whom numerous fecal leukocytes are found microscopically are often infected with an invasive bacterial pathogen (30).

Biliary tract infection

The quinolone agents hold promise for use in treating biliary tract infection. First, the quinolones are active against nearly all of the aerobic bacterial pathogens that commonly produce cholangitis and biliary sepsis (55). In one study the biliary concentration of the drugs exceeded that found in serum after oral administration (9), and therapeutic levels of drugs remained present in bile and gallbladder tissue for 6 h (16) (see chapter 4). Ciprofloxacin has been used successfully in preliminary studies to treat biliary tract infection (14, 32, 38).

Comment

Antibacterial agents play an important role in the therapy and selected prevention of certain forms of enteric infection. There are two major limitations of the commonly used drugs in the treatment of intestinal infections: prevalence or development of antimicrobial resistance and failure to achieve a satisfactory clinical response to therapy, as seen for certain forms of bacterial diarrhea (salmonellosis and campylobacteriosis). The properties of an ideal drug for therapy of bacterial diarrhea include the following: low or absence of frequency of resistance among bacterial pathogens; low potential for inducing R-factor (plasmid) resistance; activity against both *Shigella* and *Campylobacter* strains; and achievement of high levels of antibacterial drug in the gut, tissue, and blood. The newly developed drugs with perhaps the greatest potential considering these attributes are the newer quinolone derivatives, as well as bicozamycin, aztreonam, amdinocillin, and furazolidone (20). Of these agents, the quinolones may hold the greatest promise. They show a high degree of activity against enteric pathogens, they are well absorbed after oral administration, and they have large volumes of distribution with long elimination half-lives (see also chapter 4). Norfloxacin given in a dose of 400 mg to healthy volunteers led to levels in feces of between 207 and 2,716 μg/g of stool (15), while a similar study of ciprofloxacin in noninfected volunteers showed that levels in feces ranged between 185 and 2,220 μg/g (10). Ciprofloxacin therapy of traveler's diarrhea produced mean drug levels in feces exceeding 500 μg/g (20). Many of the new quinolones are well absorbed from the gastrointestinal tract, producing levels in serum well above the inhibitory concentrations of most enteric pathogens (10). The quinolones produce profound changes in intestinal flora during therapy (6, 24, 31, 45). The aerobic enterobacteria are essentially eliminated without producing resistance, and the anaerobic flora is not altered in most cases (see also chapter 10). While *C. difficile* levels have not usually been shown to increase during quinolone therapy, antibiotic-associated colitis has been reported (2). Studies are currently under way to establish the usefulness of the quinolones in the broad range of bacterial enteric infections. Clearly, the most important limitation of the quinolones in the management of diarrhea is the current lack of application to therapy in children. When one considers that diarrhea is numerically a greater problem in children, the hope remains that one or more of the quinolones will be shown to be safe for administration to infants and young children.

Undoubtedly the new quinolones will prove to play an important role in the therapy of bacterial enteric infection in adults (35). Currently they should be considered agents of choice for adults with severe diarrhea which occurs while in areas where TMP-SMX-resistant *Shigella* strains and ETEC are known to be prevalent, such as Southeast Asia, Africa, and South America (42). In any area where empiric antimicrobial therapy is indicated for clinical bacillary dysentery, where either *Shigella* or *Campylobacter* strains may be the responsible agents, a quinolone might ideally be employed. Further study of patients with intestinal and typhoidal salmonellosis is indicated. The quinolones appear to be promising in treating typhoid carriers. Finally, the quinolones should prove to be useful in the treatment of biliary tract infection and possibly combined with an agent active against anaerobes in the therapy of intra-abdominal infections where mixed organisms are suspected. Adverse reactions (detailed in chapter 14) appear to be rare when these drugs are used for short periods of time for bacterial enteric infection. The most common reactions (occurring in 2 to 3% of treated cases) are gastrointestinal complaints, insomnia, headache, and rash.

Acknowledgments. Research activities described in this chapter were supported by the National Institutes of Health (RO1 AI–72534 and RO1 AI-230491) and by grants from Merck

Sharp and Dohme Research Laboratories, West Point, Pa., and Miles Pharmaceuticals Inc., West Haven, Conn.

Dot Cowan was responsible for manuscript preparation.

Literature Cited

1. **Altwegg, M., A. Burnens, J. Zollinger-Iten, and J. L. Penner.** 1987. Problems in identification of *Campylobacter jejuni* associated with acquisition of resistance to nalidixic acid. *J. Clin. Microbiol.* **25:**1807–1808.
2. **Arcieri, G., E. Griffith, G. Grunewaldt, A. Heyd, B. O'Brien, N. Becker, and R. August.** 1987. Ciprofloxacin: an update on clinical experience. *Am. J. Med.* **82**(Suppl. 4A)**:** 381–386.
3. **Aserkoff, B., and J. V. Bennett.** 1969. Effect of antibiotic therapy in acute salmonellosis on the fecal excretion of salmonellae. *N. Engl. J. Med.* **281:**636–640.
4. **Bayerdörffer, E., G. Kasper, T. Pirlet, A. Sommer, and R. Ottenjann.** 1987. Ofloxacin in der Therapie Campylobacter-pylori-positiver Ulcera duodeni. Eine prospektive kontrollieste randomisierte Studie. *Dtsch. Med. Wochenschr.* **112:**1407–1411.
5. **Bayerdörffer, E., T. Simon, C. Bästlein, R. Ottenjann, and G. Kasper.** 1987. Bismuth/ofloxacin combination for duodenal ulcer. *Lancet* **ii:**1467–1468.
6. **Bergan, T., C. Delin, S. Johansen, I. M. Kolstad, C. E. Nord, and S. B. Thorsteinsson.** 1986. Pharmacokinetics of ciprofloxacin and effect of repeated dosage on salivary and fecal microflora. *Antimicrob. Agents Chemother.* **29:**298–302.
7. **Bhattacharya, D., G. Sen, G. B. Nair, M. K. Bhattacharya, P. Datta, and D. Datta.** 1986. Multiple drug-resistant *Shigella dysenteriae* type 1 and travelers' diarrhea. *J. Infect. Dis.* **154:**729–730.
8. **Bose, R., J. N. Nashipuri, P. K. Sen, P. Datta, S. K. Bhattacharya, D. Datta, D. Sen, and M. K. Bhattacharya.** 1984. Epidemic of dysentery in West Bengal: clinician's enigma. *Lancet* **ii:**1160.
9. **Brogard, J.-M., F. Jehl, H. Monteil, M. Adloff, J.-F. Blickle, and P. Levy.** 1985. Comparison of high-pressure liquid chromatography and microbiological assay for the determination of biliary elimination of ciprofloxacin in humans. *Antimicrob. Agents Chemother.* **28:**311–314.
10. **Brumfitt, W., I. Franklin, D. Grady, J. M. T. Hamilton-Miller, and A. Iliffe.** 1984. Changes in the pharmacokinetics of ciprofloxacin and fecal flora during administration of a 7-day course to human volunteers. *Antimicrob. Agents Chemother.* **26:**757–761.
11. **Byers, P. A., H. L. DuPont, and M. C. Goldschmidt.** 1976. Antimicrobial susceptibilities of shigellae isolated in Houston, Texas, in 1974. *Antimicrob. Agents Chemother.* **9:**288–291.
12. **Carlson, J. R., S. A. Thornton, H. L. DuPont, A. H. West, and J. J. Mathewson.** 1983. Comparative in vitro activities of ten antimicrobial agents against bacterial enteropathogens. *Antimicrob. Agents Chemother.* **24:**509–513.
13. **Chow, A. W., N. Cheng, and K. H. Bartlett.** 1985. In vitro susceptibility of *Clostridium difficile* to new β-lactam and quinolone antibiotics. *Antimicrob. Agents Chemother.* **28:**842–844.
14. **Chrysanthopoulos, C. J., A. T. Skoutelis, J. C. Starakis, E. D. Anastassiou, and H. P. Bassaris.** 1987. Use of intravenous ciprofloxacin in respiratory tract infections and biliary sepsis. *Am. J. Med.* **82**(Suppl. 4A)**:**357–359.
15. **Cofsky, R. D., L. DuBouchet, and S. H. Landesman.** 1984. Recovery of norfloxacin in feces after administration of a single oral dose to human volunteers. *Antimicrob. Agents Chemother.* **26:**110–116.
16. **Dan, M., F. Serour, A. Gorea, A. Levenbert, M. Krispin, and S. A. Berger.** 1987. Concentration of norfloxacin in human gallbladder tissue and bile after single-dose oral administration. *Antimicrob. Agents Chemother.* **31:**352–353.

17. **Delmee, M., and V. Avesani.** 1986. Comparative in vitro activity of seven quinolones against 100 clinical isolates of *Clostridium difficile. Antimicrob. Agents Chemother.* **29:** 374–375.
18. **Dirdl, G., H. Pichler, and D. Wolf.** 1986. Treatment of chronic salmonella carriers with ciprofloxacin. *Eur. J. Clin. Microbiol.* **5:**260–261.
19. **DuPont, H. L., M. L. Corrado, and J. Sabbaj.** 1987. Use of norfloxacin in the treatment of acute diarrheal disease. *Am. J. Med.* **82**(Suppl. 6B):79–83.
20. **DuPont, H. L., C. D. Ericsson, A. Robinson, and P. C. Johnson.** 1987. Current problems in antimicrobial therapy for bacterial enteric infection. *Am. J. Med.* **82**(Suppl. 4A):324–328.
21. **DuPont, H. L., and R. B. Hornick.** 1973. Adverse effect of lomotil therapy in Shigellosis. *J. Am. Med. Assoc.* **226:**1525–1528.
22. **DuPont, H. L., A. H. West, D. G. Evans, J. Olarte, and D. J. Evans, Jr.** 1978. Antimicrobial susceptibility of enterotoxigenic *Escherichia coli. J. Antimicrob. Chemother.* **4:**100–102.
23. **Ericsson, C. D., P. C. Johnson, H. L. DuPont, D. R. Morgan, J. A. Bitsura, and E. J. Cabada.** 1987. Ciprofloxacin or trimethoprim-sulfamethoxazole as initial therapy for traveler's diarrhea. *Ann. Intern. Med.* **1987:**216–220.
24. **Erizensberger, R., P. M. Shah, and H. Knothe.** 1985. Impact of oral ciprofloxacin on the faecal flora of healthy volunteers. *Infection* **13:**273–275.
25. **Glupczynski, Y., M. Labbe, A. Burette, M. Delmee, V. Avesani, and C. Bruck.** 1987. Treatment failure of ofloxacin in *Campylobacter pylori* infection. *Lancet* **i:**1096.
26. **Goodman, L. J., R. M. Fliegelman, G. M. Trenholme, and R. L. Kaplan.** 1984. Comparative in vitro activity of ciprofloxacin against *Campylobacter* spp. and other bacterial enteric pathogens. *Antimicrob. Agents Chemother.* **25:**504–506.
27. **Goodman, L. J., R. L. Kaplan, R. M. Petrak, R. M. Fliegelman, D. Taff, F. Walton, J. L. Penner, and G. M. Trenholme.** 1986. Effects of erythromycin and ciprofloxacin on chronic fecal excretion of *Campylobacter* species in marmosets. *Antimicrob. Agents Chemother.* **29:**185–187.
28. **Goossens, H., P. De Mol, H. Coignau, J. Levy, O. Grados, G. Ghysels, H. Innocent, and J. Butzler.** 1985. Comparative in vitro activities of aztreonam, ciprofloxacin, norfloxacin, ofloxacin, HR 810 (a new cephalosporin), RU 28965 (a new macrolide), and other agents against enteropathogens. *Antimicrob. Agents Chemother.* **27:**388–392.

28a. **Gotuzzo, E., J. G. Guerra, L. Benavente, J. C. Palomino, C. Carrillo, J. Lopera, F. Delgado, D. R. Nalin, and J. Sabbaj.** 1988. Use of norfloxacin to treat chronic typhoid carriers. *J. Infect. Dis.* **157:**1221–1225.

29. **Haltalin, K. C., J. D. Nelson, and H. T. Kusmiesz.** 1973. Comparative efficacy of nalidixic acid and ampicillin for severe Shigellosis. *Arch. Dis. Child.* **48:**305–312.
30. **Harris, J. C., H. L. DuPont, and R. B. Honick.** 1972. Fecal leukocytes in diarrheal illness. *Ann. Intern. Med.* **76:**697–703.
31. **Holt, H. A., D. A. Lewis, L. O. White, S. Y. Bastable, and D. S. Reeves.** 1986. Effect of oral ciprofloxacin on the fecal flora of healthy volunteers. *Eur. J. Clin. Microbiol.* **5:**201–205.
32. **Houwen, R. H. J., C. M. A. Bijleveld, and H. G. de Vries-Hospers.** 1987. Ciprofloxacin for cholangitis after hepatitic portoenterostomy. *Lancet* **i:**1367.
33. **Hudson, S. J., H. R. Ingham, and M. H. Snow.** 1985. Treatment of *Salmonella typhi* carrier state with ciprofloxacin. *Lancet* **i:**1047.
34. **Johnson, P. C., C. D. Ericsson, D. R. Morgan, H. L. DuPont, and F. J. Cabada.** 1986. Lack of emergence of resistant fecal flora during successful prophylaxis of traveler's diarrhea with norfloxacin. *Antimicrob. Agents Chemother.* **30:**671–674.
35. **Keusch, G. T.** 1988. Antimicrobial therapy for enteric infections and typhoid fever: state of the art. *Rev. Infect. Dis.* **10**(Suppl.):S199–S205.
36. **Ling, J., K. M. Kam, A. W. Lam, and G. L. French.** 1988. Susceptibilities of Hong Kong iso-

lates of multiply resistant *Shigella* spp. to 25 antimicrobial agents, including ampicillin plus sulbactam and new 4-quinolones. *Antimicrob. Agents Chemother.* **32:**20–23.
37. **Löffler, A., and H. G. von Westphalen.** 1986. Successful treatment of chronic salmonella excretor with ofloxacin. *Lancet* **i:**1206.
38. **Lonka, L., and R. S. Pedersen.** 1987. Ciprofloxacin for cholangitis. *Lancet* **ii:**212.
39. **McNulty, C. A. M., J. Dent, and R. Wise.** 1985. Susceptibility of clinical isolates of *Campylobacter pyloridis* to 11 antimicrobial agents. *Antimicrob. Agents Chemother.* **28:**837–838.
40. **Moorhead, P. J., and H. E. Parry.** 1965. Treatment of Sonne dysentery. *Br. Med. J.* **2:**913–915.
41. **Morris, J. G., Jr., J. H. Tenny, and G. L. Drusano.** 1985. In vitro susceptibility of pathogenic *Vibrio* species to norfloxacin and six other antimicrobial agents. *Antimicrob. Agents Chemother.* **28:**442–445.
42. **Murray, B. E.** 1986. Resistance of *Shigella, Salmonella,* and other selected enteric pathogens to antimicrobial agents. *Rev. Infect. Dis.* **8:**S172–S181.
43. **Oberhelman, R. A., F. J. de la Cabada, E. V. Garibay, J. M. Bitsura, and H. L. DuPont.** 1987. Efficacy of trimethoprim/sulfamethoxazole in the treatment of acute diarrhea in a Mexican pediatric population. *J. Pediatr.* **110:**960–965.
44. **O'Hare, M. D., D. Felmingham, G. L. Ridgway, and R. N. Grüneberg.** 1985. The comparative *in vitro* activity of twelve 4-quinolone antimicrobials against enteric pathogens. *Drugs Exp. Clin. Res.* **11:**253–257.
45. **Pecquet, S., A. Andremont, and C. Tancrède.** 1987. Effect of oral ofloxacin on fecal bacteria in human volunteers. *Antimicrob. Agents Chemother.* **31:**124–125.
46. **Pichler, H. E. T., G. Diridl, K. Stickler, and D. Wolf.** 1987. Clinical efficacy of ciprofloxacin compared with placebo in bacterial diarrhea. *Am. J. Med.* **82**(Suppl. 4A):329–335.
47. **Ramirez, C. A., J. L. Bran, C. R. Mejia, and J. F. Garca.** 1985. Open, prospective study of the clinical efficacy of ciprofloxacin. *Antimicrob. Agents Chemother.* **28:**128–132.
48. **Rogerie, F., D. Ott, J. Vandepitte, L. Verbist, P. Lemmens, and I. Habiyaremye.** 1986. Comparison of norfloxacin and nalidixic acid for treatment of dysentery caused by *Shigella dysenteriae* type 1 in adults. *Antimicrob. Agents Chemother.* **29:**883–886.
49. **San Joaquin, V. H., R. K. Scribner, D. A. Pickett, and D. F. Welch.** 1986. Antimicrobial susceptibility of *Aeromonas* species isolated from patients with diarrhea. *Antimicrob. Agents Chemother.* **30:**794–795.
50. **Shungu, D. L., D. R. Nalin, R. H. Gilman, H. H. Gadebusch, A. T. Cerami, C. Gill, and B. Weissberger.** 1987. Comparative susceptibilities of *Campylobacter pylori* to norfloxacin and other agents. *Antimicrob. Agents Chemother.* **31:**949–950.
51. **Shungu, D. L., E. Weinberg, and H. H. Gadebusch.** 1983. In vitro antimicrobial activity of norfloxacin (MK-0366, AM-715) and other agents against gastrointestinal tract pathogens. *Antimicrob. Agents Chemother.* **23:**86–90.
52. **Van Caekenberghe, D. L., and J. Breyssens.** 1987. In vitro synergistic activity between bismuth subcitrate and various antimicrobial agents against *Campylobacter pyloridis (C. pylori). Antimicrob. Agents Chemother.* **31:**1429–1430.
53. **Van der Auwera, P. V., and B. Scorneaux.** 1985. In vitro susceptibility of *Campylobacter jejuni* to 27 antimicrobial agents and various combinations of β-lactams with clavulanic acid or sulbactam. *Antimicrob. Agents Chemother.* **28:**37–40.
54. **Vanhoof, R., J. M. Hubrechts, E. Roebben, H. J. Nyssen, E. Nulens, J. Leger, and N. De Schepper.** 1986. The comparative activity of pefloxacin, enoxacin, ciprofloxacin and 13 other antimicrobial agents against enteropathogenic microorganisms. *Infection* **14:**294–298.
55. **Wolfson, J. S., and D. C. Hooper.** 1985. The fluoroquinolones: structures, mechanisms of action and resistance, and spectra of activity in vitro. *Antimicrob. Agents Chemother.* **28:**581–586.

Chapter 9

Treatment of Osteomyelitis and Septic Arthritis with Quinolone Antimicrobial Agents

Francis A. Waldvogel

*Clinique Médicale Thérapeutique, University Hospital,
1211 Geneva 4, Switzerland*

Introduction

The recent development of the new fluoroquinolone antimicrobial agents, the demonstration of their wide antibacterial spectrum, and the documentation of initial observations showing some remarkable clinical applications have rapidly led to the exploration of their potential efficacy in notoriously difficult, chronic or relapsing infections. Osteomyelitis represents a wide field of potential indications for the new quinolones since this disease is still characterized by a high frequency of treatment failures and recurrences, due to inadequate treatment, abscess and sequestrum formation, and other as yet unknown factors (34). These failures are also partly favored by the use of orthopedic material and prosthetic implants of various types, which have to be removed. Under certain conditions, however, the orthopedic investment is of such importance that keeping the prosthetic material may have to be achieved despite the presence of a septic focus, which is kept asymptomatic by long-term, outpatient oral antibiotic treatment (4). Thus, the development of a new group of antimicrobial agents which (i) show better activity against the commonly isolated organisms in acute and chronic osteomyelitis and arthritis, (ii) can be administered over prolonged periods of time by either the parenteral or the oral route, and (iii) have a low frequency of side effects has been the hope of many orthopedic surgeons, microbiologists, and infectious disease specialists. Whether the quinolones represent such a panacea is discussed in this chapter.

Microbiological Aspects

The spectrum of microorganisms responsible for osteomyelitis and septic arthritis has been reviewed repeatedly (23, 30, 33). In short, whether dealing with osteomyelitis of the hematogenous or the contiguous type, *Staphylococcus aureus* and *Staphylococcus epidermidis* together are still the most frequently encountered organisms. Hematogenous osteomyelitis of the long bones can be caused, in addition, by group B streptococci (neonates), *Haemophilus influenzae* (infants), and gram-negative organisms. Hematogenous disease, however, occurs mainly at an age when the use of quinolones has been contraindicated, due to previous studies showing cartilage lesions in weight-bearing joints of young animals treated with nalidixic acid and malformations due to quinolones during the development of limb buds in experimental systems (R. Stahlmann, G. Blankenburg, and D. Neubert, *Int. Symp. New Quinolones, Geneva, Switzerland*, abstr. no. 147, 1986; see also chapter 14).

Adults are often afflicted by osteomyelitis; vertebral osteomyelitis occurs frequently in this population and is due, in addition to staphylococci, to gram-negative rods (22). Osteomyelitis contiguous to a focus of infection (posttraumatic, postsurgery, etc.) is caused by not only staphylococci, but also a variety of gram-negative rods, albeit at a lower frequency. Thus, one ideal property of new agents such as the quinolones should be to show low minimal bacteriostatic concentrations (MICs) and MBCs toward all of these bacterial species. Whereas this is certainly the case for the gram-negative organisms, including *Pseudomonas aeruginosa*, as described in chapter 3, this is less so regarding *S. aureus* and *S. epidermidis*, which require MICs that are similar to, or sometimes higher than, those reported for commonly used antibacterial agents. An exception to this rule may be the methicillin-resistant *S. aureus* strains, which require quinolone MICs only slightly higher than do their susceptible congeners. The search has therefore continued in recent years for quinolones with low MICs against staphylococci, whether methicillin resistant or not, with some encouraging results (21).

Another aspect to be discussed is whether the quinolones, which seem to be highly efficient in vitro against most gram-negative organisms responsible for osteomyelitis, are active under the stringent physicochemical conditions of a focus of osteomyelitis or arthritis, characterized by a low pH (35) and low partial O_2 pressure (19). Mader et al. (20) have explored the former problem by performing kill curves assayed under aerobic (100 mm Hg) and anaerobic (10 mm Hg) conditions with difloxacin and A-56620. Both antimicrobial agents were equally effective under both conditions. Lowering of the pH increased dramatically the MICs of the compound CI-934 (21), whereas it did not affect the MICs of ciprofloxacin. Thus, it seems that the influence of pH on the MICs of the quinolones should be explored for each individual compound.

Finally, the question has to be raised whether the quinolones have adequate MICs to treat microorganisms isolated in cases of septic arthritis (13, 14). Whereas they are certainly adequate for the *Neisseria gonorrheae* strains and for gram-negative rods isolated from septic joints, the same comments made before regarding osteomyelitis are applicable to *S. epidermidis*, *S. aureus*, and streptococci. The available MIC data show no great advantage of the quinolones compared with standard antibiotics used in bone and joint infections.

Difficulties in Evaluating the Efficacy of the Quinolones

Bone, which is a heterogeneous structure, creates many problems for those who try to estimate antibiotic pharmacokinetics (2). Differences in bone matrix and crystal density between cortex and medulla, differences in blood supply between cortex, medulla, and periosteum, and differences in fluid space between infected and noninfected bone will affect the measurement of antibiotic levels in bone. This can be partly corrected by introducing correction factors such as hemoglobin (10) and myoglobin (11) assays. Extraction procedures, although well standardized, have shown either incomplete back diffusion of quinolones out of bone (11) or prolonged back diffusion requiring as many as 10 washing procedures (6). Finally, there are good reasons to believe that antibiotic distribution is not uniform in a heterogeneous structure such as bone. No data are presently available regarding this important point, but it can be hypothesized that the effects of nalidixic acid on developing bone or of quinolones on developing limb buds may be due to specific local accumulation of these compounds. In summary, quinolone levels in bone, determined mostly by high-pressure liquid chromatography, are useful indications of the intraosseous penetration of these compounds; however, any small differences found should not be attributed to the substances themselves, but rather to methodological factors. Interpretation of the data implies a careful evaluation of the extraction procedure, of the control of blood contamination, and of the standardization curve; infected bone, for instance, should not be tested against a standardization curve for which noninfected bone is used.

On a clinical basis, other problems render the evaluation of the quinolones difficult in osteomyelitis. As Arcieri et al. point out (1), all 25 studies reported by 1987 on the use of ciprofloxacin in the treatment of osteomyelitis were uncontrolled, despite the inclusion of more than 100 patients. The reasons for the lack of controlled data are evident and unavoidable: acute versus chronic disease and their respective definitions, the type of organisms involved, the mode of infection, the presence of organic or metallic foreign material, previous or concomitant surgery, and previous antibiotic therapy introduce so many variables into each individual patient that no clinical study can ever solve the statistical problem of the beta error. Therefore, only experimental infections give us clear-cut information as to whether the quinolones are a new asset in the treatment of osteomyelitis, whereas the clinical studies can be evaluated in a historical perspective only.

Experimental Osteomyelitis and Arthritis

An interesting model of hematogenous osteomyelitis in chickens (8, 9) has shown by Evans blue exclusion how rapidly a focus of osteomyelitis evolves into an area devoid of any blood supply. These studies are complementary to the important work of Norden and his associates (24, 26), who have produced a model of chronic osteomyelitis in rabbits by the injection of sodium morrhuate, a sclerosing agent, and various types of microorganisms. In a recent study Norden and Shinners (27) have shown a 95% cure rate of *P. aeruginosa* osteomyelitis after 4 weeks of treatment with ciprofloxacin versus a 6% cure rate with tobramycin. Impressive as these results are, the authors also showed that 20% of the organisms isolated at 2 weeks required for inhibition a 4- to 16-fold increase in the ciprofloxacin MIC. Under similar experimental conditions the same investigators analyzed the efficiency of ofloxacin given at a very high dosage (200 mg/kg per day either four or two times daily) (25), probably in order to achieve levels in serum and bone approximating those obtained in humans. Indeed, levels in infected serum and bone were 5 μg/ml at 2 h and 1.7 μg/ml at 6 h, respectively. These high levels led to a cure in a high proportion of cases. Whereas 94% of the rabbits had positive cultures before therapy, the positive isolation rate fell to 11% after 14 days and to 6% after 28 days of treatment. No increase in MIC against the isolated *P. aeruginosa* strain was observed in this study. These results were similar to those described previously with parenteral ciprofloxacin and better than those recorded with carbenicillin, azlocillin, and sisomicin (26). Whether the very high dosage used in this study had additional positive effects (persistently high levels in blood and bone, better compartmentalization of the quinolone in bone) and whether it may have been toxic to the animals are unanswered questions which somewhat limit the application of these results to humans.

Experimental results are less convincing with *S. aureus*. In two studies from the Mayo Clinic (16; A. L. Whitesell, R. C. Walker, M. S. Rouse, W. R. Wilson, and N. K. Henry, *Clin. Res.* **34:**55a, 1986) performed with rats, a

3-week course of treatment with ciprofloxacin led to results which were the same as those obtained with nafcillin for a methicillin-susceptible organism. A 3-week course of treatment of osteomyelitis due to methicillin-resistant organisms with vancomycin or ciprofloxacin was equally ineffective. Combination therapy with vancomycin and rifampin was less effective than the combination of ciprofloxacin with rifampin, the only effective mode of treatment. In a very careful study of experimental, chronic osteomyelitis due to *S. aureus* in rabbits, Mader et al. (20) compared the effects of nafcillin (40 mg/kg four times a day) and two aryl-fluoroquinolones, A-56619 (difloxacin) and A-56620 (15 or 20 mg/kg subcutaneously twice a day). All three treatments were started 2 weeks after infection and conducted for 4 weeks. The animals were sacrificed 2 weeks after completion of therapy. Concentrations in serum and bone were determined 1 h after administration of the quinolone and 30 min after administration of nafcillin. Whereas levels in serum were 3.1 and 3.0 μg/ml, respectively, for the two aforementioned quinolones, levels in infected and uninfected bone were 1.5 and 0.8 μg/ml for both quinolones. Nafcillin levels in serum peaked at a higher level of 21 μg/ml and contrasted with the corresponding levels in bone (2.4 and 1.2 μg/ml, respectively, in infected and uninfected bone). MICs against the *S. aureus* strain used were the same for the quinolones and for nafcillin, i.e., 0.39 μg/ml. Thus, highly comparable experimental conditions were created to compare a well-established antibiotic with the new quinolones. The best results, measured by the number of rabbits sterilized or the mean decrease in the log of CFU, were obtained with both nafcillin and difloxacin, *S. aureus* being isolated in 8 of 20 rabbits treated with the former and 6 of 20 rabbits treated with the latter compound. Taken together, the results showed that the quinolone used was equal to a beta-lactamase-stable penicillin for treatment of osteomyelitis due to methicillin-susceptible organisms; with methicillin-resistant organisms a combination of a quinolone and rifampin seemed to give the most favorable, albeit suboptimal, results.

In a recent study (3) on gram-negative experimental arthritis in rabbits, the effect of ciprofloxacin was compared with the activity of gentamicin. The offending organism was a pathogenic, serum-resistant *Escherichia coli* strain isolated from a patient with neonatal meningitis. Joints of rabbits were injected with 10^8 organisms, and animals were randomized 4 days later to receive either 80 mg of ciprofloxacin per kg per day intramuscularly or 5 mg of gentamicin per kg per day intramuscularly for 17 days. The dose of the quinolone was again high in order to generate levels in serum and joints similar to those obtained in humans. Thus, maximal ciprofloxacin levels in serum and joints were always higher than the bactericidal concentration for the *E. coli* strain, but this was especially the case for gentamicin. As expected, ciprofloxacin fared much better than gentamicin, being bactericidal in all cases at day 10 and in all but one case at day 17. In contrast, gentamicin was bactericidal in only 6 of 18 joint fluids at day 10 and 8 of 10 at day 17. Similar differences were obtained when synovial tissue was assessed. The comparison of ciprofloxacin and gentamicin showed the same beneficial effect of the quinolone when the results were expressed in terms of residual CFU. Besides the differences in dosage already discussed above, there are other factors explaining the unfavorable results obtained with gentamicin alone, such as decreased pH in infected joint fluid and binding of the aminoglycoside to purulent exudate (5) and lysed neutrophils (32).

Clinical Studies

A variety of studies have addressed the question of the efficacy of the quinolones in the treatment of osteomyelitis due to gram-negative aerobic organisms, which is—almost by definition—characterized by a chronic or recurrent evolution, with the exception of vertebral osteomyelitis. In a study of 34 patients with gram-negative osteomyelitis treated with 750 mg of oral ciprofloxacin twice a day for 6 to 24 weeks, Lesse et al. (18) reported on 23 cases, 9 of which were still on treatment. Of 23 patients, 11 had polymicrobial infection, a disputed entity; 5 of them were infected with *S. aureus*, for which they received additional antistaphylococcal treatment. Of 23 patients, 10 had received previous unsuccessful therapy. The authors reported an astonishing success rate of 23 of 23, but as mentioned above, only 14 had completed therapy, with a short mean follow-up of 6 months, while 9 were still receiving treatment. A study on gram-negative chronic osteomyelitis, with more stringent criteria for diagnosis and treatment follow-up, was carried out by Gilbert et al. (12); they used the same dosage of ciprofloxacin, i.e., 750 mg twice a day for 6 to 10 weeks, in 20 patients. Of these, 3 had osteomyelitis of the sternum and 17 had osteomyelitis of the lower extremities. Most patients had received previous unsuccessful treatment. Chronicity of the disease was defined as an infection evolving for more than 2 weeks, a definition which many would still consider including mostly acute cases. Thirteen patients had *P. aeruginosa* isolated from their infected site. No adjuvant chemotherapy was used, but 15 patients underwent debridement, and 7 were excluded from the protocol because surgical intervention was not possible. Results at a 7- to 21-month follow-up showed a 65% clinical and 70% microbiological cure rate, with slightly less satisfactory results in the case of *P. aeruginosa* infection. For the latter organism the MIC increased in four cases during therapy. Two other series with well-characterized, mostly chronic gram-negative osteomyelitides and a similar treatment regimen of ciprofloxacin (750 mg twice a day for a mean of 62 and

78 days) involved 54 cases (17, 31). Both studies encompassed together 20 cases of *P. aeruginosa* infection. In the study by Hessen et al. (17), 22 of 24 cases were evaluable; of these, 16 patients underwent surgical procedure and 8 received additional antibiotic treatment. At a follow-up of more than 6 months, 20 of 22 patients had clinical, microbiological, and radiological "cure." Of note in this series were one case of vertebral and five cases of sternotomy infections, all of which were cured. In the study by Slama et al. (31), there were 10 acute and 20 chronic cases, including eight sternal infections. Clinical and bacteriological control was achieved in 22 of 30 cases, the eight recurrences responding to a second course of ciprofloxacin and debridement. Finally, Greenberg et al. (15) performed as good a randomized study as feasible in this disease with so many confounding variables. They randomized 30 patients to receive either 750 mg of ciprofloxacin twice per day or "appropriate chemotherapy," mostly a combination of two parenteral antibiotics, including almost always an aminoglycoside. The results were slightly better with combination therapy than with the quinolones: 11 of 16 controls of infection, 4 of 16 improvements, and 1 of 16 failure with the former versus 7 of 14 controls, 3 of 14 improvements, and 4 of 14 failures with the latter. The same trend was observed in infections due to *P. aeruginosa*. Noteworthy is the high rate of drug toxicity (5 of 16) with combination therapy, which has to be weighed against the good tolerance of ciprofloxacin. Overall side effects with the quinolones in the above-mentioned studies (12, 15, 17, 18, 31) were rare and inconsequential.

Taken together, these results suggest that for gram-negative osteomyelitis, treatment with the quinolones—in particular, ciprofloxacin—achieves cure rates above 50% in most cases and represents an interesting alternative to parenteral chemotherapy with its inherent complications. Side effects are rare, but it is important to look carefully for the emergence of resistant organisms.

The interpretation of the results achieved in osteomyelitis due to gram-positive organisms or of mixed etiology is more difficult. Remarkable results were obtained with a combination of pefloxacin (400 mg twice a day) and rifampin in chronic osteomyelitides due to *S. aureus* by Desplaces et al. (7). Of 14 patients, all were cured, with a follow-up of 9 to 24 months, confirming the experimental data described previously. Similar results were reported by Dellamonica et al. (6) with pefloxacin in chronic osteomyelitis, with 5 of 15 infections being due to *S. aureus*.

An endless list of abstracts containing small numbers of cases could be added to this enumeration. For the reasons discussed previously, these studies do not help in assessing the efficacy of the quinolones in chronic osteomyelitides due to gram-positive organisms. The available data suggest that the quinolones can be used as therapeutic alternatives, if ambulatory therapy is desired. Personal experience has shown that they are effective in controlling drainage and abscess formation and in preventing hospitalization (personal data). Whether combination therapy with rifampin or other agents should be used cannot be ascertained yet from the available data.

Data on the treatment of septic arthritis with the quinolones are minimal. In an open study Ramirez et al. (28) reported the treatment and cure of three cases of *N. gonorrheae* and *Streptococcus pneumoniae* arthritis with ciprofloxacin.

Conclusions and Remaining Problems

The quinolones have many characteristics which make them suitable as effective agents in bone and joint infections due to gram-negative and gram-positive organisms. Their MICs against gram-negative pathogens are very low, often far below the levels achievable in serum by oral administration. Improved efficacy against *S. aureus* and *S. epidermidis* is, however, a desirable attribute of new compounds currently being developed. Levels in bone in experimental systems and in human bone biopsies are adequate, with the limitation of heterogeneous intraosseous distribution. Experimental models of osteomyelitis have shown their high efficacy in gram-negative bone and joint infections and suggest combined therapy—with rifampin, for instance—in gram-positive infections. Clinical studies show good efficacy of the quinolones in gram-negative bone infections, similar to or better than with parenteral combined conventional therapy. Despite many anecdotal reports as to their efficacy in staphylococcal infections, the pooled data are still inconclusive: combined therapy with another agent should still be suggested until more studies are available. Side effects of the quinolones have been limited so far, but still do not justify their use in children. Larger clinical studies have to be encouraged to further clarify the defined role of the quinolones in the treatment of bone and joint infections.

Literature Cited

1. **Arcieri, G., E. Griffith, G. Gruenwaldt, A. Heyd, B. O'Brien, N. Becker, and R. August.** 1987. Ciprofloxacin: an update on clinical experience. *Am. J. Med.* **82**(Suppl. 4A): 381–386.
2. **Auckenthaler, A., and F. A. Waldvogel.** 1984. Bone and synovial fluid, p. 505–512. *In* A. M. Ristuccia and B. A. Cunha (ed.), *Antimicrobial Therapy.* Raven Press, New York.
3. **Bayer, A. S., D. Norman, and D. Anderson.** 1985. Efficacy of ciprofloxacin in experimental arthritis caused by *Eschericia coli*—in vitro-in vivo correlations. *J. Infect. Dis.* **152:** 811–816.

4. **Bell, S. M.** 1976. Further observations on the value of oral penicillins in chronic staphylococcal osteomyelitis. *Med. J. Aust.* **2:**591–593.
5. **Bryant, R. E., and D. Hammond.** 1974. Interaction of purulent material with antibiotics used to treat *Pseudomonas* infections. *Antimicrob. Agents Chemother.* **6:**702–707.
6. **Dellamonica, P., E. Bernard, H. Etesse, and R. Garraffo.** 1986. The diffusion of pefloxacin into bone and the treatment of osteomyelitis. *J. Antimicrob. Chemother.* **17:**93–102.
7. **Desplaces, N., L. Gutmann, J. Carlet, J. Guibert, and J. F. Acar.** 1986. The new quinolones and their combinations with other agents for therapy of severe infections. *J. Antimicrob. Chemother.* **17:**25–39.
8. **Emslie, K. R., and S. Nade.** 1984. Acute hematogenous staphylococcal osteomyelitis: evaluation of cloxacillin therapy in an animal model. *Pathology* **16:**441–446.
9. **Emslie, K. R., and S. Nade.** 1986. Pathogenesis and treatment of acute hematogenous osteomyelitis: evaluation of current views with reference to an animal model. *Rev. Infect. Dis.* **8:**841–849.
10. **Fitzgerald, R. H., P. J. Kelly, R. J. Snyder, and J. A. Washington.** 1978. Penetration of methicillin, oxacillin, and cephalothin into bone and synovial tissues. *Antimicrob. Agents Chemother.* **14:**723–726.
11. **Fong, I. W., W. H. Ledbetter, A. C. Vandenbroucke, M. Simbul, and V. Rahm.** 1986. Ciprofloxacin concentrations in bone and muscle after oral dosing. *Antimicrob. Agents Chemother.* **29:**405–408.
12. **Gilbert, D. N., P. K. Marsh, P. C. Craven, and L. C. Preheim.** 1987. Oral ciprofloxacin therapy for chronic contiguous osteomyelitis caused by aerobic gram-negative bacilli. *Am. J. Med.* **82**(Suppl. 4A):254–258.
13. **Goldenberg, D. L., K. D. Brandt, E. S. Cathcart, and A. S. Cohen.** 1974. Acute arthritis caused by Gram-negative bacilli: a clinical characterization. *Medicine* (Baltimore) **53:**197–208.
14. **Goldenberg, D. L., and A. S. Cohen.** 1976. Acute infectious arthritis: a review of patients with nongonococcal joint infections (with emphasis on therapy and prognosis). *Am. J. Med.* **60:**369–377.
15. **Greenberg, R. N., A. D. Tice, P. K. Marsh, P. C. Craven, P. M. Reilly, M. Bollinger, and W. J. Weinandt.** 1987. Randomized trial of ciprofloxacin compared with other antimicrobial therapy in the treatment of osteomyelitis. *Am. J. Med.* **82**(Suppl. 4A):266–269.
16. **Henry, N. K., M. S. Rouse, A. L. Whitesell, M. E. McConnell, and W. R. Wilson.** 1987. Treatment of methicillin-resistant *Staphylococcus aureus* experimental osteomyelitis with ciprofloxacin or vancomycin alone or in combination with rifampin. *Am. J. Med.* **82**(Suppl. 4A):73–75.
17. **Hessen, M. T., M. J. Ingerman, D. H. Kaufman, P. Weiner, J. Santoro, O. M. Korzeniowski, J. Boscia, M. Topiel, L. M. Bush, D. Kaye, and M. E. Levison.** 1987. Clinical efficacy of ciprofloxacin therapy for gram-negative bacillary osteomyelitis. *Am. J. Med.* **82**(Suppl. 4A):262–265.
18. **Lesse, A. J., C. Freer, R. A. Salata, J. B. Francis, and W. M. Scheld.** 1987. Oral ciprofloxacin therapy for gram-negative bacillary osteomyelitis. *Am. J. Med.* **82**(Suppl. 4A):247–253.
19. **Mader, J. T., G. L. Brown, J. C. Guckian, C. H. Wella, and J. A. Galveston.** 1980. A mechanism for the amelioration by hyperbaric oxygen of experimental staphylococcal osteomyelitis in rabbits. *J. Infect. Dis.* **142:**915–922.
20. **Mader, J. T., L. T. Morrison, and K. R. Adams.** 1987. Comparative evaluation of A-56619, A-56620, and nafcillin in the treatment of experimental *Staphylococcus aureus* osteomyelitis. *Antimicrob. Agents Chemother.* **31:**259–263.
21. **Mandell, W., and H. C. Neu.** 1986. In vitro activity of CI-934, a new quinolone, compared

with that of other quinolones and other antimicrobial agents. *Antimicrob. Agents Chemother.* **29:**852–857.

22. **Musher, D. M., S. B. Thorsteinsson, J. N. Minuth, and R. J. Luchi.** 1976. Vertebral osteomyelitis. Still a diagnostic pitfall. *Arch. Intern. Med.* **136:**105–109.
23. **Norden, C. W.** 1985. Osteomyelitis, p. 704–711. *In* G. L. Mandell, R. G. Douglas, and J. E. Bennett (ed.), *Principles and Practice of Infectious Diseases,* 2nd ed. John Wiley & Sons, Inc., New York.
24. **Norden, C. W., and E. Keleti.** 1980. Experimental osteomyelitis caused by *Pseudomonas aeruginosa. J. Infect. Dis.* **141:**71–75.
25. **Norden, C. W., and K. Niederriter.** 1987. Ofloxacin therapy for experimental osteomyelitis caused by *Pseudomonas aeruginosa. J. Infect. Dis.* **155:**823–825.
26. **Norden, C. W., and M. A. Shafer.** 1982. Activities of tobramycin and azlocillin alone and in combination against experimental osteomyelitis caused by *Pseudomonas aeruginosa. Antimicrob. Agents Chemother.* **21:**62–65.
27. **Norden, C. W., and E. Shinners.** 1985. Ciprofloxacin as therapy for experimental osteomyelitis caused by *Pseudomonas aeruginosa. J. Infect. Dis.* **151:**291–294.
28. **Ramirez, C. A., J. L. Bran, C. R. Mejia, and J. F. Garcia.** 1985. Open, prospective study of the clinical efficacy of ciprofloxacin. *Antimicrob. Agents Chemother.* **28:**128–132.
29. **Schaad, U. B., and J. Wedgwood-Krucko.** 1987. Nalidixic acid in children: retrospective matched controlled study for cartilage toxicity. *Infection* **15:**165–168.
30. **Septimus, E. J., and D. M. Musher.** 1979. Osteomyelitis: recent clinical and laboratory aspects. *Orthop. Clin. North Am.* **10:**347–359.
31. **Slama, T. G., J. Misinski, and S. Sklar.** 1987. Oral ciprofloxacin therapy for osteomyelitis caused by aerobic gram-negative bacilli. *Am. J. Med.* **82**(Suppl. 4A)**:**259–261.
32. **Vaudaus, P., and F. A. Waldvogel.** 1980. Gentamicin inactivation in purulent exudates: role of cell lysis. *J. Infect. Dis.* **142:**586–593.
33. **Waldvogel, F. A., G. Medoff, and M. N. Swartz.** 1970. Osteomyelitis: a review of clinical features, therapeutic considerations and unusual aspects. *N. Engl. J. Med.* **282:**198–206, 260–266, 316–322.
34. **Waldvogel, F. A., and H. Vasey.** 1980. Osteomyelitis: the past decade. *N. Engl. J. Med.* **303:**360–370.
35. **Ward, T., and R. T. Steigbigel.** 1978. Acidosis of synovial fluid correlates with synovial fluid leukocytosis. *Am. J. Med.* **64:**933–936.

Chapter 10

Use of Quinolone Antimicrobial Agents in Immunocompromised Patients

Drew J. Winston

UCLA Medical Center, Los Angeles, California 90024

The profile of the new fluoroquinolones includes certain antibacterial, pharmacokinetic, and safety features which may be advantageous for the prevention and treatment of infections in immunocompromised patients. These favorable properties have been the basis for an increasing number of laboratory and clinical studies devoted to the use of the fluoroquinolones in immunocompromised patients. This chapter will review these studies and the current status of the quinolones in relation to the present epidemiology of infections in immunocompromised hosts and their prevention and treatment with alternative agents.

Prevention of Infections in Granulocytopenic Patients

Infections are frequent and almost inevitable complications in severely granulocytopenic patients. Both the level of circulating granulocytes and the duration of granulocytopenia are important. The incidence of infection increases as the granulocyte count falls below 500 cells per mm^3 (5). Most serious infections, including nearly all bacteremias, occur when the granulocyte count is less than 100 cells per mm^3 (47). Patients who remain granulocytopenic for prolonged periods (>14 days) are also more likely to develop infections. Bacterial organisms causing infections usually arise from the gastrointestinal tract of the patient. These organisms may be part of the endogenous alimentary tract flora of the patient or may be acquired from environmental sources (71).

One approach to the prevention of bacterial infections in granulocytopenic patients is the use of prophylactic oral antimicrobial agents (69). These drugs are designed not only to suppress the endogenous gastrointestinal flora of the patient but also to prevent colonization with potential pathogens from the environment. Initial prophylactic regimens consisted primarily of oral nonabsorbable antibiotics given alone or in conjunction with a protective environment (laminar airflow or similar isolation) (24, 50, 54, 65, 70, 75, 95). Common regimens were gentamicin, vancomycin, and nystatin; polymyxin, vancomycin, and nystatin; framycetin, colistin, and nystatin; and neomycin, colistin, and nystatin. These oral nonabsorbable antibiotics suppress the microbial flora of the gastrointestinal tract but usually do not completely sterilize it. Several prospective, randomized, controlled trials have shown that prophylactic oral nonabsorbable antibiotics decrease the incidence of infection in granulocytopenic patients (50, 70, 75), whereas others have not (24, 54, 95). One of the reasons for this disagreement is the variable compliance of patients for oral nonabsorbable antibiotics. These agents cause frequent nausea, vomiting, and diarrhea. They are also expensive and may be associated with the rapid acquisition of organisms after prophylaxis is discontinued (69, 70). Consequently, oral nonabsorbable antibiotics have generally been unpopular with most patients and many physicians.

Another approach to oral chemoprophylaxis is selective decontamination of the gastrointestinal tract. A series of studies by Van der Waaij and colleagues demonstrated that *Escherichia coli, Pseudomonas aeruginosa,* and *Klebsiella pneumoniae* could colonize the alimentary canal of germfree animals when only 100 organisms were given orally (80). In contrast, normal conventional animals required oral inoculation with 10^7 or more organisms for the gastrointestinal tract to be colonized. This capacity of the normal microbial flora of the alimentary canal to prevent colonization with new organisms was termed "colonization resistance." When these same workers suppressed the aerobic flora of normal animals with antibiotics while the anaerobic flora persisted, colonization resistance was maintained. An oral inoculum of 10^6 organisms was still needed to cause colonization. These findings suggested that the normal anaerobic gastrointestinal flora was responsible for preventing colonization with aerobic gram-negative bacilli, perhaps by successfully competing for nutrients in the gut (37). On the basis of these results, drugs such as trimethoprim-sulfamethoxazole and nalidixic acid, which are capable of reducing the aerobic bacterial flora of the gastrointestinal tract without suppressing the anaerobic flora, have been used for selective decontamination.

The results of studies in which either trimethoprim-sulfamethoxazole or nalidixic acid has been used for prophylaxis in granulocytopenic patients have been conflicting. Many controlled trials showed a decrease in infections in patients given prophylactic trimethoprim-sulfamethoxazole or nalidixic acid (21, 36, 38, 39, 72, 82, 85), while others found no benefit (29, 86). Moreover, certain risks are associated with the use of these agents. Trimethoprim-sulfamethoxazole may cause untoward skin rashes and may prolong the period of granulocytopenia by suppressing myelopoiesis (21, 81). Neither trimethoprim-sulfamethoxazole nor nalidixic acid is active against *P. aeruginosa*. Selection of resistant gram-negative bacilli and superinfection with fungi such as *Candida* spp. and *Aspergillus* spp. may also occur during prophylaxis (36, 81, 88). In two separate studies in which trimethoprim-sulfamethoxazole was compared with nalidixic acid, trimethoprim-sulfamethoxazole was found to be superior to nalidixic acid (6, 81). More patients receiving nalidixic acid experienced colonization and infection with gram-negative bacilli. Many of these gram-negative bacilli were resistant to nalidixic acid. These conflicting results and the risk of adverse effects and superinfections have continued to make the prophylactic use of trimethoprim-sulfamethoxazole or nalidixic acid in granulocytopenic patients controversial.

The limitations of oral nonabsorbable antibiotics, trimethoprim-sulfamethoxazole, and nalidixic acid for the prophylaxis of infection have been the impetus for evaluating the new fluoroquinolones in granulocytopenic patients. The fluoroquinolones have several favorable properties that may overcome some of the limitations of the other agents. The fluoroquinolones have excellent activity against members of the family *Enterobacteriaceae* and inhibit more than 90% of isolates at concentrations of 2 μg/ml or less (93). *P. aeruginosa* isolates are intrinsically resistant to trimethoprim-sulfamethoxazole and nalidixic acid but are susceptible to the fluoroquinolones at concentrations of 2 to 6 μg/ml. Staphylococci are also well inhibited (MIC for 90% of isolates, 1.0 to 6 μg/ml), while streptococci are less susceptible (MIC for 90% of isolates, 2 to 32 μg/ml). On the other hand, the fluoroquinolones have relatively little activity against anaerobic bacteria. In normal volunteers high concentrations of the fluoroquinolones (200 to 2,000 μg/g of stool) are achieved in the stool after oral administration and are associated with suppression of the aerobic gram-negative bowel flora with preservation of the anaerobes (7, 15, 28, 45, 60, 61, 64). Similar studies with granulocytopenic patients have shown a reduction of the enteric aerobic gram-negative flora by orally administered fluoroquinolones without a significant change in the anaerobic population (48, 67). Thus, these drugs should maintain colonization resistance. Finally, the fluoroquinolones are not associated with the plasmid-mediated resistance or myelosuppression observed with trimethoprim-sulfamethoxazole (45, 93).

Table 1 compares the effects of the oral fluoroquinolones (norfloxacin, ciprofloxacin, and ofloxacin) with those of a placebo, vancomycin plus polymyxin, or trimethoprim-sulfamethoxazole on the acquisition of gram-negative bacillary organisms during a series of clinical trials of chemoprophylaxis in granulocytopenic patients (22, 48, 92; D. J. Winston, W. G. Ho, K. Bartoni, D. A. Bruckner, R. P. Gale, and R. E. Champlin, *Program Abstr. 27th Intersci. Conf. Antimicrob. Agents Chemother.*, abstr. no. 1258, 1987). In each trial, the fluoroquinolone prevented the acquisition of members of the *Enterobacteriaceae* and *P. aeruginosa* and was more effective than the control regimen of a placebo, vancomycin plus polymyxin, or trimethoprim-sulfamethoxazole. Only colonization with non-*P. aeruginosa Pseudomonas* strains or *Acinetobacter* species occurred to a significant degree in the patients receiving the fluoroquinolones, but none of these organisms caused infection. A few of the non-*P. aeruginosa Pseudomonas* and *Acinetobacter* organisms were resistant to the fluoroquinolones. Otherwise, the acquisition of resistant organisms occurred very infrequently.

The results of different studies in which the efficacy of the fluoroquinolones was evaluated in the prevention of infections in granulocytopenic patients with hematologic malignancy are shown in Table 2. Norfloxacin has undergone the most extensive evaluation and was compared with a placebo, trimethoprim-sulfamethoxazole, or vancomycin plus polymyxin (48, 90, 92; E. J. Bow, T. J. Louie, E. Rayner, and J. Pitsanuk, *25th ICAAC*, abstr. no. 150, 1985). In more limited trials ciprofloxacin was compared with trimethoprim-sulfamethoxazole and ofloxacin was compared with vancomycin plus polymyxin (22; Winston et al., *27th ICAAC*). More than 300 patients were enrolled in these studies. Except for the study by Bow et al. (*25th ICAAC*), in which norfloxacin was compared with trimethoprim-sulfamethoxazole, patients receiving prophylaxis with an oral fluoroquinolone had fewer microbiologically documented infections than did control patients. The overall incidence of microbiologically documented infections was 38% (66 in 172 patients) in the quinolone groups versus 58% (95 in 165 patients) in the control groups. The fluoroquinolones were especially effective in the prevention of gram-negative bacteremia. Only 4 cases of gram-negative bacteremia occurred in 172 patients (2%) on quinolone prophylaxis compared with 28 cases in 165 control patients (17%). The four cases of gram-negative bacteremia reported in patients receiving fluoroquinolones for prophylaxis were caused by *P. aeruginosa* (three cases) and *Haemophilus parainfluenzae* (one case) (48). None was resistant to the quinolones. On the other hand, the number of gram-positive bacteremias was similar in the quinolone groups (31 in 172 patients [18%]) and the control groups (28 in 165 patients [17%]). The predominant gram-positive organisms causing bacteremias were coagulase-negative staphylococci

Table 1. Acquisition of gram-negative bacillary organisms in surveillance cultures of granulocytopenic patients receiving oral antimicrobial prophylaxis

Organism	Karp et al. (48)		Winston et al. (92)		Dekker et al. (22)		Winston et al. (27th ICAAC)	
	Placebo (n = 33)[a]	Norfloxacin (n = 35)	Vancomycin +polymyxin (n = 30)	Norfloxacin (n = 36)	Trimethoprim-sulfamethoxazole + colistin (n = 28)	Ciprofloxacin (n = 28)	Vancomycin + polymyxin (n = 27)	Ofloxacin (n = 25)
E. coli	8	0	0	0	12 (*E. coli* and *Klebsiella-Enterobacter-Serratia* combined)	0 (*E. coli* and *Klebsiella-Enterobacter-Serratia* combined)	0	0
Klebsiella-Enterobacter-Serratia species	14	0	19	5			8	1 (1)
Proteus species	0	0	11	1	0	0	10	0
Citrobacter species	1	0	1	2	0	0	3	0
Providencia species	0	0	3	0	0	0	1	0
P. aeruginosa	6	0	5	1	5	0	4	0
Non-*P. aeruginosa* *Pseudomonas* species	4	2 (2)[b]	10	17 (2)	4	4 (4)	0	3 (3)
Acinetobacter species	0	0	1	3	6	4 (4)	1	0
Others	0	0	6	3	0	0	0	0

[a]Number within parentheses is the number of patients.
[b]Number within parentheses is the number of isolates resistant to the fluoroquinolone.

Table 2. Controlled trials of oral fluoroquinolones for the prophylaxis of infections in granulocytopenic patients

Reference	Regimen	Dose[a]	Underlying disease	No. of patients	Micrologically documented infections	Gram-negative bacteremia	Gram-positive bacteremia	Disseminated fungal infections	Fatal infections
Karp et al. (48)	Norfloxacin	400 mg q12h	Acute leukemia and autologous marrow transplant	35	20 (57)[b]	4 (12)	4 (12)	6 (18)	6 (18)
	Placebo	Placebo		33	26 (79)	11 (33)	3 (9)	5 (15)	3 (9)
Winston et al. (90)	Norfloxacin	400 mg q8h	Hematologic malignancy	26	10 (38)	0 (0)	5 (19)	2 (8)	
	Trimethoprim-sulfamethoxazole	160 mg-800 mg q8h		28	14 (50)	1 (4)	4 (14)	1 (4)	
Bow et al. (*25th ICAAC*)	Norfloxacin	400 mg q12h	Acute leukemia	17	9 (53)	0 (0)	5 (29)	0 (0)	
	Trimethoprim-sulfamethoxazole	160 mg-800 mg q12h		14	6 (43)	1 (7)	0 (0)	0 (0)	
Winston et al. (92)	Norfloxacin	400 mg q8h	Hematologic malignancy	36	12 (33)	0 (0)	9 (25)	3 (8)	3 (8)
	Vancomycin-polymyxin	500 mg-100 mg q8h		30	16 (53)	5 (17)	4 (13)	6 (20)	4 (13)
Dekker et al. (22)	Ciprofloxacin	500 mg q12h	Acute leukemia	28	5 (18)	0 (0)	3 (11)	0 (0)	0 (0)
	Trimethoprim-sulfamethoxazole + colistin	160 mg-800 mg + 200 mg q8h		28	14 (50)	5 (18)	6 (21)	0 (0)	1 (4)
Winston et al. (*27th ICAAC*)	Ofloxacin	300 mg q12h	Hematologic malignancy	30	10 (33)	0 (0)	5 (17)	2 (7)	2 (7)
	Vancomycin-polymyxin	500 mg-100 mg q8h		32	19 (59)	5 (16)	7 (22)	2 (6)	3 (9)
	Total								
	Quinolone			172	66 (38)	4 (2)	31 (18)	13 (8)	
	Controls			165	95 (58)	28 (17)	24 (15)	14 (8)	

[a] q12h and q8h, Every 12 and 8 h, respectively.
[b] Numbers within parentheses are percentages.

and viridans group streptococci (22, 48, 90, 92; Bow et al., *25th ICAAC*; Winston et al., *27th ICAAC*). Some of the coagulase-negative staphylococcal bacteremias were related to infected central intravenous catheters, while several of the viridans group streptococci were resistant to the quinolones. The lack of efficacy of the fluoroquinolones for the prophylaxis of gram-positive infections correlated with a failure of these agents to eradicate colonizing gram-positive organisms (92; Bow et al., *25th ICAAC*; Winston et al., *27th ICAAC*). In contrast to infections with gram-negative bacteria, infections with gram-positive bacteria, however, appear to be better tolerated by patients and have been associated with a lower mortality rate. There was no increased number of fungal infections associated with the prophylactic use of the fluoroquinolones.

Chemoprophylaxis with the fluoroquinolones in granulocytopenic patients was well tolerated (Table 3). Compliance was better for norfloxacin, ciprofloxacin, or ofloxacin than for trimethoprim-sulfamethoxazole or vancomycin plus polymyxin. Gastrointestinal adverse effects were more common in patients given trimethoprim-sulfamethoxazole or vancomycin plus polymyxin and occurred infrequently in patients receiving the fluoroquinolones. Skin rashes developed in approximately 20% of the patients on trimethoprim-sulfamethoxazole but in none of the patients on the quinolones. Neurological adverse effects were also uncommon and were observed in only two patients who received ofloxacin. There were no reported cases of myelosuppression.

A concern about the use of antimicrobial agents such as the fluoroquinolones for prophylaxis is the emergence of quinolone-resistant organisms. However, colonization or infection with resistant organisms was found infrequently during the studies of prophylaxis in granulocytopenic patients (Table 4). Non-*P. aeruginosa Pseudomonas* organisms accounted for most of the colonization with resistant organisms, but none caused infection. Except for one case of colonization with an ofloxacin-resistant *Enterobacter aerogenes* strain, there were no reported cases of colonization or infection with quinolone-resistant members of the *Enterobacteriaceae* or *P. aeruginosa*. Infections caused by quinolone-resistant organisms were caused predominately by gram-positive aerobic bacteria (viridans group streptococci, coagulase-negative staphylococci, corynebacteria, and one case of methicillin-resistant *Staphylococcus aureus*). Of note, one patient developed *Pneumocystis carinii* pneumonia while receiving prophylactic ofloxacin. In contrast to trimethoprim-sulfamethoxazole (46), the fluoroquinolones have no apparent activity for *P. carinii*.

In addition to norfloxacin, ciprofloxacin, and ofloxacin, other fluoroquinolones either are currently being evaluated for prophylaxis in granulocytopenic patients or will likely be the focus of future clinical trials. These

Table 3. Compliance and adverse effects of oral antimicrobial prophylaxis in granulocytopenic patients

Reference	Regimen	No. of patients	Patients highly compliant (>90% doses taken)	Adverse effects[a]			
				GI	Rash	CNS	MYE
Karp et al. (48)	Norfloxacin	35	29 (83)[b]	1	0	0	0
	Placebo	33	29 (88)	0	0	0	0
Winston et al. (90)	Norfloxacin	26		0	0	0	0
	Trimethoprim-sulfamethoxazole	28		0	2	0	0
Winston et al. (92)	Norfloxacin	36	30 (83)	1	0	0	0
	Vancomycin-polymyxin	30	16 (53)	4	0	0	0
Dekker et al. (22)	Ciprofloxacin	28	23 (82)	6	0	0	0
	Trimethoprim-sulfamethoxazole	28	15 (54)	12	8	0	0
Winston et al. (*27th ICAAC*)	Ofloxacin	25	19 (76)	1	0	2[c]	0
	Vancomycin-polymyxin	27	14 (52)	7	0	0	0

[a]GI, Gastrointestinal; CNS, central nervous system; MYE, myelosuppression.
[b]Numbers within parentheses are percentages.
[c]One patient each with headaches and vertigo.

Table 4. Colonization and infection with quinolone-resistant organisms during oral fluoroquinolone prophylaxis in granulocytopenic patients

Reference	Prophylactic fluoroquinolone	No. of patients	Colonization with quinolone-resistant organisms	Infections with quinolone-resistant organisms
Karp et al. (48)	Norfloxacin	35	*Pseudomonas maltophilia* (3)[a]	None reported, but gram-positive infections occurred on norfloxacin
Winston et al. (92)	Norfloxacin	36	*P. maltophilia* (2)	Viridans group streptococci (3), corynebacteria (2)
Dekker et al. (22)	Ciprofloxacin	28	Non-*P. aeruginosa Pseudomonas* species (4), *Acinetobacter* species (4)	None reported, but gram-positive infections occurred on ciprofloxacin
Winston et al. (*27th ICAAC*)	Ofloxacin	30	*P. maltophilia* (1), *Pseudomonas fluorescens* (1), *Pseudomonas putida* (1), *Enterobacter aerogenes* (1)	Coagulase-negative staphylococci (1), viridans group streptococci (1), methicillin-resistant *Staphylococcus aureus* (1), *Pneumocystis carinii* (1)

[a]Numbers within parentheses are the number of isolates or organisms.

drugs include enoxacin, amifloxacin, and pefloxacin. The issue of whether any of these other agents will be more or less effective than norfloxacin, ciprofloxacin, or ofloxacin for the prevention of infections during granulocytopenia can best be addressed in comparative clinical trials. The major limitation of norfloxacin, ciprofloxacin, and ofloxacin was their lack of efficacy in the prevention of gram-positive aerobic bacterial infections, even though ciprofloxacin and ofloxacin have better in vitro activity against most gram-positive organisms than does norfloxacin (93). Thus, other fluoroquinolones with improved activity for the increasing number of resistant gram-positive pathogens emerging in granulocytopenic patients might be more effective. The safety and efficacy of using the oral fluoroquinolones for several months or longer for the prophylaxis of infections in patients with chronic granulocytopenia also need to be defined.

Treatment of Febrile Granulocytopenic Patients

Successful treatment of febrile granulocytopenic patients with suspected or documented infections requires the prompt initiation of appropriate antimicrobial therapy. In most cases this therapy has consisted of an aminoglycoside plus an antipseudomonal beta-lactam drug (44, 51). Combination therapy is aimed at providing a wide spectrum of antibacterial coverage, obtaining synergistic bactericidal activity (especially for *P. aeruginosa*), and preventing the emergence of resistant organisms. Several recent developments, however, have led to a reevaluation of this traditional approach. Due to the increasing use of nephrotoxic drugs in granulocytopenic patients (such as amphotericin B, cyclosporine, and certain chemotherapeutic agents), there is a frequent desire to avoid combination therapy with the aminoglycosides (63, 87). The incidence of *P. aeruginosa* infections has declined at many oncology centers, while the proportion of infections caused by gram-positive organisms (coagulase-negative staphylococci, viridans group streptococci, and corynebacteria) has greatly increased (16, 74, 83, 89). Many of these gram-positive pathogens are resistant to the aminoglycosides and beta-lactam antibiotics. The availability of newer beta-lactam drugs with broader and more potent antibacterial activity has also made it possible to design new approaches to therapy.

The above developments have been the basis for the recent use of alternative antimicrobial regimens in febrile granulocytopenic patients. These include double beta-lactam antibiotic combinations (a third-generation cephalosporin plus a ureidopenicillin), front-loading with an aminoglycoside (a third-generation cephalosporin plus a short course of an aminoglycoside), monotherapy with a third-generation cephalosporin or imipenem, and vancomycin plus a third-generation cephalosporin (20, 30, 44, 53, 62,

84). Each of these regimens avoids the nephrotoxicity and ototoxicity associated with the aminoglycosides but has certain limitations. Double beta-lactam antibiotic combinations are expensive and occasionally may be antagonistic (20). Front-loading with an aminoglycoside may not be as effective as the traditional combination of an aminoglycoside plus a beta-lactam antibiotic in patients with prolonged granulocytopenia (30). Monotherapy with ceftazidime does not provide adequate coverage for many gram-positive pathogens and may require the addition of vancomycin for gram-positive superinfections (53, 62, 68). Monotherapy with imipenem at high doses (4 g/day) is effective but may be associated with an increased number of seizures (91; J. C. Wade, C. Bustamante, A. Devlin, R. Finley, G. Drusano, and B. Thompson, *27th ICAAC*, abstr. no. 1251, 1987). The routine use of intravenous vancomycin as part of the initial therapy of all febrile granulocytopenic patients is extremely expensive and leads to the unnecessary treatment of many patients (68).

Clinical experience in which the fluoroquinolones have been used for the treatment of febrile granulocytopenic patients is limited. Nevertheless, a number of in vitro and animal studies have provided information suggesting a potential role for these agents in the therapy of infections in granulocytopenic patients. In vitro, the fluoroquinolones, like the aminoglycosides and newer beta-lactams, are very active against the common *Enterobacteriaceae* and *P. aeruginosa* organisms causing serious infections (93). In addition, the fluoroquinolones are active against many of the methicillin-resistant staphylococci which now cause frequent infections but are resistant to the aminoglycosides and beta-lactams. On the other hand, the fluoroquinolones are less active than the penicillins, cephalosporins, and vancomycin against viridans group and other streptococci which account for many of the gram-positive infections in granulocytopenic patients. In vitro combinations of the fluoroquinolones with the aminoglycosides or beta-lactams for *P. aeruginosa* and members of the *Enterobacteriaceae* are predominantly indifferent or additive but rarely antagonistic (11, 12, 14, 23, 31, 34, 40, 57; A. W. Chow and J. Wong, *27th ICAAC*, abstr. no. 625, 1987). If a synergistic interaction occurs, it is more often seen between a beta-lactam agent (azlocillin, mezlocillin, piperacillin, ceftazidime, cefoperazone, or imipenem) and a fluoroquinolone than between an aminoglycoside (gentamicin, tobramycin, or amikacin) and a fluoroquinolone. Similarly, the serum bactericidal activity for *P. aeruginosa, E. coli, K. pneumoniae*, and *S. aureus* of patients receiving a fluoroquinolone (pefloxacin) and amikacin indicated mostly additive or indifferent effects of the combination (79). Only occasional antagonism occurred. The bactericidal activity of ciprofloxacin plus vancomycin in serum for *S. aureus, Staphylococcus epidermidis*, and corynebacteria was also predominantly indifferent (78). There was no antagonism.

Studies with animal models

Granulocytopenic animal models evaluating the in vivo efficacy of the fluoroquinolones alone or in combination are summarized in Table 5. In most of the studies, animals were treated with dosages of the quinolones and other antimicrobial agents designed to provide levels in serum and tissue similar to those achieved in humans. While only a few of the studies actually report concentrations in serum and tissue (66; L. R. Peterson, D. N. Gerding, J. A. Moody, and C. E. Fasching, *25th ICAAC*, abstr. no. 1087, 1985; E. Ulrich, M. Trautmann, T. Weinke, and H. Hahn, *Proc. 1st Int. Symp. New Quinolones*, 1986, p. 112), an analysis of the results does provide information helpful in the design of clinical trials in humans.

For *P. aeruginosa* bacteremia in granulocytopenic mice or rats, the combination of ciprofloxacin plus gentamicin or ciprofloxacin plus azlocillin was more effective than ciprofloxacin alone in two of three studies (13; K. Jules and H. C. Neu, *24th ICAAC*, abstr. no. 27, 1984). In the third study, ciprofloxacin alone was as effective as ciprofloxacin plus azlocillin (H. G. Robson and M. Cote, *24th ICAAC*, abstr. no. 25, 1984). For all studies, ciprofloxacin alone was either as effective as or frequently more effective than single-agent therapy with gentamicin, tobramycin, azlocillin, ceftazidime, or imipenem (13; Jules and Neu, *24th ICAAC*; Robson and Cote, *24th ICAAC*; Ulrich et al., *Proc. 1st Int. Symp. New Quinolones*, 1986). In one study ciprofloxacin alone was more effective than enoxacin alone, which was more effective than ofloxacin alone, for *P. aeruginosa* bacteremia in granulocytopenic mice (13). For *P. aeruginosa* pneumonia in granulocytopenic guinea pigs, ciprofloxacin and pefloxacin each given alone were as effective as ticarcillin plus tobramycin or ceftazidime plus tobramycin and more effective than ceftazidime, ticarcillin, or tobramycin alone (35). Ciprofloxacin alone was also more effective than ceftazidime alone for *Klebsiella* pneumonia in granulocytopenic mice (66).

The model of thigh muscle infection in granulocytopenic mice has also been used to evaluate the relative efficacy of the fluoroquinolones and other agents for the treatment of *P. aeruginosa, E. coli,* and *K. pneumoniae* infections (41; M. J. Muszynski, R. K. Scribner, T. D. Lewis, and M. I. Marks, *25th ICAAC*, abstr. no. 1091, 1985; B. Vogelman, S. Gudmundsson, S. Wolz, and W. Craig, *25th ICAAC*, abstr. no. 1092, 1985). For *P. aeruginosa*, the combination of ciprofloxacin plus azlocillin was more effective than either drug alone, and ciprofloxacin alone was as effective as cefoperazone plus tobramycin and more effective than tobramycin alone. For *E. coli*, ciprofloxacin plus mezlocillin was more effective than either agent alone, and ciprofloxacin alone was more effective than cefazolin plus gentamicin or gentamicin alone. For *K. pneumoniae* infection, ciprofloxacin alone was

Table 5. Efficacy of fluoroquinolones in granulocytopenic animal models

Reference	Animal model	Organism	In vivo antibacterial potency[a]
Jules and Neu (*24th ICAAC*)	Bacteremia in granulocytopenic mice challenged i.p.[b]	*Pseudomonas aeruginosa*	Ciprofloxacin + gentamicin > ciprofloxacin > enoxacin > ofloxacin > gentamicin
Chin and Neu (13)	Bacteremia in granulocytopenic mice challenged i.p.	*Pseudomonas aeruginosa*	Ciprofloxacin + azlocillin > ciprofloxacin > azlocillin
Robson and Cote (*24th ICAAC*)	Bacteremia in granulocytopenic rats challenged i.p.	*Pseudomonas aeruginosa*	Ciprofloxacin = ceftazidime = azlocillin = tobramycin = ciprofloxacin + azlocillin
Ulrich et al.[c]	Bacteremia in granulocytopenic mice challenged i.p.	*Pseudomonas aeruginosa*	Ciprofloxacin = imipenem > ceftazidime > azlocillin = tobramycin
Gordin et al. (35)	Pneumonia in granulocytopenic guinea pigs	*Pseudomonas aeruginosa*	Ciprofloxacin = pefloxacin = ticarcillin + tobramycin = ceftazidime + tobramycin > ceftazidime = ticarcillin = tobramycin
Roosendaal et al. (66)	Pneumonia in granulocytopenic rats	*Klebsiella pneumoniae*	Ciprofloxacin > ceftazidime
Muszynski (*25th ICAAC*)	Thigh muscle infection in granulocytopenic mice	*Pseudomonas aeruginosa*	Ciprofloxacin + azlocillin > ciprofloxacin > azlocillin
Haller (41)	Thigh muscle infection in granulocytopenic mice	*Pseudomonas aeruginosa*	Ciprofloxacin + azlocillin > ciprofloxacin > azlocillin
		Escherichia coli	Ciprofloxacin + mezlocillin > ciprofloxacin > mezlocillin
Vogelman et al. (*25th ICAAC*)	Thigh muscle infection in granulocytopenic mice	*Pseudomonas aeruginosa*	Ciprofloxacin = cefoperazone + tobramycin > tobramycin
		Escherichia coli	Ciprofloxacin > cefazolin + gentamicin = gentamicin
		Klebsiella pneumoniae	Ciprofloxacin > cefazolin + gentamicin = gentamicin

Continued on following page

Table 5—*Continued*

Reference	Animal model	Organism	In vivo antibacterial potency[a]
Moody et al. (56)	Granulocytopenic chamber site infection in rabbits	*Pseudomonas aeruginosa*	Ciprofloxcin + azlocillin = azlocillin + amikacin > ceftizoxime + amikacin > ciprofloxacin + amikacin > ciprofloxacin + ceftizoxime
		Enterobacteriaceae	Ciprofloxacin + azlocillin > mezlocillin + amikacin > ceftizoxime + amikacin
		Group D streptococci	Ciprofloxacin + azlocillin > azlocillin + amikacin > penicillin + amikacin
Peterson et al. (*25th ICAAC*)	Granulocytopenic chamber site infection in rabbits	*Streptococcus pneumoniae*	Ciprofloxacin + azlocillin = azlocillin > ciprofloxacin
		Streptococcus faecalis	Ciprofloxacin + azlocillin = azlocillin + amikacin > azlocillin > ciprofloxacin
		Streptococcus avium	Ciprofloxacin + azlocillin > azlocillin + amikacin > azlocillin > ciprofloxacin

[a] >, More active than; =, of equal activity to.
[b] i.p., Intraperitoneally.
[c] *Proc. 1st Int. Symp. New Quinolones*, 1986.

more effective than the combination of cefazolin plus gentamicin or gentamicin alone. In a similar model of granulocytopenic chamber site infection in rabbits, ciprofloxacin plus azlocillin was as effective as azlocillin plus amikacin and more effective than ciprofloxacin plus amikacin for *P. aeruginosa* while ciprofloxacin plus azlocillin was more effective than either mezlocillin plus amikacin or cefotaxime plus amikacin for members of the *Enterobacteriaceae* (56). The combination of ciprofloxacin plus azlocillin was also as effective as or, in some cases, more effective than azlocillin plus amikacin and more effective than ciprofloxacin alone for streptococcal infections (Peterson et al., *25th ICAAC*).

In summary, the results from these animal studies suggest a potential synergistic role for the fluoroquinolones in combination with a broad-spectrum ureidopenicillin for the treatment of *P. aeruginosa, Enterobacteriaceae,* and streptococcal infections in granulocytopenic patients.

Clinical studies

Table 6 shows the results of a few preliminary studies in which the effectiveness of the fluoroquinolones in the therapy of febrile granulocytopenic patients was evaluated. In three uncontrolled studies intravenous ciprofloxacin was used to treat patients who had failed to respond to initial first-line therapy (73, 94; C. C. Kibbler, L. Pomeroy, R. J. Sage, P. Mannan, P. Noone, and H. G. Prentice, *Program Abstr. 3rd Eur. Congr. Clin. Microbiol.*, abstr. no. 357, 1987). Despite the high-risk character of these patients, 36 of 50 infections (72%) improved, including 15 of 28 bacteremias (54%). However, ciprofloxacin-resistant streptococcal superinfections occurred in three patients, an *S. epidermidis* causing bacteremia in another patient developed resistance to ciprofloxacin, and ciprofloxacin-resistant gram-negative organisms (including *P. aeruginosa*) developed in four patients despite the use of ciprofloxacin in combination with other antibiotics. One of two *P. aeruginosa* bacteremias treated by Berdig et al. also relapsed after treatment with ciprofloxacin plus amikacin (2). Intravenous pefloxacin alone was used in another uncontrolled trial by Beun et al. as initial empiric therapy (4). Six of seven patients (86%) with documented infections improved, but pefloxacin-resistant streptococcal superinfections occurred in two patients. In contrast, when intravenous vancomycin was combined with ciprofloxacin as first-line empiric therapy of febrile granulocytopenic patients, 21 of 32 infections (66%) improved, including 10 infections caused by gram-positive organisms (G. M. Smith, M. J. Leyland, I. Farrell, and A. M. Geddes, *Program Abstr. 3rd Eur. Congr. Clin. Microbiol.*, abstr. no. 237, 1987). No gram-positive superinfections were reported.

Two randomized comparative trials in which intravenous ciprofloxacin was used for the empiric therapy of febrile granulocytopenic patients have

Table 6. Efficacy of fluoroquinolones in febrile granulocytopenic patients with documented infections

Reference	Treatment regimen[a]	Study design	Results	Comments
Smith et al. (73)	i.v. ciprofloxacin, 200 mg q12h	Uncontrolled; therapy of febrile patients unresponsive to ceftazidime + mezlocillin	9 of 14 infections (64%) improved, including 4 of 8 bacteremias (50%)	Ciprofloxacin-resistant streptococcal superinfection in 3 patients; *S. epidermidis* in blood became resistant to ciprofloxacin
Wood and Newland (94)	i.v. ciprofloxacin, 200 mg q12h	Uncontrolled; therapy of febrile patients unresponsive to other antibiotics	11 of 14 infections (79%) improved, including 4 of 7 bacteremias (57%)	1 of 3 *P. aeruginosa* bacteremias improved
Kibbler et al.[b]	i.v. ciprofloxacin, 200 mg q12h, + other antibiotics	Uncontrolled; therapy of febrile patients unresponsive to other antibiotics	16 of 23 infections (70%) improved, including 7 of 13 bacteremias (54%)	Ciprofloxacin-resistant gram-negative organisms developed in 3 patients
Berdig et al. (2)	i.v. ciprofloxacin, 200 mg q8–12h, + amikacin	Uncontrolled; therapy of multiply resistant *P. aeruginosa* bacteremias	2 of 2 *P. aeruginosa* bacteremias improved, but 1 relapsed after therapy	Ciprofloxacin level in serum below MIC after 6 h; 8-h dosing interval recommended
Beun et al. (4)	i.v. pefloxacin, 400 mg q8h	Uncontrolled; initial empiric therapy of febrile patients	6 of 7 infections (86%) improved	Pefloxacin-resistant streptococcal superinfections in 2 patients
Smith et al.[b]	Ciprofloxacin + vancomycin	Uncontrolled; initial empiric therapy of febrile patients	21 of 32 infections (66%) improved	2 failures and 1 superinfection due to *P. aeruginosa*
Chan et al. (*27th ICAAC*)	i.v. ciprofloxacin, 200 mg q12h, + netilmicin vs piperacillin + netilmicin	Controlled, randomized; initial empiric therapy of febrile patients	41 of 76 (54%) ciprofloxacin + netilmicin vs. 37 of 69 (54%) piperacillin + netilmicin improved	Streptococcal superinfection in 3 ciprofloxacin + netilmicin vs 0 piperacillin + netilmicin
Wood and Newland (*27th ICAAC*)	i.v. ciprofloxacin, 200 mg q12h, + penicillin vs piperacillin + netilmicin	Controlled, randomized; initial empiric therapy of febrile patients	9 of 13 (69%) ciprofloxacin + penicillin vs 8 of 15 (53%) piperacillin + netilmicin improved	Renal failure in 1 ciprofloxacin + penicillin vs 7 in piperacillin + netilmicin

[a] i.v., Intravenous; q12h and q8h, every 12 and 8 h, respectively.
[b] *Program Abstr. 3rd Eur. Congr. Clin. Microbiol.*, 1987.

also been reported. In one trial, the response rates for documented infections were 54% (41 of 76) for ciprofloxacin plus netilmicin and 54% (37 of 69) for piperacillin plus netilmicin (C. C. Chan, B. A. Oppenheim, H. Anderson, and J. H. Scarffe, *27th ICAAC*, abstr. no. 1250, 1987). There were three streptococcal superinfections in the ciprofloxacin-plus-netilmicin group but none in the piperacillin-plus-netilmicin group. In a smaller trial, 9 of 13 documented infections (69%) responded to ciprofloxacin plus penicillin compared with 8 of 15 infections (53%) treated with piperacillin plus netilmicin (M. E. Wood and A. C. Newland, *27th ICAAC*, abstr. no. 1252, 1987). Of note, there were no streptococcal superinfections reported in the patients receiving ciprofloxacin in combination with penicillin, while renal failure was more frequent in the patients receiving piperacillin plus netilmicin.

Thus, the results from these preliminary studies suggest a need to use the fluoroquinolones in combination with an antistreptococcal agent such as a penicillin or vancomycin to reduce the risk for streptococcal superinfections. The use of a potentially synergistic combination such as an antipseudomonal ureidopenicillin plus a fluoroquinolone might also improve the outcome from *P. aeruginosa* infections (13, 41, 56; Muszynski et al., *25th ICAAC*). On the other hand, the coadministration of an aminoglycoside with a fluoroquinolone does not appear to prevent either streptococcal superinfections or the development of quinolone-resistant *P. aeruginosa* and *Enterobacteriaceae* organisms during therapy (1, 33; Chan et al., *27th ICAAC*; Kibbler et al., *Program Abstr. 3rd Eur. Congr. Clin. Microbiol.*, 1987).

Treatment of Specific Infections in Immunocompromised Patients

In addition to febrile granulocytopenic patients, other immunocompromised patients with *P. aeruginosa* or other gram-negative bacillary infections have been treated with the fluoroquinolones (Table 7). Follath et al. gave intravenous ciprofloxacin in combination with an aminoglycoside to 10 patients with *P. aeruginosa* infections complicating various hematologic diseases (33). Only 4 of 10 infections improved, and three *P. aeruginosa* strains developed resistance to the ciprofloxacin despite the concomitant use of an aminoglycoside. Results were somewhat better in a group of patients with solid tumors, with renal transplants, with diabetes mellitus, or receiving dialysis. Of 28 gram-negative infections in this group, 17 (61%) improved on intravenous or oral pefloxacin, including 10 of 15 *P. aeruginosa* infections (58). The development of resistance to pefloxacin was not reported.

Table 7. Efficacy of fluoroquinolones for specific infections in immunocompromised patients

Reference	Infection	Underlying disease	Regimen	Response (no. improved/ no. treated)	Comments
Follath et al. (33)	*Pseudomonas aeruginosa* bacteremia, pneumonia, and osteomyelitis	Hematologic disease	i.v.[a] ciprofloxacin + aminoglycoside	4/10	3 *Pseudomonas aeruginosa* strains developed resistance to ciprofloxacin
Morduchowicz et al. (58)	*Pseudomonas aeruginosa* and *Enterobacteriaceae* bacteremia, pneumonia, osteomyelitis, and urinary or soft tissue infection	Neoplasm, renal transplants, dialysis, diabetes	i.v. and oral pefloxacin	17/28	10 of 15 *Pseudomonas aeruginosa* infections improved
Patton et al. (59)	*Salmonella typhimurium* bacteremia	Acute leukemia	i.v. and oral ciprofloxacin	1/1	
Kiess et al. (49)	*Salmonella tennessee* osteomyelitis	Neuroblastoma	Oral ciprofloxacin	1/1	
Burns and Wallace (10)	*Salmonella typhimurium* urinary infection	Renal transplant	Oral ciprofloxacin	1/1	
Connolly et al. (18)	*Salmonella typhimurium* bacteremia	AIDS	Oral ciprofloxacin	1/1	
Klein et al. (52)	*Salmonella* bacteremia	AIDS	Oral ciprofloxacin	2/2	
Heseltine et al. (42)	*Salmonella* enteritis and bacteremia	AIDS	Oral norfloxacin	7/7	
Heseltine and Corrado (43)	*Salmonella* bacteremia	Chronic granulomatous disease	Oral norfloxacin	0/1	
Heseltine and Corrado (43)	*Shigella* or *Campylobacter* enteritis	AIDS	Oral norfloxacin	1/2	

[a] i.v., Intravenous.

The major bacterial pathogens causing diarrhea (toxigenic *E. coli*, *Shigella* spp., *Salmonella* spp., *Campylobacter* spp., *Aeromonas* spp., and *Vibrio* spp.) are all inhibited by the fluoroquinolones at concentrations of less than 1 μg/ml (93; see also chapters 3 and 8). Some *Salmonella* strains resistant to ampicillin or chloramphenicol are inhibited by the fluoroquinolones, which are capable of penetrating phagocytic cells and killing intracellular bacteria (9, 26, 27). In murine models ciprofloxacin reduces mortality from systemic *Salmonella* infections in mice with no effective immunity as well as in mice with normal immunity (8, 25). For these reasons ciprofloxacin and norfloxacin have been used for the treatment of *Salmonella* infections complicating malignancy, acquired immunodeficiency syndrome (AIDS), renal transplantation, or chronic granulomatous disease (10, 18, 42, 43, 49, 52, 59). Of 14 *Salmonella* infections, 13 (93%) responded to ciprofloxacin or norfloxacin (Table 7). Many of these patients had multiple prior clinical episodes of salmonellosis which had failed to be eradicated with ampicillin or trimethoprim-sulfamethoxazole. One of two AIDS patients with *Shigella* or *Campylobacter* gastroenteritis also improved on norfloxacin (43). The pathogens were eradicated from the stool in all cases.

Mycobacteria and quinolones

Serious infections caused by the *Mycobacterium avium* complex have been increasing and are especially common in patients with AIDS (97). The therapy of *M. avium* complex infections is limited by the resistance of these organisms to individual standard agents and by the toxicity of combined drug regimens. Among the fluoroquinolones, ciprofloxacin has the greatest in vitro activity for *M. avium*, including isolates from AIDS patients (32, 96; B. G. Yangco, C. S. Lackman, K. Halkias, and H. Chmel, *27th ICAAC*, abstr. no. 1362, 1987; S. P. Naik and R. E. Ruck, *27th ICAAC*, abstr. no. 1365, 1987). However, only one-fourth to one-third of these isolates are inhibited by ciprofloxacin alone. Combinations of ciprofloxacin plus imipenem and amikacin, ciprofloxacin plus rifampin and ethambutol, and ciprofloxacin plus ansamycin are synergistic (96; Yangco et al., *27th ICAAC*; Naik and Ruck, *27th ICAAC*). In the beige and C57 black immunodeficient mouse models of disseminated *M. avium* complex infections, ciprofloxacin alone has little in vitro activity (C.B. Inderlied, L. S. Young, P. T. Kolonoski, S. Yadagar, and J. K. Yamada, *25th ICAAC*, abstr. no. 1118, 1985). However, the combination of ciprofloxacin, imipenem, and amikacin produced a 90% reduction in infection of the liver and spleen and a 50-fold decrease in bacteremia. Mortality was decreased from 100 to 20% in the black mice and from 75 to 30% in the beige mice.

There are no clinical studies documenting the efficacy of ciprofloxacin in the treatment of *M. avium* complex infections, but the in vitro and animal model results provide a rational basis for using ciprofloxacin as part of a combination regimen in future clinical trials. Ciprofloxacin and the other fluoroquinolones also inhibit most *Mycobacterium tuberculosis* strains at achievable concentrations in serum and may be useful agents for the treatment of tuberculosis in AIDS and other immunocompromised patients (3, 17, 19, 55, 76). Ofloxacin was given in combination with other antimycobacterial drugs to 19 nonimmunocompromised patients with chronic cavitary pulmonary tuberculosis caused by organisms resistant to many conventional agents (77). These patients had all previously failed other therapies. A decrease in the number of tubercle bacilli in the sputum occurred in most patients, and negative conversion was observed in five patients. Ofloxacin-resistant tubercle bacilli appeared in patients who did not show negative sputum conversion.

Summary

The new fluoroquinolones are effective agents for the prevention of gram-negative bacillary colonization and infection in granulocytopenic patients. They are better tolerated than oral nonabsorbable antibiotics or trimethoprim-sulfamethoxazole and have few side effects. For the empiric therapy of febrile granulocytopenic patients, the use of the fluoroquinolones has been limited and requires additional study in controlled comparative trials in which proven standard regimens are used. On the basis of in vitro synergy studies, granulocytopenic animal models, and preliminary clinical results, a combination of fluoroquinolone plus a ureidopenicillin may be more effective than a fluoroquinolone alone by reducing the risk for streptococcal superinfections and the emergence of quinolone-resistant *P. aeruginosa* strains. The fluoroquinolones are effective alternative drugs for the treatment of enteric bacterial infections in AIDS and other immunocompromised patients, including patients with hypersensitivity to trimethoprim-sulfamethoxazole or infections with organisms resistant to ampicillin. There are no clinical data establishing the efficacy of the fluoroquinolones for the treatment of mycobacterial diseases in immunocompromised patients, but results from in vitro and animal model studies suggest a possible role for these agents as part of combination therapy of *M. avium* and *M. tuberculosis* infections. As with patients with normal immunity, the oral administration of the fluoroquinolones should also facilitate the outpatient management of immunocompromised patients with other specific infections caused by susceptible organisms.

Literature Cited

1. **Azadian, B. S., J. W. A. Bendig, and D. M. Samson.** 1986. Emergence of ciprofloxacin-resistant *Pseudomonas aeruginosa* after combined therapy with ciprofloxacin and amikacin. *J. Antimicrob. Chemother.* **18:**771.
2. **Berdig, J. W. A., P. W. Kyle, P. L. F. Giangrande, D. M. Samson, and B. S. Azadian.** 1987. Two neutropenic patients with multiple resistant *Pseudomonas aeruginosa* septicemia treated with ciprofloxacin. *J. R. Soc. Med.* **80:**316–317.
3. **Berlin, O. G. W., L. S. Young, and D. A. Bruckner.** 1987. In vitro activity of six fluorinated quinolones against *Mycobacterium tuberculosis*. *J. Antimicrob. Chemother.* **19:**611–615.
4. **Beun, G. D. M., L. L. Debrus-Palmans, M. S. M. Daniels-Bosman, and G. H. Blijham.** 1988. Therapy with pefloxacin in febrile neutropenic patients. *Rev. Infect. Dis.* **10**(Suppl. 1)**:**S236.
5. **Bodey, G. P., M. Buckley, Y. S. Sathe, and E. J. Freireich.** 1966. Quantitative relationship between circulating leukocytes and infection in patients with acute leukemia. *Ann. Intern. Med.* **64:**328–340.
6. **Bow, E. J., E. Rayner, B. A. Scott, and T. J. Jouie.** 1987. Selective gut decontamination with nalidixic acid or trimethoprim-sulfamethoxazole for infection prophylaxis in neutropenic cancer patients: relationship of efficacy to antimicrobial spectrum and timing of administration. *Antimicrob. Agents Chemother.* **31:**551–557.
7. **Brumfitt, W., I. Franklin, P. Grady, J. M. T. Hamilton-Miller, and A. Iliffe.** 1984. Changes in the pharmacokinetics of ciprofloxacin and fecal flora during administration of a 7-day course to human volunteers. *Antimicrob. Agents Chemother.* **26:**757–761.
8. **Brunner, H., and H. J. Zeiler.** 1988. Oral ciprofloxacin treatment for *Salmonella typhimurium* infection of normal and immunocompromised mice. *Antimicrob. Agents Chemother.* **32:**57–62.
9. **Bryan, J. P., H. Rocha, and W. M. Scheld.** 1986. Problems in salmonellosis: rationale for clinical trials with newer β-lactam agents and quinolones. *Rev. Infect. Dis.* **8:**189–207.
10. **Burns, B. J., and M. R. Wallace.** 1987. Treatment of *Salmonella typhimurium* infection in a renal transplant patient with ciprofloxacin. *N.Z. Med. J.* **100:**190.
11. **Bustamante, C. I., G. L. Drusano, R. C. Wharton, and J. C. Wade.** 1987. Synergism of the combinations of imipenem plus ciprofloxacin and imipenem plus amikacin against *Pseudomonas aeruginosa* and other bacterial pathogens. *Antimicrob. Agents Chemother.* **31:**632–634.
12. **Chalkley, L. J., and H. J. Koornhof.** 1985. Antimicrobial activity of ciprofloxacin against *Pseudomonas aeruginosa, Escherichia coli,* and *Staphylococcus aureus* determined by the killing curve method: antibiotic comparison and synergism interactions. *Antimicrob. Agents Chemother.* **28:**331–342.
13. **Chin, N. X., and H. C. Neu.** 1986. Synergy of ciprofloxacin and azlocillin in vitro and in a neutropenic mouse model of infection. *Eur. J. Clin. Microbiol.* **5:**23–28.
14. **Chin, N. X., and H. C. Neu.** 1987. Synergy of imipenem—a novel carbapenem, and rifampin and ciprofloxacin against *Pseudomonas aeruginosa, Serratia marcescens* and *Enterobacter* species. *Chemotherapy* (Basel) **33:**183–188.
15. **Cofsky, R. D., L. DuBouchet, and S. H. Landesman.** 1984. Recovery of norfloxacin in feces after administration of a single oral dose to human volunteers. *Antimicrob. Agents Chemother.* **26:**110–111.
16. **Cohen, J., J. P. Donnelly, A. M. Worsley, D. Catovsky, J. M. Goldman, and D. A. G. Galton.** 1983. Septicemia caused by viridans streptococci in neutropenic patients with leukemia. *Lancet* **ii:**1452–1454.
17. **Collins, C. H., and A. H. G. Uttley.** 1985. In vitro susceptibility of mycobacteria to ciprofloxacin. *J. Antimicrob. Chemother.* **16:**575–580.

18. **Connolly, M. J., M. H. Snow, and H. R. Ingham.** 1986. Ciprofloxacin treatment of recurrent *Salmonella typhimurium* septicemia in a patient with acquired immune deficiency syndrome. *J. Antimicrob. Chemother.* **18:**647–648.
19. **Davies, S., P. D. Sparham, and R. C. Spencer.** 1987. Comparative in vitro activity of five fluoroquinolones against mycobacteria. *J. Antimicrob. Chemother.* **19:**605–609.
20. **Dejace, P., and J. Klastersky.** 1986. Comparative review of combination therapy: two beta-lactams versus beta-lactam plus aminoglycoside. *Am. J. Med.* **80**(Suppl. 6B)**:**29–38.
21. **Dekker, A. W., M. Rozenberg-Arska, J. J. Sixma, and J. Verhoef.** 1981. Prevention of infection by trimethoprim-sulfamethoxazole plus amphotericin B in patients with acute nonlymphocytic leukemia. *Ann. Intern. Med.* **95:**555–559.
22. **Dekker, A. W., M. Rozenberg-Arska, and J. Verhoef.** 1987. Infection prophylaxis in acute leukemia: a comparison of ciprofloxacin with trimethoprim-sulfamethoxazole and colistin. *Ann. Intern. Med.* **106:**7–12.
23. **Desplaces, N., L. Gutmann, J. Carlet, J. Guibert, and J. F. Acar.** 1986. The new quinolones and their combination with other agents for therapy of severe infections. *J. Antimicrob. Chemother.* **17**(Suppl. A)**:**25–39.
24. **Dietrich, M., W. Gaus, J. Vossen, D. Van der Waaik, and F. Wendt.** 1977. Protective isolation and antimicrobial decontamination in patients with high susceptibility to infection: a prospective cooperative study of gnotobiotic care in acute leukemia patients. I. Clinical results. *Infection* **5:**107–114.
25. **Easmon, C. S. F.** 1987. Protective effects of ciprofloxacin in a murine model of Salmonella infection. *Am. J. Med.* 82(Suppl. 4A)**:**71–72.
26. **Easmon, C. S. F., and J. P. Crane.** 1985. Uptake of ciprofloxacin by human neutrophils. *J. Antimicrob. Chemother.* **16:**67–73.
27. **Easmon, C. S. F., and J. P. Crane.** 1985. Uptake of ciprofloxacin by macrophages. *J. Clin. Pathol.* **38:**442–444.
28. **Edlund, C., A. Lidbeck, L. Kagen, and C. E. Nord.** 1987. Effect of enoxacin on colonic microflora of healthy volunteers. *Eur. J. Clin. Microbiol.* **6:**298–300.
29. **EORTC International Antimicrobial Therapy Project Group.** 1984. Trimethoprim-sulfamethoxazole in the prevention of infection in neutropenic patients. *J. Infect. Dis.* **150:**372–379.
30. **EORTC International Antimicrobial Therapy Cooperative Group.** 1987. Ceftazidime combined with a short or long course of amikacin for empirical therapy of gram-negative bacteremia in cancer patients with granulocytopenia. *N. Engl. J. Med.* **317:**1692–1698.
31. **Farrag, N. N., J. W. A. Bendig, C. Talboys, and B. S. Azadian.** 1986. In vitro study of the activity of ciprofloxacin combined with amikacin or ceftazidime against *Pseudomonas aeruginosa*. *J. Antimicrob. Chemother.* **18:**770.
32. **Fenlon, C. H., and M. H. Cynamon.** 1986. Comparative in vitro activities of ciprofloxacin and other 4-quinolones against *Mycobacterium tuberculosis* and *Mycobacterium intracellulare*. *Antimicrob. Agents Chemother.* **29:**386–388.
33. **Follath, F., M. Bindschedler, M. Wenk, R. Frei, H. Stalder, and H. Reber.** 1986. Use of ciprofloxacin in the treatment of *Pseudomonas aeruginosa* infections. *Eur. J. Clin. Microbiol.* **5:**236–240.
34. **Giamarellou, H., and G. Petrikkos.** 1987. Ciprofloxacin interactions with imipenem and amikacin against multiresistant *Pseudomonas aeruginosa*. *Antimicrob. Agents Chemother.* **31:**958–961.
35. **Gordin, F. M., C. J. Hackbarth, K. G. Scott, and M. A. Sande.** 1985. Activities of pefloxacin and ciprofloxacin in experimentally induced *Pseudomonas* pneumonia in neutropenic guinea pigs. *Antimicrob. Agents Chemother.* **27:**452–454.

36. **Gualtier, R. J., G. R. Donowitz, D. L. Kaiser, C. E. Hess, and M. A. Sande.** 1983. Double-blind randomized study of prophylactic trimethoprim-sulfamethoxazole in granulocytopenic patients with hematologic malignancies. *Am. J. Med.* **74:**934–940.
37. **Guiot, H. F. L.** 1982. Role of competition for substrate in bacterial antagonism in the gut. *Infect. Immun.* **38:**887–892.
38. **Guiot, H. F. L., P. J. Van den Broek, J. W. M. Van der Meer, and R. Van Furth.** 1983. Selective antimicrobial modulation of the intestinal flora of patients with acute nonlymphocytic leukemia: a double-blind, placebo-controlled study. *J. Infect. Dis.* **147:**615–623.
39. **Gurwith, M. J., J. L. Braunton, B. A. Lank, G. K. M. Harding, and A. R. Ronald.** 1979. A prospective controlled investigation of prophylactic trimethoprim-sulfamethoxazole in hospitalized granulocytopenic patients. *Am. J. Med.* **66:**248–256.
40. **Haller, I.** 1985. Comprehensive evaluation of ciprofloxacin-aminoglycoside combinations against *Enterobacteriaceae* and *Pseudomonas aeruginosa* strains. *Antimicrob. Agents Chemother.* **28:**663–666.
41. **Haller, I.** 1986. Comprehensive evaluation of ciprofloxacin in combination with β-lactam antibiotics against Enterobacteriaceae and *Pseudomonas aeruginosa. Drug Res.* **26:**226–229.
42. **Heseltine, P. N. R., D. M. Causey, M. D. Appleman, M. L. Corrado, and J. M. Leedom.** 1988. Norfloxacin in the eradication of enteric infections in AIDS patients. *Eur. J. Cancer Clin. Oncol.* **24**(Suppl. 1):S25–S28.
43. **Heseltine, P. N. R., and M. L. Corrado.** 1987. Compassionate use of norfloxacin. *Am. J. Med.* **82**(Suppl. 6B):88–92.
44. **Ho, W. G., and D. J. Winston.** 1986. Infection and transfusion therapy in acute leukemia. *Clin. Hematol.* **15:**873–904.
45. **Hooper, D. C., and J. S. Wolfson.** 1985. The fluoroquinolones: pharmacology, clinical uses, and toxicities in humans. *Antimicrob. Agents Chemother.* **28:**716–721.
46. **Hughes, W. T., S. Kuhn, and S. Chaudhary.** 1987. Successful prophylaxis for *Pneumocystic carinii* pneumonia. *N. Engl. J. Med.* **297:**1419–1426.
47. **Joshi, J. H., and S. C. Schimpff.** 1985. Infections in the compromised host, p. 1644–1648. *In* G. L. Mandell, R. C. Douglas, and J. E. Bennett (ed.), *Principles and Practice of Infectious Diseases*, 2nd ed. John Wiley & Sons, Inc., New York.
48. **Karp, J. E., W. G. Menz, C. Hendricksen, B. Laughon, T. Redden, B. Bamberger, J. G. Bartlett, R. Saral, and P. J. Burke.** 1986. Oral norfloxacin for prevention of gram-negative bacterial infections in patients with acute leukemia and granulocytopenia. A randomized double-blind, placebo-controlled trial. *Ann. Intern. Med.* **106:**1–7.
49. **Kiess, W., R. Haas, and W. Marget.** 1984. Chloramphenicol-resistant *Salmonella tennessee* osteomyelitis. *Infection* **12:**359.
50. **Klastersky, J., L. Debusscher, and D. Daneau.** 1974. Use of oral antibiotics in protected environment units: clinical effectiveness and role in the emergence of antibiotic-resistant strains. *Pathol. Biol.* **22:**5–12.
51. **Klastersky, J., S. H. Zinner, T. Calandra, H. Gaya, M. P. Glauser, F. Meunier, M. Rossi, S. C. Schimpff, M. Tattersal, C. Viscoli, and EORTC Antimicrobial Therapy Cooperative Group.** 1988. Empiric antimicrobial therapy for febrile granulocytopenic cancer patients: lessons from four EORTC trials. *Eur. J. Cancer Clin. Oncol.* **24**(Suppl. 1):S35–S45.
52. **Klein, E., M. Trautman, and H. G. Hoffmann.** 1986. Ciprofloxacin bei Salmonellen infektion und Typhus abdominalis. *Dtsch. Med. Wochenschr.* **111:**1599–1602.
53. **Kramer, B. S., R. Ramphal, and K. H. Rand.** 1986. Randomized comparison between two ceftazidime-containing regimens and cephalothin-gentamicin-carbenicillin in febrile granulocytopenic cancer patients. *Antimicrob. Agents Chemother.* **30:**64–68.
54. **Levine, A. S., S. E. Siegel, A. D. Schreiber, J. Hauser, H. Preisler, I. M. Goldstein, F. Seidler, R. Simon, S. Perry, J. E. Bennett, and E. S. Henderson.** 1973. Protected environ-

ments and prophylactic antibiotics. A prospective controlled study of their utility in the therapy of acute leukemia. *N. Engl. J. Med.* **288:**477–483.

55. **Marinis, E., and N. J. Legakis.** 1985. In vitro activity of ciprofloxacin against clinical isolates of mycobacteria resistant to antimycobacterial drugs. *J. Antimicrob. Chemother.* **16:**527–530.
56. **Moody, J. A., D. N. Gerding, and L. R. Peterson.** 1987. Evaluation of ciprofloxacin's synergism with other agents by multiple in vitro methods. *Am. J. Med.* **82**(Suppl. 4A)**:**44–54.
57. **Moody, J. A., L. R. Peterson, and D. N. Gerding.** 1985. In vitro activity of ciprofloxacin combined with azlocillin. *Antimicrob. Agents Chemother.* **28:**849–850.
58. **Morduchowicz, G., C. S. Block, M. Drucker, J. B. Rosenfeld, and S. D. Pitlik.** 1988. Pefloxacin for the treatment of various gram-negative infections in immunocompromised patients. *Rev. Infect. Dis.* **10**(Suppl. 1)**:**S237–S238.
59. **Patton, W. N., G. M. Smith, M. J. Leyland, and A. M. Geddes.** 1985. Multiply resistant *Salmonella typhimurium* septicemia in an immunocompromised patient successfully treated with ciprofloxacin. *J. Antimicrob. Chemother.* **16:**667–669.
60. **Pecquet, S., A. Andremont, and C. Tancrède.** 1986. Selective antimicrobial modulation of the intestinal tract by norfloxacin in human volunteers and in gnotobiotic mice associated with a human fecal flora. *Antimicrob. Agents Chemother.* **29:**1047–1052.
61. **Pecquet, S., A. Andremont, and C. Tancrède.** 1987. Effect of oral ofloxacin on fecal bacteria in human volunteers. *Antimicrob. Agents Chemother.* **31:**124–125.
62. **Pizzo, P. A., J. W. Hathorn, J. Hiemenz, M. Browne, J. Cummers, D. Cotton, J. Gress, D. Longo, D. Marshall, J. McKnight, M. Rubin, J. Skelton, M. Thaler, and R. Wesley.** 1986. A randomized trial comparing ceftazidime alone with combination antibiotic therapy in cancer patients with fever and neutropenia. *N. Engl. J. Med.* **315:**552–558.
63. **Pizzo, P. A., K. J. Robichaud, F. A. Gill, and F. G. Witebsky.** 1982. Empirical antibiotic and antifungal therapy for cancer patients with prolonged fever and granulocytopenia. *Am. J. Med.* **72:**101–111.
64. **Reeves, D. S.** 1986. The effect of quinolone antibacterials on the gastrointestinal flora compared with that of other antibacterials. *J. Antimicrob. Chemother.* **18**(Suppl. D)**:**89–112.
65. **Rodriguez, V., G. P. Bodey, E. J. Freireich, K. B. McCredie, J. U. Gutterman, M. J. Keating, T. L. Smith, and E. A. Gehan.** 1978. Randomized trial of protected environment prophylactic antibiotics in 145 adults with acute leukemia. *Medicine* (Baltimore) **57:**253–266.
66. **Roosendaal, R., I. A. J. M. Bakker-Woudenberg, M. Van Den Berghe-Van Raffe, J. C. Vink-van Den Berg, and M. F. Michel.** 1987. Comparative activities of ciprofloxacin and ceftazidime against *Klebsiella pneumoniae* in vitro and in experimental pneumonia in leukopenic rats. *Antimicrob. Agents. Chemother.* **31:**1809–1815.
67. **Rozenberg-Arska, M., A. W. Dekker, and J. Verhoef.** 1985. Ciprofloxacin for selective decontamination of the alimentary tract in patients with acute leukemia during remission induction treatment: the effect on fecal flora. *J. Infect. Dis.* **152:**104–107.
68. **Rubin, M., J. W. Hathorn, D. Marshall, J. Gress, S. M. Steinberg, and P. A. Pizzo.** 1988. Gram-positive infections and the use of vancomycin in 550 episodes of fever and neutropenia. *Ann. Intern. Med.* **108:**30–35.
69. **Schimpff, S. C.** 1980. Infection prevention during profound granulocytopenia. New approaches to alimentary canal microbial suppression. *Ann. Intern. Med.* **93:**358–361.
70. **Schimpff, S. C., W. H. Greene, V. M. Young, C. L. Fortner, L. Jepsen, N. Cusack, J. B. Block, and P. H. Wiernik.** 1975. Infection prevention in acute nonlymphocytic leukemia: laminar air flow room reverse isolation with oral, nonabsorbable antibiotic prophylaxis. *Ann. Intern. Med.* **82:**351–358.
71. **Schimpff, S. C., V. M. Young, W. H. Greene, G. P. Vermeulen, M. R. Moody, and P. H.**

Wiernik. 1982. Origin of infection in acute nonlymphocytic leukemia. Significance of hospital acquisition of potential pathogens. *Ann. Intern. Med.* **79:**707–714.

72. **Sleijfer, D. T., N. H. Mulder, H. G. De Vries-Hospers, V. Fidler, H. O. Nieweg, D. Van der Waaij, and H. K. F. Van Saeve.** 1980. Infection prevention in granulocytopenic patients by selective decontamination of the digestive tract. *Eur. J. Cancer* **16:**859–869.
73. **Smith, G. M., M. J. Leyland, I. D. Farrell, and A. M. Geddes.** 1986. Preliminary evaluation of ciprofloxacin, a new 4-quinolone antibiotic in the treatment of febrile neutropenic patients. *J. Antimicrob. Chemother.* **18**(Suppl. D):165–174.
74. **Stamm, W. E., L. E. Tompkins, K. F. Wagner, G. W. Counts, E. D. Thomas, and J. D. Meyers.** 1979. Infection due to *Corynebacterium* species in marrow transplant patients. *Ann. Intern. Med.* **91:**167–173.
75. **Storring, R. A., B. Jameson, T. J. McElwain, E. Wittshaw, A. S. P. Spies, and H. Gaya.** 1977. Oral nonabsorbable antibiotics prevent infection in acute nonlymphoblastic leukemia. *Lancet* **ii:**837–840.
76. **Tsukamura, M.** 1985. In vitro antituberculosis activity of a new antibacterial substance, ofloxacin (DL8280). *Am. Rev. Respir. Dis.* **131:**348–351.
77. **Tsukamura, M., E. Nakamura, S. Yashii, and H. Amano.** 1985. Therapeutic effect of a new antibacterial substance, ofloxacin (DL8280), on pulmonary tuberculosis. *Am. Rev. Respir. Dis.* **131:**352–356.
78. **Van Der Auwera, P., and J. Klastersky.** 1986. Bactericidal activity and killing rate of serum in volunteers receiving ciprofloxacin alone or in combination with vancomycin. *Antimicrob. Agents Chemother.* **30:**892–895.
79. **Van Der Auwera, P., J. Klastersky, S. Lieppe, M. Husson, D. Lauzon, and A. P. Lopez.** 1986. Bactericidal activity and killing rate of serum from volunteers receiving pefloxacin alone or in combination with amikacin. *Antimicrob. Agents Chemother.* **29:**230–234.
80. **Van der Waaij, D., J. M. Berghuis, and J. E. C. Lekkerkerk.** 1971. Colonization resistance of the digestive tract in conventional and antibiotic-treated mice. *J. Hyg.* **69:**405–411.
81. **Wade, J. C., C. A. DeJongh, K. A. Newman, J. Crowley, P. H. Wiernik, and S. C. Schimpff.** 1983. Selective antimicrobial modulation as prophylaxis against infection during granulocytopenia: trimethoprim-sulfamethoxazole vs. nalidixic acid. *J. Infect. Dis.* **147:**624–634.
82. **Wade, J. C., S. C. Schimpff, M. T. Hargadon, C. L. Fortner, V. M. Young, and P. H. Wiernik.** 1981. A comparison of trimethoprim-sulfamethoxazole plus nystatin with gentamicin plus nystatin in the prevention of infections in acute leukemia. *N. Engl. J. Med.* **304:**1057–1062.
83. **Wade, J. C., S. C. Schimpff, K. A. Newman, and P. H. Wiernik.** 1982. *Staphylococcus epidermidis*: an increasing cause of infection in patients with granulocytopenia. *Ann. Intern. Med.* **97:**503–508.
84. **Wade, J. C., H. C. Standiford, G. L. Drusano, D. E. Johnson, M. R. Moody, C. I. Bustamante, J. H. Joshi, C. DeJongh, and S. C. Schimpff.** 1985. Potential of imipenem as single-agent empiric antibiotic therapy of febrile neutropenic patients with cancer. *Am. J. Med.* **78**(Suppl. 5A):62–72.
85. **Watson, J. G., B. Jameson, H. R. L. Powler, T. J. McElwain, D. N. Lawson, I. Judson, G. R. Morgenstern, H. Lumley, and H. E. M. Kay.** 1982. Co-trimoxazole vs. nonabsorbable antibiotics in acute leukemia. *Lancet* **i:**6–9.
86. **Weiser, B., M. Lange, M. S. Fialk, C. Singer, T. H. Szatrowski, and D. Armstrong.** 1981. Prophylactic trimethoprim-sulfamethoxazole during consolidation chemotherapy for acute leukemia: a controlled trial. *Ann. Intern. Med.* **95:**436–438.
87. **Whiting, P. H., I. S. Simpson, and A. W. Thomson.** 1983. Nephrotoxicity of cyclosporine

in combination with aminoglycoside and cephalosporin antibiotics. *Transplant. Proc.* **15:**2702–2706.

88. **Wilson, J. M., and D. G. Guiney.** 1982. Failure of oral trimethoprim-sulfamethoxazole prophylaxis in acute leukemia. Isolation of resistant plasmids from strains of Enterobacteriaceae causing bacteremia. *N. Engl. J. Med.* **306:**16–20.
89. **Winston, D. J., D. V. Dudnick, M. Chapin, W. G. Ho, R. P. Gale, and W. J. Martin.** 1983. Coagulase-negative staphylococcal bacteremia in patients receiving immunosuppressive therapy. *Arch. Intern. Med.* **143:**32–36.
90. **Winston, D. J., W. G. Ho, R. E. Champlin, J. Karp, J. Bartlett, R. S. Finley, J. H. Joshi, G. Talbot, L. Levitt, S. Deresinski, and M. Corrado.** 1987. Norfloxacin for prevention of bacterial infection in granulocytopenic patients. *Am. J. Med.* **82**(Suppl. 6B)**:**40–46.
91. **Winston, D. J., W. G. Ho, D. A. Bruckner, R. P. Gale, and R. E. Champlin.** 1988. Controlled trials of double beta-lactam therapy with cefoperazone plus piperacillin in febrile granulocytopenic patients. *Am. J. Med.* **85**(Suppl. 1A)**:**21–30.
92. **Winston, D. J., W. G. Ho, S. L. Nakao, R. P. Gale, and R. E. Champlin.** 1986. Norfloxacin versus vancomycin/polymyxin for prevention of infections in granulocytopenic patients. *Am. J. Med.* **80:**884–890.
93. **Wolfson, J. S., and D. C. Hooper.** 1985. The fluoroquinolones: structures, mechanisms of action and resistance, and spectra of activity in vitro. *Antimicrob. Agents Chemother.* **28:**581–586.
94. **Wood, M. E., and A. C. Newland.** 1986. Intravenous ciprofloxacin in the treatment of infection in immunocompromised patients. *J. Antimicrob. Chemother.* **18**(Suppl. D)**:**175–178.
95. **Yates, J. W., and J. F. Holland.** 1973. A controlled study of isolation and endogenous microbial suppression in acute myelocytic leukemia. *Cancer* **32:**1490–1498.
96. **Young, L. S., O. G. W. Berlin, and C. B. Inderlied.** 1987. Activity of ciprofloxacin and other fluorinated quinolones against Mycobacteria. *Am. J. Med.* **82**(Suppl. 4A)**:**23–26.
97. **Young, L. S., C. B. Inderlied, O. G. Berlin, and M. S. Gottlieb.** 1986. Mycobacterial infections in AIDS patients, with an emphasis on the *Mycobacterium avium* complex. *Rev. Infect. Dis.* **8:**1024–1033.

Chapter 11

Treatment of Experimental and Human Bacterial Endocarditis and Meningitis with Quinolone Antimicrobial Agents

Arnold S. Bayer

UCLA School of Medicine, Los Angeles, California 90024, and
Division of Infectious Diseases, LAC Harbor-UCLA Medical Center,
Torrance, California 90509

Introduction

The in vivo efficacy of newly developed antimicrobial agents in relevant animal infection models has been an important transitional link between the in vitro activity of an agent and human utility. A review of the in vitro activity spectrum of the newer quinolones suggested that these agents would be potential candidates for the therapy of two important human infections, infective endocarditis and bacterial meningitis.

The newer quinolones possess in vitro activity spectra encompassing several organisms associated with particularly recalcitrant forms of bacterial endocarditis, *Staphylococcus aureus* (both methicillin susceptible and resistant [MSSA and MRSA, respectively]), *Pseudomonas aeruginosa*, and other gram-negative rods (e.g., *Enterobacter* species). These forms of endocarditis have been associated with relatively poor clinical outcomes when treated with standard antimicrobial regimens in humans (35, 39).

The newer quinolones are very active against the aerobic gram-negative bacilli, such as *Escherichia coli*, *Enterobacter* species, and *P. aeruginosa* (1, 23). Such bacterial strains are common causes of nosocomially acquired meningitis, particularly in the setting of recent neurosurgical procedures with placement of indwelling intraventricular devices for pressure monitoring or treatment of hydrocephalus. Moreover, gram-negative bacillary meningitis may be community acquired, occurring spontaneously in patients with cirrhosis or underlying lymphoreticular or hematologic malignancies (18). Traditionally, meningitis caused by the aerobic gram-negative bacilli has responded relatively poorly to regimens featuring aminoglycoside and older β-lactam agents, with high morbidity and mortality rates (16, 36), although clinical outcomes have substantially improved recently with extended-spectrum cephalosporin therapy.

Both the experimental animal models of endocarditis and bacterial meningitis provide a severe test of the efficacy of any antimicrobial regimen. For example, in experimental bacterial endocarditis the infection is induced in rabbits or rats by the placement of a polyethylene catheter across a heart valve (e.g., aortic or tricuspid) to induce marantic endocarditis, followed by the seeding of the sterile vegetation with a large bacterial challenge administered either intravenously or through the catheter. The catheter is generally secured in place for the duration of the study. This procedure, after the methods of Freedman and Valone (24), reliably induces endocarditis in catheterized animals, with infected vegetations containing $>10^7$ to 10^9 bacteria per g. Thus, antimicrobial regimens must sterilize vegetations containing high bacterial densities in the presence of an indwelling foreign body (intraventricular catheter). Moreover, many of the organisms in the interstices of the experimental valvular vegetation appear to be in a metabolically inactive phase of growth (stationary [22]).

Experimental gram-negative bacillary meningitis presents many of the same therapeutic difficulties as experimental endocarditis. The infection is generally initiated by the classic model of Dacey and Sande in which rabbits are immobilized in a stereotactic frame, with meningitis being induced by the direct intracisternal injection of a large bacterial inoculum ($\sim10^5$ to 10^7 CFU [19]). At 12 to 24 h after bacterial inoculation, the antimicrobial regimens are initiated intravenously, with direct access into the basilar cistern providing serial measurements of bacterial counts and cerebrospinal fluid (CSF) chemistries, cell counts, and antimicrobial levels. As with experimental endocarditis, meningitis provides a rigorous test of antimicrobial efficacy because of the high bacterial densities achieved in the subarachnoid space before therapy ($\sim10^6$ to 10^8 CFU/ml) and the low CSF pH, diminishing the activities of certain antimicrobial classes (e.g., aminoglycosides [45]).

The newer quinolone antibiotics appear to be particularly promising for the treatment of experimental bacterial endocarditis and meningitis because these agents possess features in vitro that potentially may overcome the therapeutic problems presented in these infection model systems. First, the newer quinolones do not exhibit a pronounced inoculum effect in vitro (an effect seen with many aminoglycosides and β-lactams [48]), remaining bactericidal at challenge inocula of $\sim10^6$ to 10^8 CFU/ml, depending upon the organism (3, 8, 41). Second, although strain specific, the newer quinolones have in some circumstances bactericidal activity against stationary-phase bacteria in vitro (51). Third, these agents are usually relatively active at acidic pHs (2, 3). Last, the quinolones often exhibit concentration-dependent killing kinetics in vivo, such that increases in administered doses result in roughly proportional increases in bacterial killing in vivo (27).

Since there is little clinical information concerning the efficacies of the newer quinolones in the therapy of human bacterial endocarditis and meningitis at this time, this chapter will concentrate predominantly on a detailed analysis of the results of quinolone treatment protocols in the discriminative animal models outlined above. This chapter will emphasize the lessons learned from these animal models that may be relevant to the treatment of human endocarditis and meningitis.

Endocarditis

Tissue penetration of newer quinolones in endocarditis

There is little information on the clinical efficacy of the newer quinolones in human endocarditis. Data from the experimental endocarditis model, as well as from human pharmacokinetic studies, have, however, suggested that these agents may be potentially useful in the treatment or prophylaxis of human valvular infections caused by selected organisms. Using pefloxacin as a model, Contrepois et al. (17) showed that this agent achieved peak levels of ~20 and 40 μg/g in rabbit aortic valves of normal animals and those with *E. coli* endocarditis, respectively, after an intravenous dose of 15 mg/kg. Moreover, this dosage given repeatedly at 15 mg/kg every 12 h was effective in significantly reducing intravegetation *E. coli* titers in animals with endocarditis, as compared with untreated controls. Work from our laboratory examined the ability of pefloxacin to penetrate vegetations of animals with aortic endocarditis due to MRSA; we showed that after single intravenous doses of pefloxacin at 20 or 40 mg/kg, mean peak concentrations within aortic vegetations were only ~1.5 and 3 μg/g, respectively (4); in addition, these levels of pefloxacin within vegetations were significantly lower than those achieved after single intravenous doses of vancomycin (15 mg/kg). These latter differences in the ability of pefloxacin and vancomycin to penetrate MRSA-infected aortic vegetations were roughly mirrored in vivo, as vancomycin caused a more rapid decline in intravegetation MRSA densities than did pefloxacin at both dose regimens. The disparities in the levels of drug achievable in vegetations in our study as compared with that of Contrepois et al. may relate either to differences in the route of administration (intravenous infusion [17] versus intravenous bolus [4]) or to methods of preparing vegetations for tissue drug assays. In our study, for example, vegetations after removal from the animals were carefully dried to evaporate extravascular fluid contamination.

As a corollary to the above investigations, Brion et al. (14) studied the penetrability of pefloxacin into abnormal cardiac valves of humans undergoing open heart surgery for prosthetic valve insertion. They showed that a single dose of intravenous pefloxacin at ~10 to 15 mg/kg (800 mg) achieved mean peak levels of ~2 to 9 μg/g in abnormal aortic or mitral valves for 4 to 24 h after infusion; additionally, the mean ratios of valve to plasma pefloxacin concentrations were often >1.0, suggesting complete penetrance from the vascular to the tissue compartment. In addition, the levels of pefloxacin achieved in vegetations were well above the MBC for 90% of strains for most important valvular pathogens, including the viridans and enterococcal streptococci, MRSA and MSSA, and *P. aeruginosa.* These data from human endocarditis and the experimental endocarditis models suggest that in selected circumstances newer quinolones with long half-lives (such as pefloxacin) may be useful in the prophylaxis or treatment of clinical endocarditis. However, formal chemoprophylactic studies in animals have not been reported to date, and there is little published information on utilizing such agents in humans with confirmed bacterial endocarditis.

Experimental endocarditis

P. aeruginosa. Infective endocarditis due to *P. aeruginosa* remains an important infection among parenteral drug abusers in the United States, particularly in Detroit and Chicago (39, 43). Recent experiences in treating this infection in humans and experimental animals have emphasized the difficulties in achieving cures with aminoglycoside–β-lactam regimens alone, especially with left-sided valve involvement (9, 39). Such regimens have been limited by primary drug failures, bacteriologic relapses, and the development of antibiotic resistance in vivo (9, 33, 39). Our laboratory and others have recently reported on the efficacies of newer quinolones in experimental *P. aeruginosa* endocarditis.

Ingerman et al. examined the relative efficacies of ciprofloxacin versus ceftazidime or an investigational β-lactam (BMY-28142), with and without gentamicin, in the rat model of experimental aortic endocarditis due to *P. aeruginosa* (31). In vitro, ciprofloxacin exhibited a significantly faster onset of bactericidal action against the infecting pseudomonal strain than did either β-lactam (with or without gentamicin added). For example, after 2 h of incubation, ciprofloxacin (1 μg/ml) lowered the initial bacterial inoculum from ~8 to ~4 $\log_{10}$ CFU/ml; in contrast, all β-lactam regimens lowered the initial pseudomonal inoculum by less than 10-fold. Similarly, when infected aortic valvular vegetations were exposed to drugs in vitro, ciprofloxacin produced bacterial killing more rapidly or to a greater extent

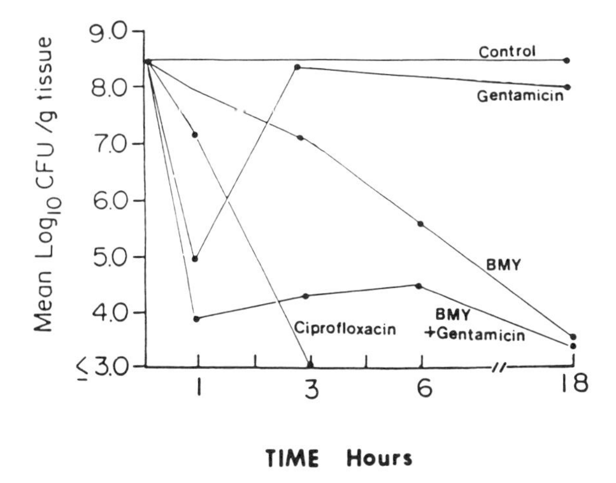

Figure 1. Fate of *P. aeruginosa* in vegetations suspended in broth containing 60 μg of BMY-28142 per ml, 6 μg of gentamicin per ml, 2 μg of ciprofloxacin per ml, BMY-28142 plus gentamicin, or broth without drugs. (Reproduced with permission of the University of Chicago Press [31].)

than did β-lactam regimens (Fig. 1). Of importance, these investigators confirmed that ciprofloxacin (but not the β-lactams) exerted a persistent suppression of pseudomonal growth in vitro ("postantibiotic effect") for ≥2.5 h after brief exposure (2 h) to the drug (Fig. 2). In vivo, ciprofloxacin therapy produced significantly greater reductions in bacterial density in vegetations and more sterile vegetations than did any of the β-lactam regimens tested. This therapeutic difference occurred despite the β-lactam agents being present within infected vegetations at concentrations above the MBC for periods of time similar to those in the ciprofloxacin-treated animals. This study confirmed the importance of pharmacodynamic parameters such as postantibiotic effect and time above MBC at the tissue infection site as dual determinants of β-lactam therapeutic outcome in experimental infections (25).

Our laboratory studied the efficacy of ciprofloxacin in the rabbit models of experimental tricuspid and aortic endocarditis caused by *P. aeruginosa* (3, 6, 7). The comparative drug regimen for the quinolone featured a combination of an aminoglycoside (amikacin or netilmicin) plus an antipseudomonal penicillin (azlocillin) synergistically active against the infecting bacterial strain. In tricuspid valve endocarditis, ciprofloxacin and a combination of amikacin and azlocillin were equally effective in reducing mortality, preventing pulmonary infarction, and reducing mean pseudomonal densities in vegetations as compared with untreated controls. Also, both regimens were equivalent in preventing bacteriologic relapse after therapy

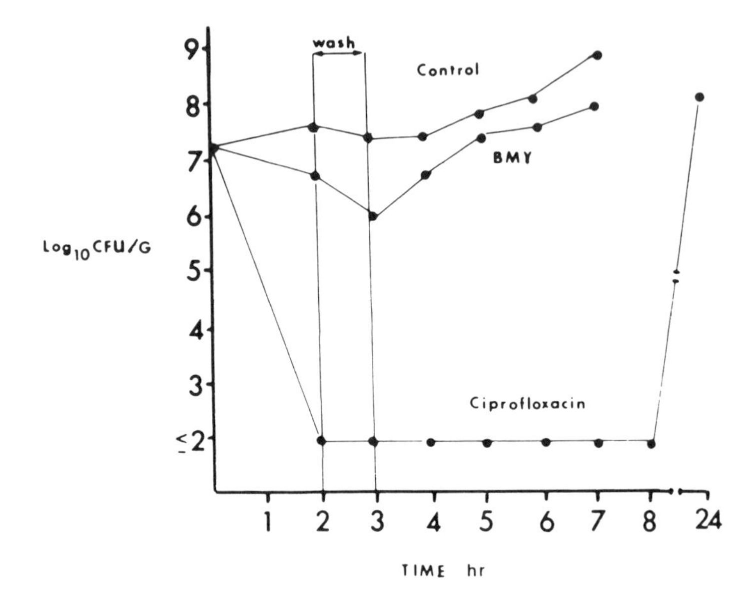

Figure 2. Postantibiotic effect of *P. aeruginosa* after 2-h exposure to BMY-28142 (15 μg/ml) or ciprofloxacin (0.5 μg/ml). Drugs were removed by repeated washing. (Reproduced with permission of the University of Chicago Press [31].)

was discontinued, a substantial problem in the therapy of human pseudomonal endocarditis (6). Moreover, no development of resistance to ciprofloxacin, amikacin, or azlocillin was observed in vivo. The addition of amikacin did not enhance the outcome of ciprofloxacin therapy alone in vivo, despite the frequently additive bactericidal effects of such combinations in vitro (28). In aortic valve endocarditis, ciprofloxacin was significantly more effective than a combination of netilmicin and azlocillin in sterilizing vegetations, reducing pseudomonal densities within vegetations, and preventing bacteriologic relapses after therapy (Table 1). In contrast, both the ciprofloxacin and netilmicin-plus-azlocillin regimens were equally effective in sterilizing renal abscesses. Resistance to azlocillin, but not to ciprofloxacin or netilmicin, was occasionally observed in vivo among *P. aeruginosa* strains isolated from vegetations during week 2 of treatment. Development of resistance to β-lactams (e.g., ceftazidime) in week 2 of therapy has been previously noted in this model and has been associated with the selection of stably derepressed mutants with constitutive overproduction of chromosomal β-lactamase (9). The ability of the aminoglycoside and β-lactam combination to sterilize renal, but not endocardial, infection in this

Table 1. Mortality rates of antibiotic-treated and control rabbits with aortic valve endocarditis due to *P. aeruginosa*[a]

Treated group	Mortality (%)
Control[b, c]	12/19 (63)
Netilmicin + azlocillin[c]	11/23 (47)
Ciprofloxacin[b]	8/23 (34)

[a]Reproduced with permission of the *Journal of Antimicrobial Chemotherapy* and the British Society of Chemotherapy (3).
[b]$P < 0.05$.
[c]Not significantly different.

model suggested differences in the penetration of these agents into renal parenchyma and cardiac vegetations. Strunk et al. (46) also confirmed the efficacy of ciprofloxacin in experimental aortic endocarditis due to *P. aeruginosa*. These investigators showed that ciprofloxacin was as effective as the synergistically active combination of tobramycin and azlocillin in reducing intrarenal and intravegetation pseudomonal densities, as well as rendering these tissues sterile. Moreover, none of the surviving bacteria from renal tissue or vegetations was found to be resistant to the pertinent study drug. There was also a suggestion that relapses were less likely to occur after ciprofloxacin therapy than after therapy with the combination of tobramycin and azlocillin, although this difference did not reach statistical significance. The equal efficacy of the quinolone and the aminoglycoside–β-lactam regimens in this latter study is in contrast to the inferior outcome with aminoglycosides and β-lactams in our study of aortic endocarditis cited above. This disparity may relate to the differences in the model utilized. In the study of Strunk et al. (46), the catheter was removed approximately 1 h after its placement across the aortic valve and the injection of the pseudomonal inoculum through the catheter; in our study the catheter remained within the left ventricle for the duration of the study. It seems likely that the persistent presence of the foreign body adversely affected the ability of the aminoglycoside–β-lactam regimen to eradicate *P. aeruginosa* from vegetations.

Our laboratory has recently reported on the efficacy of pefloxacin, a newer quinolone with a prolonged half-life, in experimental aortic valve *P. aeruginosa* endocarditis (5). Pefloxacin was compared with a combination of high doses of amikacin and ceftazidime which exhibited bactericidal synergy against the infecting pseudomonal strain in vitro. Pefloxacin and the combination regimen both significantly reduced bacterial densities within vegetations as compared with untreated controls. As seen in our previous models of aortic pseudomonal endocarditis treated with β-lactams, bacteria isolated from vegetations after 2 weeks of therapy exhibited ceftazidime resistance related to the constitutive overproduction of β-lactamase (3, 9). However, as opposed to experiences with ciprofloxacin therapy in this model, intravegetation isolates from pefloxacin recipients showed significant increases in pefloxacin MICs as early as day 4 of therapy. These increases ranged from four- to eightfold. For these pefloxacin-resistant variants increases in ciprofloxacin MICs, as well as pleiotropic resistance to ticarcillin and chloramphenicol, but not to amikacin, ceftazidime, or tetracycline, were also exhibited (Table 2). Because of the similarity in the profile of MIC increases for our strain with that induced for *E. coli* with the *cfxB, nfxB,* or *norB* gene mutation associated with decreased porin protein OmpF (29, 30), it appeared possible that altered drug permeability may be the underlying mechanism of resistance in our strain. We have recently shown that the major mechanism of quinolone resistance in our endocarditis variants was "target insensitivity" (altered DNA gyrase). Although a drug permeability defect was also noted, MICs correlated closely with IC_{50} for inhibition of DNA synthesis (S. Chamberland, A. S. Bayer, T. Schollardt, S. A. Wong, and L. E. Bryan, *Antimicrob. Agents Chemother.*, in press).

Over the last decade a number of β-lactamase-stable, extended-spectrum cephalosporins have been developed (e.g., cefotaxime and ceftazidime). Recently, increasing numbers of reports have documented the emergence of resistance to these agents by selection of mutants with derepressed β-lactamase overproduction (33, 38, 40, 42); in addition, several of these multiply β-lactam-resistant strains showed cross resistance to aminoglycosides (38). This resistance phenomenon has been most commonly seen among bacterial genera within the expanded spectrum of these newer β-lactams, especially *Enterobacter, Serratia,* and *Pseudomonas* (40); the quinolones generally retain good in vitro efficacy against these β-lactam-resistant mutants. We utilized the experimental aortic endocarditis model to evaluate the efficacy of ciprofloxacin against such strains in vivo (7). The infecting strain was a multiply β-lactam-resistant *P. aeruginosa* variant derepressed for constitutive β-lactamase overproduction (10). Ciprofloxacin significantly lowered pseudomonal densities within vegetations and rendered significantly more animals abacteremic and with sterile vegetations as compared with recipients of ceftazidime or no therapy. This study suggested that the newer quinolones warrant further evaluation in the treatment of multiply drug-resistant, gram-negative bacillary infections due to quinolone-susceptible strains.

Enterobacter aerogenes. *Enterobacter* species rarely cause endocarditis in humans. Such strains, however, are relatively common causes of serious nosocomial infections, especially in patients residing in intensive care units (13). Boscia et al. (11, 12) performed two studies to evaluate newer oral

Table 2. Cross and pleotropic resistances among selected intravegetation pseudomonal variants isolated during pefloxacin therapy for experimental aortic endocarditis[a]

Agent	MIC (μg/ml) for:			
	PA96 (parent)	22_V[b]	$22V_3$[b]	$23V_1$[c]
Pefloxacin	0.19	1.56	0.78	1.56
Ciprofloxacin	0.19	1.56	1.56	1.56
Ceftazidime	2	2	2	2
Amikacin	2	2	2	2
Chloramphenicol	5	1,600	800	800
Tetracycline	50	50	100	50
Ticarcillin	16	16	64	32

[a]Reproduced with permission of the American Society for Microbiology (5).
[b]Pseudomonal variant isolated on pefloxacin-containing agar (2 μg/ml) after 4 days of pefloxacin therapy.
[c]Pseudomonal variant isolated on pefloxacin-containing agar after 10 days of pefloxacin therapy.

quinolone agents in *E. aerogenes* aortic endocarditis as a severe test of drug efficacies in this difficult model of bacteremic infection. These investigators first compared oral enoxacin with parenteral cefoperazone in *Enterobacter* endocarditis (11). In vitro both enoxacin and cefoperazone exerted a rapid and substantial bactericidal effect against the infecting *Enterobacter* strain. However, high-dose enoxacin (100 mg/kg, given every 6 h by the oral syringe method) was significantly better at reducing intravegetation bacterial densities than was lower-dose enoxcacin (25 mg/kg, given every 6 h) or cefoperazone (60 mg/kg, given every 6 h intramuscularly). The data of the authors suggested that the longer half-life of enoxacin (~3 h) versus cefoperazone (~1 h) was probably important in the therapeutic differences observed. In a similar study Boscia et al. (12) compared the efficacy of two oral quinolones, enoxacin and difloxacin, with that of parenterally administered cefoperazone in experimental aortic endocarditis caused by *E. aerogenes*. As in their previous study above, all three agents were active in vitro against the infecting *Enterobacter* strain, although difloxacin exhibited the most rapid and complete killing of this strain in time-kill experiments. Difloxacin (100 mg/kg, given orally every 12 h) was significantly better at reducing intravegetation *Enterobacter* densities than was enoxacin given at the same dose regimen or cefoperazone. As before, the authors ascribed the better efficacy of difloxacin in this model to the longer half-life of this agent (~3.5 h) as compared with enoxacin (~2.3 h) and cefoperazone (0.6 h). They also concluded that the relatively inferior outcome of enoxacin therapy in this study as compared with their prior report was related to differences in the enoxacin dosing intervals utilized in the two studies (every 6 versus every 12 h).

E. coli. Contrepois et al. (17) reported on the efficacy of 3 days of pefloxacin given at 30 mg/kg per day intramuscularly in experimental *E. coli* aortic endocarditis. This regimen significantly lowered intravegetation *E. coli* densities as compared with untreated controls; unfortunately, these investigators did not include a comparative therapy group for pefloxacin.

MSSA and MRSA. Several studies in the animal model of experimental endocarditis have evaluated the efficacy of several newer quinolones against both MSSA and MRSA strains. Sullam et al. (47) compared pefloxacin with cephalothin in MSSA endocarditis and with vancomycin in MRSA endocarditis. Cephalothin and pefloxacin were equally effective in reducing mortality and intravegetation MSSA densities as compared with untreated controls. Likewise, both vancomycin and pefloxacin were significantly better at reducing mortality and vegetation MRSA densities versus untreated controls. In a very similar study Carpenter et al. (15) confirmed that ciprofloxacin was as effective as nafcillin in experimental MSSA aortic endocarditis and as effective as vancomycin in experimental MRSA aortic endocarditis in reducing intravegetation staphylococcal densities. Kaatz et al. (34) also confirmed the equivalent efficacies of a new quinolone (in this case intravenously administered ciprofloxacin) and vancomycin in reducing intravegetation MRSA densities as compared with untreated controls. In addition, this study confirmed a significant reduction in endocarditis-related intrarenal and intrasplenic MRSA infection by both ciprofloxacin and vancomycin (Table 3). Moreover, Kaatz et al. confirmed that the multiple-dose pharmacokinetics of ciprofloxacin in infected animals were markedly different from those which were predicted from single-dose pharmacokinetics in uninfected rabbits. Of particular importance was the higher-than-predicted peak ciprofloxacin levels in the serum of rabbits with endocarditis after multiple drug doses. This difference might have resulted from a decrease in clearance of this agent, but the contribution of endocarditis to renal dysfunction and higher ciprofloxacin levels was not addressed. The authors found no development of ciprofloxacin resistance (MIC, >5 μg/ml) among surviving MRSA cells in cardiac vegetations.

Gilbert et al. studied the comparative efficacies of an orally administered quinolone (enoxacin) versus intravenously administered vancomycin in MRSA aortic endocarditis in rabbits (26). They observed that both agents significantly lowered intravegetation MRSA densities over the 5-day treatment period as compared with untreated controls, although the bactericidal effect of enoxacin in vivo was seen earlier than that of vancomycin (3 versus 5 days of treatment). We have recently examined the efficacies of intravenously administered pefloxacin (40 or 80 mg/kg per day) and vancomycin (30 mg/kg per day) in experimental MRSA aortic endocarditis (4). Our results with pefloxacin differ from those of Gilbert et al. As in their study, both the higher-dose pefloxacin regimen and vancomycin significantly lowered intravegetation MRSA densities after 6 days of treatment as compared with

Table 3. Counts of MRSA 494 in vegetations and tissues[a]

Treatment group[b]	Mean ± SD $\log_{10}$ CFU/g[c]		
	Vegetation	Kidney	Spleen
Ciprofloxacin (27)	3.37 ± 1.58 (14)	2.01 ± 0.93 (21)	1.56 ± 0.23 (24)
Vancomycin (22)	3.56 ± 1.67 (12)	1.91 ± 1.01 (18)	1.79 ± 0.97 (21)
No treatment (controls; 17)	8.45 ± 0.70	5.27 ± 1.55	5.15 ± 0.58

[a]Reproduced with permission of the American Society for Microbiology (34).
[b]Numbers within parentheses are the number of rabbits.
[c]Numbers within parentheses are the number of rabbits rendered culture negative. No significant difference in bacterial counts between the ciprofloxacin and vancomycin treatment groups was observed. Significant differences were noted between the ciprofloxacin-treated animals and the controls and between the vancomycin-treated animals and the controls in all cases ($P < 0.001$).

untreated controls. We, however, observed that the onset of bactericidal activity in vivo was more rapid with vancomycin than with pefloxacin therapy. Significant reductions in intravegetation MRSA densities were seen by day 3 of treatment only in vancomycin recipients. We examined various pharmacokinetic and pharmacodynamic parameters that explain in part the superior effect of vancomycin as compared with pefloxacin in this model. The in vitro and ex vivo (i.e., intravegetation) postantibiotic effects were virtually identical for pefloxacin and vancomycin against the infecting MRSA strain. Moreover, the trough bactericidal titers in serum were significantly greater in pefloxacin recipients than in animals given vancomycin, reflecting the longer half-life of pefloxacin in serum. Of interest, despite the superior pharmacokinetics of pefloxacin in serum, the penetration of vancomycin in vegetations exceeded that of pefloxacin by 7- to 11-fold. This finding supports the concept of Gengo et al. (25) and suggests that levels of antimicrobial agents achievable in vegetations may correlate better with therapeutic outcome in experimental endocarditis than the levels achievable in serum.

Boscia et al. (J. Boscia, W. Kobasa, and D. Kaye, *Program Abstr. 27th Intersci. Conf. Antimicrob. Agents Chemother.*, abstr. no. 934, 1987) reported on the comparative efficacies of oral difloxacin and enoxacin versus parenteral cefazolin in experimental MSSA aortic endocarditis in rabbits. As with their studies in experimental *Enterobacter* endocarditis, difloxacin therapy yielded significantly greater reductions of intravegetation bacterial densities than did enoxacin treatment, presumably reflecting the longer elimination half-life and higher achievable levels for difloxacin in serum.

Staphylococcus epidermidis. One study of newer quinolones in the treatment of experimental methicillin-resistant *S. epidermidis* endocarditis was cited by Rouse et al. (M. S. Rouse, R. W. Wilcox, M. J. Engler, N. K. Henry, J. E. Geraci, and W. R. Wilson, *26th ICAAC*, abstr. no. 289, 1986). These investigators evaluated ciprofloxacin (alone or with rifampin) in comparison with teichoplanin or vancomycin (alone or in combination with gentamicin, rifampin, or both). Their data showed that ciprofloxacin alone was more effective at reducing intravegetation methicillin-resistant *S. epidermidis* densities in vivo than was vancomycin alone or combined with gentamicin; ciprofloxacin combined with rifampin and vancomycin combined with gentamicin and rifampin were equally effective in reducing intravegetation methicillin-resistant *S. epidermidis* densities and were the two most active regimens studied.

Human endocarditis

Daikos et al. (20) have recently reported on experiences with oral quinolone therapy of two patients with infective endocarditis caused by *P. aeruginosa*. One patient was an intravenous drug addict with mitral valve endocarditis refractory to two mitral valve replacements and three courses of combination parenteral antipseudomonal regimens. Oral ciprofloxacin therapy with daily doses of between 1 and 2 g was continued for 3.5 months (total dose, 150 g). Blood cultures were sterilized during therapy, and fever abated. The patient, however, expired with fulminant endocarditis approximately 1 month after a reduction in the dosage of ciprofloxacin necessitated by drug-induced transaminitis. At autopsy, mitral vegetations contained enormous densities of *P. aeruginosa* ($\sim 10^8$ CFU/g of tissue).

The second patient with a permanent cardiac pacemaker developed mural endocarditis due to *P. aeruginosa*. He failed treatment with parenteral antipseudomonal agents and was treated with 1.5 g of oral ciprofloxacin per day for nearly 2 years, with sterilization of blood cultures and partial abatement of fever. The patient, however, failed a test of cure after 1 year of therapy during a planned discontinuation of the drug. There occurred a rapid return of pseudomonal bacteremia and fever, and progressive biventricular heart failure and anemia developed. At postmortem examination, right ventricular mural endocarditis, as well as tricuspid and aortic valvular endocarditis, with vegetation bacterial densities of $\sim 10^5$ CFU was found. Of interest, for pseudomonal isolates after quinolone therapy in both patients, selective increases in the MIC of ciprofloxacin (four- to eightfold higher than for the pretreatment strain) were demonstrated, without increases in the MICs of other classes of antibiotics.

Summary

The newer quinolones, especially ciprofloxacin, have performed well in the experimental animal models of *P. aeruginosa* endocarditis. Most studies have documented that ciprofloxacin is equally or more effective than synergistically active combinations of aminoglycosides and β-lactams in the treatment of this infection. Reports documenting the development of increases in the MICs of the newer quinolone agents during the therapy of experimental pseudomonal endocarditis are disturbing, and such studies need to be extended to define the relative prevalence of this problem. The oral quinolones with extended elimination half-lives (e.g., difloxacin) look promising for the therapy of bacteremic *Enterobacter* infections, based on the results of the experimental endocarditis treatment studies.

The newer quinolones also appear to be highly active in vivo against both MSSA and MRSA strains in experimental endocarditis infection models. Their efficacy in general appears to be roughly equivalent to that of standard antimicrobial regimens, such as vancomycin for MRSA and semisynthetic penicillins or cephalosporins for MSSA endocarditis. Among the quinolones evaluated to date, parenteral ciprofloxacin and pefloxacin and oral difloxacin appear to be the most active agents in experimental *S. aureus* endocarditis. The development of quinolone resistance in vivo does not seem to be a major problem in the therapy of experimental MSSA or MRSA endocarditis, although few of the studies have systematically looked for this phenomenon. Further evaluation is needed. Moreover, there is little, if any, experimental information concerning the utility of the newer quinolone antibiotics in the prophylaxis of MRSA or MSSA endocarditis; such data need to be generated to assess the potential role of these agents in human endocarditis. Information on the efficacy of the newer quinolones in preventing experimental *S. epidermidis* endocarditis remains scanty to date.

Meningitis

Penetrance of newer quinolones into CSF of humans

Studies of pefloxacin penetration into CSF in the presence of uninflamed and inflamed meninges indicate that this agent penetrates well into CSF under both circumstances. Dow et al. (21) studied the transfer kinetics of a single intravenous dose of pefloxacin (400 mg) into the ventricular CSF of nine patients with hydrocephalus, but without overt meningitis. The apparent half time of pefloxacin transfer from plasma to ventricular CSF was 1.26 h, while the elimination half-life from ventricular CSF was 13.4 h (similar to the elimination half-life of pefloxacin from plasma). Mean pefloxacin concentrations in CSF exceeded the MIC for most pefloxacin-susceptible gram-negative and gram-positive pathogens for at least 12 h after a dose. Wolff et al. (50) studied the penetration of orally or intravenously administered pefloxacin into CSF of 15 patients with bacterial meningitis or ventriculitis receiving other antibiotics for their infection; 2 patients eventually received pefloxacin therapeutically after failure of the primary antibiotic regimen. With intravenously administered pefloxacin, mean peak and trough levels in CSF were 1.28 and 3.54 μg/ml, respectively, well above the MIC for 90% of pefloxacin-susceptible strains; similarly, when administered orally, the mean peak concentrations in CSF were ~10 μg/ml in recipients of the higher-dose oral regimen (15 mg/kg). Thus, pefloxacin appears to penetrate intact and inflamed meninges quite well, whether administered orally or intravenously. This agent also achieves over the full dosing interval levels in all compartments of CSF that are above the MIC for most of the important meningeal pathogens (except pneumococci).

Wolff et al. also recently studied the penetration of intravenously administered ciprofloxacin into CSF of patients with purulent meningitis and ventriculitis (49). Twenty-three adult patients received three successive doses of ciprofloxacin (200 mg intravenously), 12 h apart. During the most active phase of the meningitis (based on CSF formulas), mean ciprofloxacin levels in lumbar CSF ranged from 0.35 to 0.56 μg/ml over an 8-h period after the dose; these concentrations in CSF are above the MICs and MBCs for most members of the family *Enterobacteriaceae*. During the convalescent phase of the meningitis, when the meninges were presumably less inflamed, mean ciprofloxacin levels in CSF ranged from 0.15 to 0.27 μg/ml. The penetrance of CSF by ciprofloxacin ranged from 6.5 to 16% during the acute meningitis phase and 4 to 9.9% during the later stages of infection. Peak ciprofloxacin levels in ventricular CSF ranged from ~0.27 to 0.43 μg/ml. These data suggest that, similar to pefloxacin, ciprofloxacin diffuses well into the lumbar and ventricular CSF of patients with bacterial meningitis, during both the acute and the convalescent phase of infection.

Newer quinolones in experimental meningitis

Most of the experimental meningitis studies in which the newer quinolones have been evaluated have involved either *E. coli* or *P. aeruginosa* meningitis in rabbits. Shibl et al. (44) examined the comparative efficacies of pefloxacin, cefotaxime, and chloramphenicol in experimental *E. coli* meningitis. The mean penetrance of pefloxacin from plasma to CSF was ~50%, similar to data generated from the human studies cited above; in contrast, the mean penetrances of cefotaxime and chloramphenicol were ~11 and 22%, respectively (Table 4). Cefotaxime and pefloxacin (30 mg/kg per h) were equally effective in diminishing intracisternal bacterial densities of *E. coli*, with both agents being superior to chloramphenicol (Table 5). Kill curves of the infecting *E. coli* strain determined in vitro confirmed that the optimal bactericidal effect was achieved in broth or infected CSF when pefloxacin concentrations were ~200 times above the MBC for the strain (0.125 μg/ml). Of interest, this phenomenon was mirrored in vivo, as maxi-

Table 4. Concentrations of pefloxacin, cefotaxime, and chloramphenicol in serum and CSF of rabbits with *E. coli* meningitis[a]

Dosage (mg/kg per h)	Concn[b] (μg/ml) in: Serum	CSF	% Penetration
Pefloxacin			
30.0	45.8 ± 19.7	19.1 ± 4.1	37.4 ± 9.8
15.0	23.5 ± 7.0	12.7 ± 3.4	55.0 ± 17.7
5.0	10.0 ± 2.9	4.8 ± 0.8	53.6 ± 19.1
2.5	5.6 ± 0.4	2.7 ± 0.4	48.4 ± 7.2
1.0	1.9 ± 0.5	0.9 ± 0.2	46.3 ± 8.6
Cefotaxime (100.0)	142.8 ± 37.6	16.4 ± 6.1	11.1 ± 1.0
Chloramphenicol	52.4 ± 8.6	11.3 ± 1.5	22.3 ± 1.5

[a]Reproduced with permission of the American Society for Microbiology (44).
[b]Mean ± standard deviation.

mal killing of intracisternal *E. coli* was achieved at pefloxacin concentrations in CSF of ~20 μg/ml. Sobieski and Scheld (M. W. Sobieski and W. M.Scheld, *26th ICAAC,* abstr. no. 542, 1986) evaluated antimicrobial penetrance into CSF and early intracisternal bacterial killing after single parenteral doses of ofloxacin, ciprofloxacin, and cefotaxime in experimental *E. coli* meningitis and ciprofloxacin and ceftazidime in experimental *P. aeruginosa* meningitis. In experimental *E. coli* meningitis ciprofloxacin and ofloxacin penetrated equally well into cisternal CSF from plasma (~30%). Mean peak concentrations of ofloxacin and ciprofloxacin in CSF were ~30 and 100 times the MBC for the infecting strain; this value for cefotaxime was

Table 5. Bactericidal activity of pefloxacin, cefotaxime, and chloramphenicol in CSF of rabbits with *E. coli* meningitis[a]

Dosage (mg/kg per h)	Killing rate (log_{10} CFU/ml per h)[b]
Pefloxacin	
30.0 (n = 4)	−0.77 ± 0.18
15.0 (n = 7)	−0.31 ± 0.11
5.0 (n = 5)	−0.31 ± 0.14
2.5 (n = 4)	−0.53 ± 0.10
1.0 (n = 4)	−0.88 ± 0.15
Cefotaxime (100.0 [n = 4])	−0.88 ± 0.23
Chloramphenicol (60.0 [n = 2])	−0.10 ± 0.14
Untreated controls (n = 8)	0.22 ± 0.10

[a]Reproduced with permission of the American Society for Microbiology (44).
[b]Mean ± standard deviation.

~500 times the MBC. Both quinolone agents and cefotaxime were highly active in reducing mean intracisternal *E. coli* densities at 8 h after the single-dose administrations (reductions of ~3 to 5 $\log_{10}$ CFU/ml). In *P. aeruginosa* experimental meningitis, ceftazidime appeared to have a better pharmacokinetic profile within CSF than did ciprofloxacin after single-dose administrations. The mean penetrances of ceftazidime and ciprofloxacin into CSF were ~38 and 20%, respectively; the mean peak concentrations of ceftazidime and ciprofloxacin in CSF were ~10 and 1.2 times the MBC for the infecting strain, respectively. Despite this difference in pharmacokinetic profiles, both ciprofloxacin and ceftazidime were equally effective in the early reduction of intracisternal *P. aeruginosa* densities after single-dose drug administrations. In a similar model system Hackbarth et al. (27) studied the relative efficacies of ciprofloxacin and the synergistic combination of ceftazidime and tobramycin in experimental *P. aeruginosa* meningitis. Both regimens penetrated the cisternal CSF well (relative to plasma, ~11 to 40% for ciprofloxacin, depending on dosage; ~23% for ceftazidime and ~33% for tobramycin); mean peak ciprofloxacin concentrations in CSF exceeded the MBC for the infecting strain when the drug was administered at the higher doses (15 to 30 mg/kg per h). Both the quinolone and the aminoglycoside–β-lactam regimens were equally efficacious at reducing mean intracisternal *P. aeruginosa* counts (reductions of ~0.5 $\log_{10}$ CFU/ml per h).

Efficacy of newer quinolones in human bacterial meningitis

There have been several brief descriptions of patients with purulent meningitis treated with the newer quinolones. Wolff et al. (49) mentioned two patients with *Enterobacter cloacae* meningitis who received oral pefloxacin combined with intrathecal gentamicin after failing treatment with parenteral cefotaxime and gentamicin; no delineation of treatment outcome was provided by the authors, however. Millar et al. (37) described a patient with neurosurgically related *P. aeruginosa* meningitis who was cured with 14 days of intravenous ciprofloxacin and tobramycin. Lastly, Isaacs et al. have reported the successful treatment of an infant with *P. aeruginosa* ventriculitis with intravenously administered ciprofloxacin (32) and of a patient developing *Morganella morganii* meningitis after laminectomy with intravenously administered pefloxacin (R. D. Isaacs and R. B. Ellis-Peglar, Letter, *J. Antimicrob. Chemother.* **20**:769–770, 1987).

Summary

Data on the pharmacokinetics of the newer quinolones in CSF of humans with and without inflamed meninges are relatively limited, but appear to indicate that agents with long half-lives in plasma (e.g., pefloxacin) penetrate CSF well and achieve levels above the MBCs for susceptible strains for the entire dosing interval. There are few detailed clinical data on treating patients with bacterial meningitis with these agents. However, on the basis of the data generated in the animal model of meningitis, gram-negative bacillary meningitis caused by *E. coli* or *P. aeruginosa* would appear to be a prime candidate for initial clinical studies of the efficacy of the newer quinolones in human bacterial meningitis.

Literature Cited

1. **Auckenthaler, R. N., M. Michea-Hamzehpour, and J. C. Perchere.** 1986. *In vitro* activity of newer quinolones against aerobic bacteria. *J. Antimicrob. Chemother.* **17**(Suppl.): 29–39.
2. **Bauernfeind, A., and C. Petermuller.** 1983. *In vitro* activity of ciprofloxacin, norfloxacin and nalidixic acid. *Eur. J. Clin. Microbiol.* **2:**111–115.
3. **Bayer, A. S., I. K. Blomquist, and K. S. Kim.** 1986. Ciprofloxacin in experimental aortic valve endocarditis due to *Pseudomonas aeruginosa. J. Antimicrob. Chemother.* **17:**641–649.
4. **Bayer, A. S., D. P. Greenberg, and J. Yih.** 1988. Correlates of therapeutic efficacy in experimental methicillin-resistant *Staphylococcus aureus* endocarditis. *Chemotherapy* (Basel) **34:**46–55.
5. **Bayer, A. S., L. Hirano, and J. Yih.** 1988. Development of β-lactam resistance and increased quinolone MICs during therapy of experimental *Pseudomonas aeruginosa* endocarditis. *Antimicrob. Agents Chemother.* **32:**231–235.
6. **Bayer, A. S., K. Lam, D. Norman, K. S. Kim, and J. Morrison.** 1985. *In vivo* efficacy of azlocillin and amikacin versus ciprofloxacin with and without amikacin in experimental right-sided endocarditis due to *Pseudomonas aeruginosa. Chemotherapy* (Basel) **32:** 364–373.
7. **Bayer, A. S., P. Lindsay, J. Yih, L. Hirano, D. Lee, and I. K. Blomquist.** 1986. Efficacy of ciprofloxacin in experimental aortic valve endocarditis caused by a multiply β-lactam-resistant variant of *Pseudomonas aeruginosa* stably derepressed for β-lactamase production. *Antimicrob. Agents Chemother.* **30:**528–531.
8. **Bayer, A. S., D. Norman, and D. Anderson.** 1985. Efficacy of ciprofloxacin in experimental arthritis caused by *Escherichia coli—in vitro-in vivo* correlations. *J. Infect. Dis.* **152:**811–816.
9. **Bayer, A. S., D. Norman, and K. S. Kim.** 1985. Efficacy of amikacin and ceftazidime in experimental aortic valve endocarditis due to *Pseudomonas aeruginosa. Antimicrob. Agents Chemother.* **28:**781–785.
10. **Bayer A. S., J. Peters, T. R. Parr, L. Chan, and R. E. W. Hancock.** 1987. Role of β-lactamase in in vivo development of ceftazidime resistance in experimental *Pseudomonas aeruginosa* endocarditis. *Antimicrob. Agents Chemother.* **31:**253–258.
11. **Boscia, J. A., W. D. Kobasa, and D. Kaye.** 1985. Enoxacin compared with cefoperazone for the treatment of experimental *Enterobacter aerogenes* endocarditis. *Antimicrob. Agents Chemother.* **27:**708–711.
12. **Boscia, J. A., W. D. Kobasa, and D. Kaye.** 1987. Comparison of difloxacin, enoxacin, and cefoperazone for treatment of experimental *Enterobacter aerogenes* endocarditis. *Antimicrob. Agents Chemother.* **31:**458–460.

13. **Bouza, E., M. Garcia de la Torre, A. Erice, E. Loza, J. M. Diaz-Bnrrega, and L. Buzon.** 1985. *Enterobacter* bacteremia—an analysis of 50 episodes. *Arch. Intern. Med.* **145:**1024–1027.
14. **Brion, N., A. Lessana, F. Mosset, J. J. Lefevre, and G. Montay.** 1986. Penetration of pefloxacin in human heart valves. *J. Antimicrob. Chemother.* **17**(Suppl. B)**:**89–92.
15. **Carpenter, T. C., C. J. Hackbarth, H. F. Chambers, and M. A. Sande.** 1986. Efficacy of ciprofloxacin for experimental endocarditis caused by methicillin-susceptible or -resistant strains of *Staphylococcus aureus. Antimicrob. Agents Chemother.* **30:**382–384.
16. **Cherubin, C. E., J. Marrm, M. F. Sierra, and S. Becker.** 1981. *Listeria* and gram-negative bacillary meningitis in New York City, 1972–1979: frequent causes of meningitis in the adult. *Am. J. Med.* **71:**199–209.
17. **Contrepois, A., C. Daldoss, B. Pangon, J. J. Garaud, M. Kecir, C. Sarrazin, J. M. Valois, and C. Carbon.** 1984. Pefloxacin in rabbits: protein binding, extravascular diffusion, urinary excretion and bactericidal effect in experimental endocarditis. *J. Antimicrob. Chemother.* **14:**51–57.
18. **Crane, L. R., and A. M. Lerner.** 1978. Non-traumatic gram-negative bacillary meningitis in the Detroit Medical Center, 1964–1974. *Medicine* (Baltimore) **57:**197–210.
19. **Dacey, R. G., and M. A. Sande.** 1974. Effect of probenicid on cerebrospinal fluid concentrations of penicillin and cephalosporin derivatives. *Antimicrob. Agents Chemother.* **6:**437–441.
20. **Daikos, G. L., S. B. Kathpalia, V. T. Lolans, G. G. Jackson, and E. Fosslein.** 1988. Long-term oral ciprofloxacin: experience in the treatment of incurable infective endocarditis. *Am. J. Med.* **84:**786–790.
21. **Dow, J., J. Chazal, A. M. Frydman, P. Janny, R. Woehrle, F. Djebbar, and J. Gaillot.** 1986. Transfer kinetics of pefloxacin into cerebrospinal fluid after one hour iv infusion of 500 mg in man. *J. Antimicrob. Chemother.* **17**(Suppl.)**:**81–87.
22. **Durack, D. T., and P. B. Beeson.** 1972. Experimental bacterial endocarditis. II. Survival of endocardial vegetations. *Br. J. Exp. Pathol.* **53:**50–53.
23. **Eliopoulos, G. M., A. Gardella, and R. C. Moellering, Jr.** 1984. In vitro activity of ciprofloxacin, a new carboxyquinolone antimicrobial agent. *Antimicrob. Agents Chemother.* **25:**331–335.
24. **Freedman, L. R., and J. Valone, Jr.** 1979. Experimental infective endocarditis. *Prog. Cardiovasc. Dis.* **22:**169–180.
25. **Gengo, F. M., T. W. Mannion, C. H. Nightingale, and J. J. Schentag.** 1984. Integration of pharmacokinetics and pharmacodynamics of methicillin in curative treatment of experimental endocarditis. *J. Antimicrob. Chemother.* **14:**619–631.
26. **Gilbert, M., J. A. Boscia, W. Kobasa, and D. Kaye.** 1986. Enoxacin compared with vancomycin for the treatment of experimental methicillin-resistant *Staphylococcus aureus* endocarditis. *Antimicrob. Agents Chemother.* **29:**461–463.
27. **Hackbarth, C. J., H. F. Chambers, F. Stella, A. M. Shibl, and M. A. Sande.** 1986. Ciprofloxacin in experimental *Pseudomonas aeruginosa* meningitis in rabbits. *J. Antimicrob. Chemother.* **18**(Suppl.)**:**65–69.
28. **Haller, I.** 1985. Comprehensive evaluation of ciprofloxacin-aminoglycoside combinations against *Enterobacteriaceae* and *Pseudomonas aeruginosa* strains. *Antimicrob. Agents Chemother.* **28:**663–666.
29. **Hooper, D. C., J. S. Wolfson, E. Y. Ng, and M. N. Swartz.** 1987. Mechanism of action of and resistance to ciprofloxacin. *Am. J. Med.* **82**(Suppl.)**:**12–20.
30. **Hooper, D. C., J. S. Wolfson, K. S. Souza, C. Tung, G. L. McHugh, and M. N. Swartz.** 1986. Genetic and biochemical characterization of norfloxacin resistance in *Escherichia coli. Antimicrob. Agents Chemother.* **29:**639–644.

31. **Ingerman, M. J., P. K. Pitsakis, A. F. Rosenberg, and M. E. Levinson.** 1986. The importance of pharmacodynamics in determining the dosing interval in therapy for experimental *Pseudomonas* endocarditis in the rat. *J. Infect. Dis.* **153:**707–714.
32. **Isaacs, D., M. P. E. Slack, A. R. Wilkinson, and A. W. Westwood.** 1986. Successful treatment of *Pseudomonas* ventriculitis with ciprofloxacin. *J. Antimicrob. Chemother.* **17:**535–538.
33. **Jimenez-Lucho, V. E., L. D. Saravolatz, A. A. Medeiros, and D. Pohlod.** 1986. Failure of therapy in *Pseudomonas* endocarditis—selection of resistant mutants. *J. Infect. Dis.* **154:**64–68.
34. **Kaatz, G. W., S. L. Barriere, D. R. Schaberg, and R. Fekety.** 1987. Ciprofloxacin versus vancomycin in the therapy of experimental methicillin-resistant *Staphylococcus aureus* endocarditis. *Antimicrob. Agents Chemother.* **31:**527–530.
35. **Karchmer, A. W.** 1985. Staphylococcal endocarditis—laboratory and clinical basis for antibiotic therapy. *Am. J. Med.* **78**(Suppl. 6B):116–127.
36. **Landesman, S. H., M. L. Corrado, P. M. Shah, M. Armenguad, M. Barza, and C. E. Cherubin.** 1981. Past and current roles for cephalosporin antibiotics in treatment of meningitis—emphasis on use in gram-negative bacillary meningitis. *Am. J. Med.* **71:**693–703.
37. **Millar, M., M. A. Bransby-Zachary, D. S. Tompkins, P. M. Hawkey, and R. M. Gibson.** 1986. Ciprofloxacin for *Pseudomonas aeruginosa* meningitis. *Lancet* **i:**1325.
38. **Preheim, L. C., R. G. Penn, C. C. Sanders, R. V. Goering, and D. K. Giger.** 1982. Emergence of resistance to β-lactam and aminoglycoside antibiotics during moxalactam therapy of *Pseudomonas aeruginosa* infections. *Antimicrob. Agents Chemother.* **22:**1037–1041.
39. **Reyes, M. P., and A. M. Lerner.** 1983. Current problems in the treatment of infective endocarditis due to *Pseudomonas aeruginosa. Rev. Infect. Dis.* **5:**414–421.
40. **Sanders, C. C., and W. E. Sanders.** 1983. Emergence of resistance during therapy with the new β-lactam antibiotics: role of inducible β-lactamases and implications for the future. *Rev. Infect. Dis.* **5:**639–648.
41. **Sanders, C. C., W. E. Sanders, and R. V. Goering.** 1987. Overview of preclinical studies with ciprofloxacin. *Am. J. Med.* **82**(Suppl.):196–201.
42. **Sanders, C. C., W. E. Sanders, Jr., R. V. Goering, and V. Werner.** 1984. Selection of multiple antibiotic resistance by quinolones, β-lactams, and aminoglycosides with special reference to cross-resistance between unrelated drug classes. *Antimicrob. Agents Chemother.* **26:**797–801.
43. **Shekar, R., T. W. Rice, C. H. Zierdt, and C. A. Kallick.** 1985. Outbreak of endocarditis caused by *Pseudomonas aeruginosa* serotype O11 among pentazocine and tripelennamine abusers in Chicago. *J. Infect. Dis.* **151:**203–208.
44. **Shibl, A. M., C. J. Hackbarth, and M. A. Sande.** 1986. Evaluation of pefloxacin in experimental *Escherichia coli* meningitis. *Antimicrob. Agents Chemother.* **29:**409–411.
45. **Strasbaugh, L., and M. A. Sande.** 1978. Factors influencing the therapy of experimental *Proteus mirabilis* meningitis in rabbits. *J. Infect. Dis.* **137:**251–260.
46. **Strunk, R. W., J. C. Gratz, R. Maserati, and W. M. Scheld.** 1985. Comparison of ciprofloxacin with azlocillin plus tobramycin in the therapy of experimental *Pseudomonas aeruginosa* endocarditis. *Antimicrob. Agents Chemother.* **28:**428–432.
47. **Sullam, P. M., M. Tauber, C. J. Hackbarth, H. F. Chambers, K. G. Scott, and M. A. Sande.** 1985. Pefloxacin therapy for experimental endocarditis caused by methicillin-susceptible or methicillin-resistant strains of *Staphylococcus aureus. Antimicrob. Agents Chemother.* **27:**685–687.
48. **Thrupp, L. D.** 1980. Susceptibility testing of antibiotics in liquid media, p. 73–113. *In* V. Lorian (ed.), *Antibiotics in Laboratory Medicine.* The Williams & Wilkins Co., Baltimore.
49. **Wolff, M., L. Boutron, E. Singlas, B. Clair, J. M. Decazes, and B. Regnier.** 1987. Penetra-

tion of ciprofloxacin into cerebrospinal fluid of patients with bacterial meningitis. *Antimicrob. Agents Chemother.* **31:**899–902.

50. **Wolff, M., B. Regnier, C. Daldoss, M. Nkam, and F. Vachon.** 1984. Penetration of pefloxacin into cerebrospinal fluid of patients with meningitis. *Antimicrob. Agents Chemother.* **26:**289–291.
51. **Zeiler, H. J.** 1985. Evaluation of the in vitro bactericidal action of ciprofloxacin in cells of *Escherichia coli* in the logarithmic and stationary phases of growth. *Antimicrob. Agents Chemother.* **28:**524–527.

Chapter 12

Treatment of Skin and Soft Tissue Infections with Quinolone Antimicrobial Agents

David C. Hooper and John S. Wolfson

Infectious Disease Unit, Medical Services, Massachusetts General Hospital, Boston, Massachusetts 02114

Introduction

The newer fluoroquinolone antimicrobial agents have a broad spectrum of activity and excellent penetration into tissues, including skin structures, as determined by maximum drug levels in the inflammatory exudate of experimentally induced skin blisters in humans. These levels have ranged from 1 to 5 μg/ml, representing 47 to 81% of levels achievable in serum (Table 1) (1, 3, 14, 28; see chapter 4). In one study of patients with skin infections, levels in tissues were reported to be 175% of levels in serum (13).

Primary pyogenic infections of the skin and skin structures are most frequently caused by *Staphylococcus aureus* and streptococci which may colonize and then infect the skin. These organisms are considered susceptible to several of the newer fluoroquinolones, but the MICs for 90% of strains are close to the levels achievable in skin (see also Table 2). These infections include impetigo, erysipelas, furuncles, carbuncles, and cellulitis. Occasionally group A streptococci alone or in combination with *S. aureus* will also produce a necrotizing fasciitis.

Aerobic gram-negative bacilli tend to be more susceptible to the newer fluoroquinolones than are aerobic gram-positive cocci. Such gram-negative bacteria may also be involved in skin and soft tissue infections in certain circumstances and patients. Infections of traumatic wounds may be infected by multiple organisms, and surgical wound infections may be caused by gram-negative bacilli, particularly after surgery on the gastrointestinal or genitourinary tract, body areas often heavily contaminated with these

Table 1. Penetration of fluoroquinolones into cutaneous tissues of humans

Drug (reference)	Dose (mg)	Route	Drug concn[a] (% of concn in serum)		
			Blister fluid[b] (μg/ml)	Skin (μg/g)	Subcutaneous fat (μg/g)
Ciprofloxacin (3, 4, 13)	500	Oral	1.0 (67)		
	750	Oral		4.0 (170)[c]	
	100	i.v.[d]	0.6 (28)	0.23 (27)	0.4 (47)
Ofloxacin (14)	600	Oral	5.2 (47)		
Enoxacin (15, 28)	600	Oral	3.0 (81)	2.2 (81)	<1.0 (<39)
Pefloxacin (27)	400	Oral	3.9 (59)		

[a] Peak values for tissue, fluid, and serum.
[b] Inflammatory exudate from experimentally induced skin blisters.
[c] Determined at the site of infection.
[d] i.v., Intravenous.

organisms. Decubitus ulcers and ischemic ulcers of the lower extremities are also frequently colonized and may become secondarily infected with a mixture of anaerobic, gram-positive, and gram-negative bacteria. Cellulitis with or without ulceration in the feet of patients with diabetes mellitus or ischemic vascular disease is particularly likely to be caused by a mixture of organisms. Uncommonly, cellulitis involving the face, neck, and upper extremities in adults (as in children) with upper respiratory tract disease may result from infection with *Haemophilus influenzae*, an organism highly susceptible to the newer fluoroquinolones.

Table 2. Potency of fluoroquinolones against skin pathogens in vitro[a]

Organism	MIC for 90% of strains (μg/ml)			
	Ciprofloxacin	Ofloxacin	Pefloxacin	CI-934
Staphylococcus aureus	0.5	0.5	0.5	0.12
Streptococci groups A, C, and G	2.0	2.0	8.0	0.5
Enterococci	2.0	4.0	4.0	0.5
Escherichia coli	<0.06	0.12	0.12	1.0
Enterobacter cloacae	0.12	0.5	0.5	0.5
Serratia spp.	0.25	2.0	1.0	2.0
Pseudomonas aeruginosa	0.25–1	2–8	2–8	8.0–16
Bacteroides fragilis	8.0	8.0	32	8.0–16
Anaerobic gram-positive cocci	2.0	4.0	8.0	0.5–1.0
Clostridium sp.	8.0	8.0	64	4.0–8.0

[a] See chapter 3 for details and a complete list of references for activity in vitro.

Anaerobic bacteria have been recognized to be involved in some of the mixed infections of the skin stemming from ulcers and in diabetics, as well as in the entities of necrotizing anaerobic cellulitis (due to *Clostridium perfringens, Bacteroides* spp., or anaerobic cocci), necrotizing fasciitis (due to *Bacteroides* spp. in combination with facultative gram-negative aerobes), and synergistic necrotizing gangrene (due to anaerobic cocci in combination with facultative gram-negative aerobes). The currently available fluoroquinolone agents lack sufficient activity against these anaerobes to be useful alone for the treatment of these infections.

Assessment of the outcome of treatment of skin and soft tissue infections can often be made readily on clinical grounds because of the accessibility of the infected site. Obtaining reliable bacteriologic data for some of these infections, however, may be difficult in the absence of purulent drainage or material obtained at surgery (biopsy or debridement). In particular, in many cases of cellulitis an etiologic organism is not readily cultured, even with aspiration of sterile saline injected in or ahead of the area of local inflammation, and in cases of chronic skin ulcers it may be difficult to distinguish by surface cultures colonizing from infecting bacteria.

With these limitations in mind, we summarize below the current clinical studies assessing the treatment of skin and skin structure infections with the newer fluoroquinolones. Comparative studies are considered first, followed by a discussion of open studies and then conclusions about the role of fluoroquinolones in relation to other agents available for therapy of these common infections.

Comparative Studies of Treatment of Skin Infections

Fluoroquinolone treatment of skin and soft tissue infections has been studied principally with ciprofloxacin. There have been four studies, all prospective, randomized, and double blind in design, in which ciprofloxacin (750 mg given orally twice daily) was compared with cefotaxime (2 g given intravenously every 8 h) (18, 19, 22, 25) (Table 3). Each treatment group received a placebo formulation of the comparison drug. In one study (19) clindamycin was also given to any patients with documented anaerobic infection in either treatment group.

Parish and Asper (18) studied 60 patients, most of whom were elderly women with infected decubitus ulcers believed to have sufficiently severe infection to warrant hospitalization. A total of 24 patients receiving ciprofloxacin and 32 patients receiving cefotaxime were evaluable; the reasons for excluding patients from evaluation were not given. One patient in the ciprofloxacin group and four patients in the cefotaxime group had more

Table 3. Prospective, randomized, double-blind studies in which ciprofloxacin and cefotaxime were compared for the treatment of skin and soft tissues infections

Study (reference)	Cures/total (%)	
	Ciprofloxacin[a]	Cefotaxime[b]
Parish and Asper (18)	22/25 (88)	25/36 (69)
Self et al. (25)	34/38 (89)	33/35 (93)
Pérez-Ruvalcaba et al. (19)	24/31 (77)	22/29 (75)
Ramirez-Ronda et al. (22)	22/28 (78)	18/28 (64)

[a] Dosage of 750 mg orally twice daily.
[b] Two grams intravenously every 8 h.

than one site of skin infection, and about half of the patients had multiple organisms at the infected site. After 5 to 21 days of treatment, clinical resolution occurred at 22 of 25 sites (88%) in patients given ciprofloxacin and at 25 of 36 sites (69%) in patients given cefotaxime. Clinical failures were seen at only one site treated with ciprofloxacin and at two sites treated with cefotaxime; the remainder of sites were classified as improved. Thus, there was no statistical difference in the clinical outcomes in the two groups. Infecting organisms were roughly equally distributed in the two treatment groups and included the following: *Proteus mirabilis,* 30; *S. aureus,* 25; *Escherichia coli,* 17; *Pseudomonas aeruginosa,* 15; and other gram-negative bacteria, 15. It is not clear from the study protocol, however, the manner in which colonizing and infecting organisms were distinguished. Eradication of these organisms was accomplished in most cases in both groups. In patients treated with ciprofloxacin organisms persisted at three sites (12%) (*S. aureus,* 2; group B streptococci, 1), and in patients treated with cefotaxime organisms persisted at eight sites (22%) (*P. mirabilis,* 4; *P. aeruginosa,* 2; *S. aureus,* 2; *E. coli,* 1; group B streptococci, 1), differences that were not statistically significant.

Pérez-Ruvalcaba and associates (19), using a similar study design, treated hospitalized patients with a variety of skin and soft tissue infections. The protocol differed from that in the study by Parish and Asper in that patients with organisms resistant to either agent were excluded, some patients had additional debridement, and some patients (eight treated with ciprofloxacin and six treated with cefotaxime) received concurrent therapy with clindamycin for coverage of anaerobic infections. Of the infections, 30% were cellulitis, 28% were infected skin ulcers, 15% were abscesses, 13% were wound infections, and 13% were unspecified skin structure infections. Half of the patients were diabetics. Clinical resolution occurred in 24 of 31 (77%) infections treated with ciprofloxacin and 22 of 28 (76%) treated with cefotaxime. Only one clinical failure occurred, in the cefotaxime group. The predominant pathogens were the following: *S. aureus,* 24; *P. mirabilis,* 13; *E.*

coli, 11; *Klebsiella pneumoniae*, 6; and *Proteus vulgaris*, 5. *P. aeruginosa* was cultured from only three patients. The pathogens were eradicated in 90% of sites in both treatment groups; persisting organisms included one isolate each of *S. aureus* and *P. mirabilis* in the ciprofloxacin group and two isolates of *P. aeruginosa* and one of *S. aureus* in the cefotaxime group. In patients infected with *S. aureus* who also received clindamycin, the role of the test agent in eliminating this pathogen is uncertain.

Ramirez-Ronda and associates (22) also studied a mixture of skin infections in hospitalized patients similar to that studied by Pérez-Ruvalcaba et al. In this study, however, there were significantly more clinical failures in the group treated with cefotaxime (6 of 28 [21%]) than in the group treated with ciprofloxacin (1 of 28 [4%]) ($P < 0.05$). Infecting organisms were again dominated by enteric gram-negative bacilli (27 isolates), but included the following: *S. aureus*, 15; *P. aeruginosa*, 6; and group B streptococci, 5. Bacteriological eradication was equivalent in the two groups, with small numbers of cases in which organisms persisted (two for ciprofloxacin [*S. aureus* and *Citrobacter diversus*] and four for cefotaxime [one each for *K. pneumoniae*, *Pseudomonas magnus*, *Citrobacter freundii*, and *P. aeruginosa*]).

Finally, in the study conducted by Self and co-workers (25), a similar mixture of infections sufficiently severe to require hospitalization (cellulitis, 42%; abscesses, 26%; wound infections, 16%; miscellaneous infections, 16%) was treated, but in contrast to the other three studies, in these infections gram-positive cocci were the predominant pathogens. *S. aureus*, streptococci, or combinations of the two accounted for 76% of the 73 infections. Clinical and bacteriological outcomes were analyzed together, with cures defined as clinical resolution in addition to eradication of the pathogen, improvement as clinical resolution without elimination of the pathogen, and failure as lack of clinical resolution or bacteriologic eradication. Cures were effected in 34 of 38 patients (89%) treated with ciprofloxacin and 33 of 35 patients (94%) treated with cefotaxime. The one failure, in the ciprofloxacin group, was caused by a combination of *S. aureus* and beta-hemolytic streptococci.

Thus, in each of these four studies orally administered ciprofloxacin was comparable to parenteral cefotaxime in the treatment of skin infections caused by both gram-negative bacilli and gram-positive cocci. Overall, cures were effected in 101 of 120 patients (84%) treated with ciprofloxacin and in 90 of 120 patients (75%) treated with cefotaxime. For the more common skin infections caused by *S. aureus* and streptococci, however, cefotaxime is not the preferred currently available agent and, thus, is not the best standard of comparison. Needed are studies in which ciprofloxacin is compared with semisynthetic penicillins (or possibly first-generation cephalosporins) in the treatment of skin infections caused by these aerobic gram-positive cocci.

Open Studies of Treatment of Skin Infections

There have been at least 14 open studies in the English literature that have reported the use of ciprofloxacin for the treatment of skin and skin structure infections (2, 7–13, 16, 20, 21, 24, 26, 29). In nine of these studies patients with skin infections were reported in numbers sufficient to constitute a substantial fraction of the patients reported (2, 7, 8, 12, 13, 20, 24, 26, 29) (Table 4).

In these nine studies a mixture of infections was treated with ciprofloxacin orally (500 to 750 mg twice daily) for 5 to 181 days. A total of 197 patients were studied, and where specified (in 138 cases; see references 7, 8, 20, 24, and 26) the infections consisted of the following (in percentages): cellulitis, 33; wound infections, 30; infected ulcers, 20; abscesses, 10; and burn infections, 6. One study recorded clinical responses for each type of infection (8); cures were recorded in eight of eight patients (100%) with severe cellulitis, seven of eight patients (87%) with postoperative wound infections, and three of five patients (60%) with infected decubitus or diabetic foot ulcers. In the nine studies overall clinical cures were reported in 136 patients (69%). An additional 42 patients (21%) were classified as improved based on clinical resolution with persistence of the organism. Nineteen patients (10%) failed treatment, and diabetic foot ulcers accounted for at least eight of these (7, 8, 24, 26).

The bacteriologic response of skin infections to ciprofloxacin treatment has been analyzed by the infecting organism (6). Organism-specific rates of eradication were as follows (in percentages): *P. aeruginosa*, 62 (n = 111); *Enterobacter* spp., 75 (n = 40); *Klebsiella* spp., 71 (n = 41); *E. coli*, 78 (n = 67); *S. aureus*, 75 (n = 188); streptococci, 64 (n = 88); and anaerobic bacteria, 56 (n = 18). Failures were seen in 11% of *P. aeruginosa* infections, 6% of streptococcal infections, and 5% of *S. aureus* infections. Noteworthy are the results of one study in which seven cases of infections caused by methicillin-resistant *S. aureus* (MRSA) were cured with ciprofloxacin (12), raising the possibility of oral treatment for certain of these infections which usually would otherwise require parenteral vancomycin therapy. Single cases of *Pasteurella multocida* and *Eikenella* sp. infection have been cured with ciprofloxacin (8).

Eradication of Colonization by MRSA

Colonization of the skin and mucous membranes of hospitalized patients with MRSA has been associated with perpetuation of outbreaks of in fections with these organisms. In an effort to interrupt the spread of these

Table 4. Results of open studies of the use of ciprofloxacin in the treatment of skin and soft tissue infections

Study (reference)	No. of patients[a]				Characteristics of failures[b]
	Total	Clinical cures	Improved	Failed	
Berman and Schwartz (2)	23	15	6	2	Renal transplant recipient, *P. aeruginosa*
Eron et al. (7)	26	16	5	5	Diabetes mellitus, 3; *P. aeruginosa*, 3; *S. aureus*, 1
Fass (8)	21	18	2	1	Diabetes mellitus, 1
Greenberg et al. (12)	20	18	0	2	*S. aureus*, 1
Licitra et al. (13)	21	16	5	0	
Pien and Yamane (20)	19	9	10	0	
Scully et al. (24)	15	11	0	4	Diabetes mellitus, 1; *P. aeruginosa*, 1
Valainis et al. (26)	32	19	9	5	Diabetes mellitus, 3
Wood and Logan (29)	20	14	5	0	

[a] Of a total of 197 patients, 136 (69%) were clinical cures, 42 (21%) improved, and 19 (10%) failed.
[b] Characteristics of some patients who failed therapy could not be determined in many studies.

organisms to hospital staff and patients, a number of antimicrobial regimens have been used in attempts to eliminate colonization with MRSA (5, 23, 30). Only limited success has been achieved, however.

There has as yet been only one open study assessing the efficacy of ciprofloxacin in eliminating MRSA colonization (17). Most of the 20 patients enrolled had colonization at multiple sites, including the nares, perineum, and open skin lesions. Ciprofloxacin was given orally (750 mg twice daily) for 7 to 28 days. Excluding the 8 patients in whom the drug was stopped prematurely for possible adverse effects, MRSA was eradicated from all sites in 11 of 14 courses (78%, in 12 patients) of treatment. Generally a 2- to 3-week course of therapy was necessary before eradication occurred, and the three failures were associated with the development of bacterial resistance to ciprofloxacin. Colonization with susceptible organisms also recurred in 4 of the 12 successfully treated patients within a month after cessation of therapy; it was not possible to distinguish relapse from recolonization in these patients. Thus, overall ciprofloxacin produced long-term eradication of MRSA in only 8 of 20 patients (40%) in whom therapy was undertaken. In retrospective controls MRSA was eliminated from none of 17 patients receiving no therapy or therapy with vancomycin alone.

The optimal regimen for eradication of MRSA colonization remains uncertain. Comparisons of regimens containing ciprofloxacin alone or in combination with regimens containing trimethoprim-sulfamethoxazole, rifampin, or both are needed.

Development of Bacterial Resistance

Development of bacterial resistance associated with persistence of the infecting organism was quantitated in seven studies (8–13, 16). Two other studies also mentioned that resistance occurred, but the frequency was not quantitated (24, 26). Increases in MIC were detected for persisting bacteria in 13 of 98 cases (13%) in these seven studies. *P. aeruginosa* was the most common (or the exclusive) pathogen reported in several studies and thus predominated among the organisms developing ciprofloxacin resistance. Increases in MIC for this organism have ranged from 4- to 80-fold. For *S. aureus* and *Enterococcus faecium* 4- and 64-fold increases in MIC have also been reported to develop during therapy (8, 17). Treatment of MRSA colonization has been associated with a particularly high rate of development of resistance in *S. aureus* (7 of 22 episodes [32%]) (17). Infections in the feet of patients with diabetes mellitus seem particularly likely to develop bacterial resistance when *P. aeruginosa* is the pathogen (6, 24), probably resulting from poor tissue perfusion and drug delivery. Infections in limbs made ischemic by atherosclerosis are also likely to be similarly predisposed.

Conclusions

Ciprofloxacin appears to have moderate efficacy in the treatment of skin and soft tissue infections and is comparable to cefotaxime given in moderate dosages for treatment of these infections, particularly when caused by aerobic gram-negative bacilli. Mixed infections with both gram-positive cocci and gram-negative bacilli may also respond, allowing the use of a single oral agent. Infections known or likely to be caused by staphylococci or streptococci alone are still best treated with penicillin, semisynthetic penicillins, first-generation cephalosporins, erythromycin, clindamycin, or vancomycin until comparative studies suggest otherwise. In selected patients in whom oral agents are indicated, it may be possible to treat skin infections caused by MRSA with oral ciprofloxacin. Attempts to eradicate colonization with this organism, however, should not be attempted with ciprofloxacin alone, at least until additional comparative data are available. Infections in which anaerobic bacteria are likely to be involved will need treatment with an additional agent, such as clindamycin or metronidazole, which may also be given orally.

On the basis of potency or pharmacokinetics, several other fluoroquinolones undergoing clinical testing, such as ofloxacin and pefloxacin, may also have efficacy in the treatment of skin and soft tissue infections. In addition, certain fluoroquinolones undergoing preclinical testing appear to have enhanced potency against staphylococci, streptococci, and in some cases anaerobic bacteria. These congeners may well prove to be more generally suited for the treatment of skin and soft tissue infections. The results of clinical testing are awaited with interest.

Literature Cited

1. **Adhami, Z. N., R. Wise, D. Weston, and B. Crump.** 1984. The pharmacokinetics and tissue penetration of norfloxacin. *J. Antimicrob. Chemother.* **13:**87–92.
2. **Berman, S. J., and S. M. Schwartz.** 1987. Clinical evaluation of ciprofloxacin in patients with moderately severe bacterial infections. *Am. J. Med.* **82**(Suppl. 4A):233–235.
3. **Crump, B., R. Wise, and J. Dent.** 1983. Pharmacokinetics and tissue penetration of ciprofloxacin. *Antimicrob. Agents Chemother.* **24:**784–786.
4. **Daschner, F. D., M. Westenfelder, and A. Dalhoff.** 1986. Penetration of ciprofloxacin into kidney, fat, muscle, and skin tissue. *Eur. J. Clin. Microbiol.* **5:**212–213.
5. **Ellison, R. T., III, F. N. Judson, L. C. Peterson, D. L. Cohn, and J. M. Ehret.** 1984. Oral rifampin and trimethoprim/sulfamethoxazole therapy in asymptomatic carriers of methicillin-resistant *Staphylococcus aureus* infections. *West. J. Med.* **140:**735–740.
6. **Eron, L. J.** 1987. Therapy of skin and skin structure infections with ciprofloxacin: an overview. *Am. J. Med.* **82**(Suppl. 4A):224–226.
7. **Eron, L. J., L. Harvey, D. L. Hixon and D. M. Poretz.** 1985. Ciprofloxacin therapy of infections caused by *Pseudomonas aeruginosa* and other resistant bacteria. *Antimicrob. Agents Chemother.* **27:**308–310.
8. **Fass, R. J.** 1986. Treatment of skin and soft tissue infections with oral ciprofloxacin. *J. Antimicrob. Chemother.* **18**(Suppl. D):153–157.
9. **Follath, F., M. Bindschedler, M. Wenk, R. Frei, H. Stalder, and H. Reber.** 1986. Use of ciprofloxacin in the treatment of *P. aeruginosa* infections. *Eur. J. Clin. Microbiol.* **5:**236–240.
10. **Giamarellou, H., E. Daphnis, C. Dendrinos, and G. K. Daikos.** 1985. Experience with ciprofloxacin in the treatment of various infections caused mainly by *Pseudomonas aeruginosa. Drugs Exp. Clin. Res.* **11:**351–356.
11. **Giamarellou, H., N. Galanakis, C. Dendrinos, J. Stefanou, E. Daphnis, and G. K. Daikos.** 1986. Evaluation of ciprofloxacin in the treatment of *Pseudomonas aeruginosa* infections. *Eur. J. Clin. Microbiol.* **5:**232–235.
12. **Greenberg, R. N., D. J. Kennedy, P. M. Reilly, K. L. Luppen, W. J. Weinandt, M. R. Bollinger, F. Aguirre, F. Kodesch, and A. M. K. Saeed.** 1987. Treatment of bone, joint, and soft-tissue infections with oral ciprofloxacin. *Antimicrob. Agents Chemother.* **31:**151–155.
13. **Licitra, C. M., R. G. Brooks, and B. E. Sieger.** 1987. Clinical efficacy and levels of ciprofloxacin in tissue in patients with soft tissue infection. *Antimicrob. Agents Chemother.* **31:**805–807.
14. **Lockley, M. R., R. Wise, and J. Dent.** 1984. The pharmacokinetics and tissue penetration of ofloxacin. *J. Antimicrob. Chemother.* **14:**647–652.
15. **Malmborg, A. S., and S. Rannikko.** 1988. Enoxacin distribution in human tissues after multiple oral administration. *J. Antimicrob. Chemother.* **21**(Suppl. B):57–60.

16. **Mehtar, S., Y. Drabu, and P. Blakemore.** 1986. Ciprofloxacin in the treatment of infections caused by gentamicin-resistant gram-negative bacteria. *Eur. J. Clin. Microbiol.* **5:**248–251.
17. **Mulligan, M. E., P. J. Ruane, L. Johnston, P. Wong, J. P. Wheelock, K. MacDonald, J. F. Reinhardt, C. C. Johnson, B. Statner, I. Blomquist, J. McCarthy, W. O'Brien, S. Gardner, L. Hammer, and D. M. Citron.** 1987. Ciprofloxacin for eradication of methicillin-resistant *Staphylococcus aureus* colonization. *Am. J. Med.* **82**(Suppl. 4A):215–219.
18. **Parish, L. C., and R. Asper.** 1987. Systemic treatment of cutaneous infections: a comparative study of ciprofloxacin and cefotaxime. *Am. J. Med.* **82**(Suppl. 4A):227–229.
19. **Pérez-Ruvalcaba, J. A., N. P. Qurntero-Pérez J. J. Morales-Reyes, J.A. Huitrón-Ramirez, J. J. Rodríguez-Chagollán, and E. Rodríguez-Noriega.** 1987. Double-blind comparison of ciprofloxacin with cefotaxime in the treatment of skin and skin structure infections. *Am. J. Med.* **82**(Suppl. 4A):242–246.
20. **Pien, F. D., and K. K. Yamane.** 1987. Ciprofloxacin treatment of soft tissue and respiratory infections in a community outpatient practice. *Am. J. Med.* **82**(Suppl. 4A):236–238.
21. **Ramirez, C. A., J. L. Bran, C. R. Mejia, and J. F. Garcia.** 1985. Open, prospective study of the clinical efficacy of ciprofloxacin. *Antimicrob. Agents Chemother.* **28:**128–132.
22. **Ramirez-Ronda, C. H., S. Saavedra, and C. R. Rivera-Vázquez.** 1987. Comparative, double-blind study of oral ciprofloxacin and intravenous cefotaxime in skin and skin structure infections. *Am. J. Med.* **82**(Suppl. 4A):220–223.
23. **Sande, M. A., and G. L. Mandell.** 1975. Effect of rifampin on nasal carriage of *Staphylococcus aureus. Antimicrob. Agents Chemother.* **7:**294–297.
24. **Scully, B. E., H. C. Neu, M. F. Parry, and W. Mandell.** 1986. Oral ciprofloxacin therapy of infections due to *Pseudomonas aeruginosa. Lancet* **i:**819–822.
25. **Self, P. L., B. A. Zeluff, D. Sollo, and L. O. Gentry.** 1987. Use of ciprofloxacin in the treatment of serious skin and skin structure infections. *Am. J. Med.* **82**(Suppl. 4A):239–241.
26. **Valainis, G. T., G. A. Pankey, H. P. Katner, L. M. Cortez, and J. R. Dalovisio.** 1987. Ciprofloxacin in the treatment of bacterial skin infections. *Am. J. Med.* **82**(Suppl. 4A):230–232.
27. **Webberley, J. M., J. M. Andrews, J. P. Ashby, A. McLeod, and R. Wise.** 1987. Pharmacokinetics and tissue penetration of orally administered pefloxacin. *Eur. J. Clin. Microbiol.* **6:**521–524.
28. **Wise, R., R. Lockley, J. Dent, and M. Webberly.** 1984. Pharmacokinetics and tissue penetration of enoxacin. *Antimicrob. Agents Chemother.* **26:**17–19.
29. **Wood, M. J., and M. N. Logan.** 1986. Ciprofloxacin for soft tissue infections. *J. Antimicrob. Chemother.* **18**(Suppl. D):159–164.
30. **Yu, V. L., A. Goetz, M. Wagener, P. B. Smith, J. D. Rihs, J. Hanchett, and J. J. Zuravleff.** 1986. *Staphylococcus aureus* nasal carriage and infection in patients on hemodialysis: efficacy of antibiotic prophylaxis. *N. Engl. J. Med.* **315:**91–96.

Chapter 13

Quinolone Antimicrobial Agents and Ophthalmologic Infections

John S. Wolfson and David C. Hooper

Infectious Disease Unit, Medical Services,
Massachusetts General Hospital, Boston, Massachusetts 02114

Introduction

Bacterial infections of the eye can be conveniently grouped into three categories: conjunctivitis, keratitis, and endophthalmitis (11–13). The purpose of this chapter is to review the limited number of studies with quinolones relevant to treatment of these three categories of bacterial infections of the eye.

Bacterial Species Causing Ophthalmologic Infections

Bacterial species causing conjunctivitis include commonly *Streptococcus pneumoniae, Haemophilus influenzae, Staphylococcus aureus,* and *Staphylococcus epidermidis* and less commonly *Neisseria gonorrhoeae, Haemophilus aegyptius, Neisseria meningitidis,* viridans streptococci, *Proteus vulgaris, Moraxella lacunata, Corynebacterium diphtheriae, Francisella tularensis,* and other bacterial species. *Chlamydia trachomatis* is also an important cause of conjunctivitis.

Bacterial species causing keratitis in the United States include most commonly *S. aureus, S. pneumoniae, Pseudomonas aeruginosa,* and *Moraxella* spp. Additional species that may cause keratitis include *S. epidermidis,* streptococci, *Morganella morganii, Klebsiella pneumoniae, Serratia marcescens, Escherichia coli, N. gonorrhoeae,* and other bacterial species.

Bacterial endophthalmitis generally occurs in the setting of ocular surgery or, much less commonly, after nonsurgical trauma to the eye or bacteremic spread from another site of infection. Bacterial species causing these infections include *S. aureus, P. aeruginosa, Proteus* spp., *E. coli,* and *S.*

epidermidis and less commonly *Streptococcus* spp., *H. influenzae, Clostridium perfringens, N. meningitidis, S. marcescens, K. pneumoniae, Bacillus cereus,* and other bacterial species.

Shungu and associates (18) and Bywater and associates (4) evaluated the potency of norfloxacin in vitro against bacterial species that cause ophthalmologic infection, using isolates from sites other than the eye, and demonstrated excellent activity against most species, particularly members of the family *Enterobacteriaceae, H. influenzae,* and *N. gonorrhoeae,* but also *S. aureus, S. epidermidis, P. aeruginosa,* and *Moraxella* spp. Potency against streptococci appeared to be less favorable. The study suggested that norfloxacin had potential use in the treatment of superficial bacterial infections of the eye. This finding was subsequently confirmed and extended by Goldstein and associates (10), who carried out similar studies with bacterial strains isolated from patients with eye infections.

On a molar basis ciprofloxacin, ofloxacin, and other new quinolone agents are in general more potent than or equal in potency in vitro to norfloxacin against a similar spectrum of bacterial species (21; for specific MICs, see chapter 3). In addition, certain new quinolone agents, including CI-934 and temafloxacin, exhibit enhanced potency in vitro against gram-positive bacterial species, including both streptococci and staphylococci. Thus, the spectrum of potency of new quinolones in addition to norfloxacin suggests their possible use in the treatment of bacterial infections of the eye.

The quinolones also have potency in vitro against *C. trachomatis,* the agent of trachoma and of inclusion conjunctivitis (see chapter 2).

Pharmacokinetic Considerations

The delivery of an antimicrobial agent in the treatment of bacterial conjunctivitis or keratitis is uncomplicated, as the drug can be applied in either drops or an ointment at a high concentration directly to the involved area. The treatment of bacterial endophthalmitis is more complex, however, because of problems of delivery of the drug to deep sites of infection. A drug can be delivered by topical, subconjunctival, intravitreal, and intravenous administration.

Data on the penetration of systemically administered quinolones into the eye are only just being reported (2, 3, 8, 15, 16, 20). Findings of four studies with pefloxacin (16), ofloxacin (3, 8), and ciprofloxacin (2) are presented in Table 1. In these studies uninfected patients undergoing cataract extraction received a single dose of quinolone, and drug concentrations in serum, aqueous humor, and (in two of the four studies) lens tissue were determined at specific times after dosing. The data indicate that concentrations of pefloxacin and ofloxacin achieved in aqueous humor appear to have

Table 1. Quinolone concentrations in serum, aqueous humor, and lens in uninfected patients undergoing cataract extraction

Drug (reference)	Dose[a] (mg), route	No. of patients	Time after dose (h)	Drug concn		
				Serum (μg/ml)	Aqueous humor (μg/ml)	Lens (μg/g)
Pefloxacin (16)	400, i.v.[b]	20	2	4.4	0.8	0.09
			6	3.5	1.5	0.20
			12	2.5	1.0	0.26
			24	1.2	0.9	0.43
Ofloxacin (8)	400, oral	20	2	5.0	1.0	0.04
			6	3.2	1.0	0.04
			12	2.0	1.0	0.02
			24	0.5	0.3	<0.01
Ofloxacin (3)	200, oral	30	2	3.0	0.3	ND[c]
			6	2.6	0.6	ND
Ciprofloxacin (2)	200, i.v.	16	1	1.8	0.2	ND
			3	0.9	0.1	ND

[a] Patients received a single dose.
[b] i.v., Intravenous.
[c] ND, Not done.

peak values in the range of from 0.6 to 1.5 μg/ml and thus appear to be sufficient to inhibit certain of the bacterial species that cause endophthalmitis. Concentrations of pefloxacin and ofloxacin achieved in lens tissue, in contrast, were substantially lower.

Additional studies in which concentrations of quinolones in aqueous humor, vitreous humor, and lens tissue in patients receiving single and especially multiple doses of quinolone agents are evaluated seem warranted. Similarly, studies evaluating the concentrations of quinolones in the aqueous and vitreous humors after subconjunctival administration seem needed.

Therapeutic Efficacy

Topical norfloxacin was compared with topical tobramycin for the therapy of purulent conjunctivitis in a prospective, randomized study with 30 patients in each group (J. A. Jacobson, N. B. Call, and E. M. Kasworm, *Program Abstr. 27th Intersci. Conf. Antimicrob. Agents Chemother.* abstr. no. 94, 1987). Pathogens included *H. influenzae, S. pneumoniae, S. aureus,* and other bacterial species. Clinical cure rates (79% for norfloxacin and 90% for tobramycin) and elimination of pathogens (70% for topical norfloxacin and 63% for tobramycin) did not differ statistically. Toxicity occurred in three patients in each group and was mild. Thus, in this study norfloxacin appeared to be as efficacious as tobramycin for the treatment of purulent conjunctivitis.

Data on the therapy of bacterial keratitis have not been published.

Data on the therapy of bacterial endophthalmitis in humans are also not available. Ciprofloxacin, however, has been compared with gentamicin and imipenem in the treatment of *P. aeruginosa* infections in the rabbit model of bacterial endophthalmitis (7). A single intravitreal dose of drug was administered either 24 or 48 h after infection. For all three drugs treatment at 24 h resulted in a dose-dependent decrease in bacterial counts, while, in contrast, treatment at 48 h did not significantly decrease bacterial counts. Poor efficacy with therapy at 48 h represented neither altered drug clearance nor emergence of bacterial resistance but may have resulted in part from a decreased pH or slowed bacterial growth due to the exhaustion of nutrients, both conditions that antagonize the killing of bacteria in vitro by quinolones and other agents. This study thus documented in an animal model the ability of ciprofloxacin to reduce bacterial counts when administered 24 but not 48 h after infection.

Ocular Toxicity of Quinolones

A comment on possible adverse effects of systemically administered quinolones on the eye seems warranted because quinolones have been associated with ocular toxicity in animals, although published data are minimal. Pefloxacin caused subcapsular cataracts in rats, and rosoxacin may cause subcapsular cataracts in dogs (5, 17). Schlüter (17) evaluated ciprofloxacin in monkeys and found no accumulation of the drug in lens tissue and no change in lens density in animals receiving the drug for 6 months. Changes in electrophysiology and histology of the retina have been documented in cats receiving nalidixic acid but not norfloxacin (6) or ciprofloxacin (17).

In humans, ocular toxicity appears to be uncommon. More than 800 patients who received ciprofloxacin underwent fundoscopy, visual acuity testing, and color testing both before and after treatment, with no significant drug-related ocular toxicity (1, 17). Three cases of serous macular detachment in patients with chronic renal failure who received flumequine, however, have been reported (14). These patients presented with scotoma, and on ophthalmologic examination bilateral symmetrical macular bullae were seen. Lesions resolved within 2 days of cessation of drug therapy.

Conclusion

Data are limited but suggest some promise for quinolones in the topical treatment of bacterial conjunctivitis. The spectrum of potency in vitro of quinolones and the preliminary results for the therapy of bacterial conjunctivitis suggest that the quinolones should be evaluated in the treatment of bacterial keratitis. The spectrum of potency in vitro, several studies of penetration of systemically administered quinolones into aqueous humor, and the one animal model study described above also suggest that the drugs merit further study to define the penetration of quinolones into the vitreous humor after systemic administration, penetration into the aqueous and vitreous humor after subconjunctival administration, and efficacy in the therapy of bacterial endophthalmitis.

In considering the role of quinolones in the therapy of bacterial and chlamydial conjunctivitis in neonates, it is noteworthy that the drugs are currently not recommended for systemic administration to children because of drug-induced cartilage damage in juvenile animals (17). Although no data exist for quinolones, topical administration could possibly lead to systemic absorption and toxicity, as has occurred with other agents (9, 19).

Literature Cited

1. **Arcieri, G., E. Griffith, G. Gruenwaldt, A. Heyd, B. O'Brien, N. Becker, and R. August.** 1987. Ciprofloxacin: an update on clinical experience. *Am. J. Med.* **82**(Suppl. 4A): 381–386.
2. **Behrens-Baumann, W., and J. Martell.** 1987. Ciprofloxacin concentrations in human aqueous humor following intravenous administration. *Chemotherapy* (Basel) **33:**328–330.
3. **Bron, A., D. Talon, B. Delbosc, J. M. Estavoyer, G. Kaya, and J. Royer.** 1987. La pénétration intracamérulaire de l'ofloxacine chez l'homme. *J. Fr. Ophtalmol.* **10:**443–446.
4. **Bywater, M. J., H. A. Holt, and D. S. Reeves.** 1987. In vitro activity of norfloxacin in comparison with 11 topical antimicrobial agents against 142 ocular pathogens. *Rev. Infect. Dis.* **10**(Suppl. 1):S248–S250.
5. **Christ, W., T. Lehnert, and B. Ulbrichi.** 1988. Specific toxicologic aspects of the quinolones. *Rev. Infect. Dis.* **10**(Suppl. 1):S141–S146.
6. **Corrado, M. L., W. E. Struble, C. Peter, V. Hoagland, and J. Sabbaj.** 1987. Norfloxacin: review of safety studies. *Am. J. Med.* **82**(Suppl. 4A):22–26.
7. **Davey, P. G., M. Barza, and M. Stuart.** 1987. Dose response of experimental *Pseudomonas* endophthalmitis to ciprofloxacin, gentamicin, and imipenem: evidence for resistance to "late" treatment of infections. *J. Infect. Dis.* **155:**518–523.
8. **Fisch, A., C. Lafaix, A. Salvanet, M. Cherifi, and A. Meulemans.** 1987. Ofloxacin in human aqueous humour and lens. *J. Antimicrob. Chemother.* **20:**453–454.
9. **Fraunfelder, F. T., and G. C. Bagby, Jr.** 1983. Ocular chloramphenicol and aplastic anemia. *N. Engl. J. Med.* **308:**1536.
10. **Goldstein, E. J. G., D. M. Citron, L. Bendon, A. E. Vagvolgyi, M. D. Trousdale, and M. D. Appleman.** 1987. Potential of topical norfloxacin therapy. Comparative in vitro activity against clinical ocular bacterial isolates. *Arch. Ophthalmol.* **105:**991–994.
11. **Hirst, L. W., J. V. Thomas, and W. R. Green.** 1985. Conjunctivitis, p. 749–754. *In* G. L. Mandell, R. G. Douglas, Jr., and J. E. Bennett (ed.), *Principles and Practice of Infectious Diseases.* John Wiley & Sons, Inc., New York.
12. **Hirst, L. W., J. V. Thomas, and W. R. Green.** 1985. Keratitis, p. 754–760. *In* G. L. Mandell,

R. G. Douglas, Jr., and J. E. Bennett (ed.), *Principles and Practice of Infectious Diseases.* John Wiley & Sons, Inc., New York.

13. **Hirst, L. W., J. V. Thomas, and W. R. Green,** 1985. Endophthalmitis, p. 760–767. *In* G. L. Mandell, R. G. Douglas, Jr., and J. E. Bennett (ed.), *Principles and Practice of Infectious Diseases.* John Wiley & Sons, Inc., New York.
14. **Hurault de Ligny, B., D. Sirbat, M. Kessler, P. Trechot, and J. Chanliau.** 1984. Effets secondaires oculaires de la fluméquine. Trois observations d'atteintes maculaires. *Thérapie* **39:**595–600.
15. **Lüthy, R., B. Joos, and F. Gassman.** 1986. Penetration of ciprofloxacin into the human eye, p. 192–196. *In* H. C. Neu and H. Weuta (ed.), *Proceedings of the First International Ciprofloxacin Workshop.* Excerpta Medica, Amsterdam.
16. **Salvanet, A., A. Fisch, C. Lafaix, G. Montay, P. Dubayle, F. Forestier, and G. Haroche.** 1986. Pefloxacin concentrations in human aqueous humour and lens. *J. Antimicrob. Chemother.* **18:**199–201.
17. **Schlüter, G.** 1987. Ciprofloxacin: review of potential toxicologic effects. *Am. J. Med.* **82**(Suppl. 4A):91–93.
18. **Shungu, D. L., V. K. Tutlane, E. Weinberg, and H. H. Gadebusch.** 1985. In vitro antibacterial activity of norfloxacin and other agents against ocular pathogens. *Chemotherapy* (Basel) **31:**112–118.
19. **Williams, T., and W. H. Ginther.** 1982. Hazard of optic timolol. *N. Engl. J. Med.* **306:**1485–1486.
20. **Wise, R., and I. A. Donovan.** 1987. Tissue penetration and metabolism of ciprofloxacin. *Am. J. Med.* **82**(Suppl. 4A):103–107.
21. **Wolfson, J. S., and D. C. Hooper.** 1985. The fluoroquinolones: structures, mechanisms of action and resistance, and spectra of activity in vitro. *Antimicrob. Agents Chemother.* **28:**581–586.

Chapter 14

Adverse Effects of Quinolone Antimicrobial Agents

David C. Hooper and John S. Wolfson

Infectious Disease Unit, Medical Services, Massachusetts General Hospital, Boston, Massachusetts 02114

Introduction

It is difficult to assess accurately drug-related adverse experiences in humans. To establish rigorously in an individual patient a cause-and-effect relationship between the drug administered and the adverse experience reported or observed, the putative toxic effect should improve with cessation of therapy and return reproducibly on rechallenge with the drug but not a placebo. It is obviously seldom possible (or desirable to attempt) to fulfill these criteria in humans, for ethical reasons and because some toxic effects may not be reversible.

The best estimates of the frequency and types of drug-related adverse experiences come from placebo-controlled, double-blind, randomized trials, which in the population studied allow a determination of the statistical likelihood that an observed effect is drug related. Such studies could be performed with healthy volunteers, but the results might not be extrapolated reliably to other groups of patients with infections for which the drug is to be tested for efficacy or used clinically. In addition, the use of placebo controls in studies of treatment of infections is limited to those infections that are nonfatal, resolve spontaneously, and have little residual morbidity. Crossover designs of drug trials may overcome some of these problems.

Because of the limitations cited above, information on adverse experiences for clinically relevant populations comes most commonly from open studies or studies in which fluoroquinolones are compared with another antimicrobial agent. Assessments of adverse experiences are particularly difficult in open studies because the frequency of reporting varies with the method of ascertainment and with the patient population studied. In all studies the investigators must make clinical judgments as to whether the symptom or sign reported is possibly, probably, or definitely drug related. Even in comparative studies, which should preferably be double-blind to eliminate possible bias in these clinical judgments, the information gained is, of course, relative and must be assessed in the context of the general clinical experience with the tolerability of the comparison agent.

Clinical Adverse Effects

Frequencies of adverse experiences in open and comparative clinical studies

Because the largest clinical experience with the newer fluoroquinolones has to date been with norfloxacin, ciprofloxacin, and ofloxacin, the most extensive safety profiles are available for these agents. The safety profiles of other newer fluoroquinolone agents may differ, as was the case with some of the older quinolone analogs (41, 58).

In the English literature there have thus far been nine double-blind studies in which fluoroquinolone therapy has been compared with therapy with other antimicrobial agents or a placebo (Table 1) (28, 67, 72, 75, 77, 87, 101, 103, 107). Some double-blind comparative studies from Japan have also been reviewed (Table 1) (61). Adverse experiences thought to be either possibly, probably, or definitely drug related ranged from 2 to 36%. In two studies adverse experiences were significantly fewer in the fluoroquinolone group, when norfloxacin was compared with trimethoprim-sulfamethoxazole (103) and when ciprofloxacin was compared with ampicillin (107). In the remaining studies the frequency of adverse experiences in the fluoroquinolone group did not differ significantly from that with ampicillin, amoxicillin, cefaclor, cefotaxime, doxycycline, or a placebo, but in many studies the numbers of cases entered were small.

The types and overall frequency of clinical adverse experiences encountered with norfloxacin, ciprofloxacin, ofloxacin, enoxacin, and pefloxacin, as tabulated from clinical studies (both open and comparative, including double-blind studies) in the English literature, are shown in Table 2. Apparent differences among drugs in total adverse effects may not be meaningful because of the different mixes of patients studied. In two double-blind direct comparisons of two fluoroquinolones, ciprofloxacin and ofloxacin had similar frequencies of side effects in small numbers of patients with cystic fibrosis (51) or complicated urinary tract infections (55).

The frequency of adverse experiences when arranged by types of infections treated showed few consistent trends. Bone and joint infections (treatment of which has principally been reported with ciprofloxacin or

Table 1. Adverse effects of fluoroquinolones in comparison with other agents or a placebo in double-blind studies

Quinolone agent	%	Comparison agent	%	Study (reference)
Norfloxacin	34[a]	Trimethoprim-sulfamethoxazole	49	Urinary Tract Infection Study Group (103)
Ofloxacin	5	Amoxicillin	4	Sasaki, 1984 (61)
	6	Amoxicillin	5	Takase, 1984 (61)
	2	Cefaclor	5	Fujita, 1984 (61)
	8	Cefaclor	8	Fujimora, 1984 (61)
Ciprofloxacin	4	Ampicillin	6	Roddy et al., 1986 (77)
	17[a]	Ampicillin	35	Wollschlager et al., 1987 (107)
	5	Placebo	5	Renkonen et al., 1987 (75)
	11	Placebo	12	Pichler et al., 1986 (67)
	16	Doxycycline	23	Fong et al., 1987 (28)
	24	Cefotaxime	17	Self et al., 1987 (87)
	36	Cefotaxime	36	Ramirez-Ronda et al., 1987 (72)
Enoxacin	28	Placebo	33	Tsuei et al., 1984 (101)

[a]$P < 0.05$ relative to the comparison agent.

pefloxacin) had strikingly higher rates of reported adverse experiences (21 and 73%) than did other infections treated, possibly reflecting the usage of higher doses and longer durations of therapy. With the exception of norfloxacin, the treatment of sexually transmitted diseases with fluoroquinolones tended to have lower frequencies of adverse experiences relative to the treatment of other infections, probably because in most studies only one or two doses of drug were administered. Dose-related toxicity has been noted with norfloxacin (103).

Additional published summaries of the adverse effects reported to the manufacturers were available for some drugs (Table 3) (6, 10, 12, 15a, 17, 25, 44, 54, 61, 104). These studies provide the most comprehensive source of information on the types of adverse reactions encountered with each of several fluoroquinolones. These values are likely to be an upper estimate of drug-related adverse effects because experiences thought possibly, probably, and definitely drug related were included. Similar frequencies of overall adverse experiences were seen from these data as were seen in the tabulation of data from published studies. An additional group of 140 granulocytopenic cancer patients given norfloxacin as prophylaxis had a very high rate of total adverse experiences (63%), but in only 2 patients (1.4%) were the effects thought to be related to norfloxacin; many of the adverse effects were likely related to antitumor chemotherapy (104).

Table 2. Clinical adverse effects of fluoroquinolones as tabulated from published clinical studies

Drug and type of study[a]	No. of patients	No. of adverse experiences by organ system (%)			
		Total	Gastro-intestinal	Central nervous system	Skin-allergic
Norfloxacin					
UTI	2,591	135	84	41	10
STD	418	52	24	28	0
GI[b]	414	9	9		
Total	3,423	196 (5.7)	117 (3.4)	69 (2.0)	10 (0.3)
Ciprofloxacin					
Mixed	388	42	32	1	9
UTI	1,063	80	40	32	8
STD	476	27	19	8	0
GI	241	10	7	0	3
RTI[c]	1,117	107	67	28	12
BJ	145	30	20	4	6
SK-ST	232	31	23	5	3
Total	3,662	327 (8.9)	208 (5.6)	78 (2.1)	41 (1.1)
Ofloxacin					
STD	320	2	0	2	0
RTI	49	5	3	1	1
Total	369	7 (1.9)	3 (0.8)	3 (0.8)	1 (0.3)
Enoxacin					
UTI	48	13	5	7	1
STD	286	8	5	2	1
RTI	20	1	0	0	1
Total	354	22 (6.2)	10 (2.8)	9 (3)	3 (0.8)
Pefloxacin					
UTI	104	27	17	5	5
RTI	62	1	1	0	0
BJ	15	11	6	1	4
Total	181	39 (21)	24 (13)	6 (3)	9 (5)

[a]UTI, Urinary tract infections; STD, sexually transmitted diseases; GI, gastrointestinal infections; RTI, respiratory tract infections; BJ, bone and joint infections; SK-ST, skin and soft tissue infections.
[b]Includes bacterial diarrheas and prophylaxis in neutropenic patients.
[c]Includes patients with pneumonia and bronchitis.

In only a small percentage (<1.0 to 2.6%) of patients treated with norfloxacin (106), ciprofloxacin (6), or ofloxacin (61) were the adverse experiences sufficiently severe to prompt cessation of treatment.

Table 3. Clinical adverse effects of fluoroquinolones as reported to the manufacturers

Category[a]	No. of patients (%)				
	Norfloxacin (104)[b]	Ciprofloxacin (10)	Ofloxacin (12, 54)	Pefloxacin (40)	Enoxacin (25)
Total studied	2,206	5,512	15,641	1,181	2,407
Total with AE	202 (9.1)	354 (6.4)	704 (4.5)	94 (8.0)	149 (6.2)
Total requiring cessation of therapy	(<1.0)	25 (2.0)[c]	231 (1.5)	26 (2.2)[d]	132 (2.6)[d]
Types of AE					
GI	85 (3.9)[e]	230 (4.2)[f]	464 (3.0)	66 (5.6)	91 (3.8)
Nausea or vomiting	44 (2.0)	73 (2.5)	160 (1.0)[g]	42 (3.6)	
Abdominal discomfort	26 (1.2)	10 (0.6)[h]	171 (1.1)[g]	15 (1.3)	
Diarrhea	9 (0.4)	49 (1.7)	73 (0.5)	4 (0.3)	
Other	6 (0.3)	7 (0.4)	60 (0.4)		
CNS	103 (4.4)	84 (1.5)	152 (1.0)	11 (0.9)	29 (1.2)
Headache	34 (1.4)	7 (0.4)[h]	29 (0.2)	1 (0.08)	
Dizziness	29 (1.2)	10 (0.6)[h]	28 (0.2)	3 (0.2)	
Sleep disorder	13 (0.6)		48 (0.3)	3 (0.2)	
Mood change	13 (0.6)		18 (0.1)		
Seizures	4 (0.2)	2 (0.4)			
Other	10 (0.4)		29 (0.2)		
Skin-allergic	11 (0.5)[e]	51 (0.9)	71 (0.4)	26 (2.2)	14 (0.6)
Rash	11 (0.5)	19 (0.6)[i]	48 (0.3)	15 (1.3)	
Pruritis		7 (0.4)[f]	14 (0.1)	0 (0)	

[a]AE, Adverse effects; GI, gastrointestinal; CNS, central nervous system.
[b]Reference number.
[c]Data from reference 6, including a total of 1,241 patients.
[d]Data from reference 40, based on denominators of 1,181 patients for enoxacin and 5,088 patients for pefloxacin.
[e]Granulocytopenic patients not included, leaving a total of 2,206 patients assessed for gastrointestinal or skin symptoms.
[f]Japanese studies not subcategorized, but included in the total; denominator for individual symptoms is 2,934.
[g]Of 87 cases listed as having either nausea, vomiting, or abdominal discomfort, 43 were assigned to "nausea or vomiting" and 44 were assigned to "abdominal discomfort."
[h]Denominator is 1,693.
[i]Denominator is 2,934.

Gastrointestinal reactions

In general, gastrointestinal symptoms were the most frequent symptoms reported, occurring in 3.0 to 5.6% of patients overall. These symptoms have included nausea, vomiting, abdominal discomfort, and anorexia and in many instances likely resulted from gastric irritation. In some cases, however, nausea and vomiting may have resulted from toxicity to the central nervous system or theophylline toxicity (see below). Patients should be cautioned not to take antacids for these symptoms because those antacids containing magnesium or aluminum salts impair fluoroquinolone absorption from the gastrointestinal tract (see below). Diarrhea developing during fluoroquinolone therapy has been infrequent, and the development of antibiotic-associated colitis, as detected by the presence of *Clostridium difficile* cytotoxin in the stool, has been rare, having been documented in only one or two cases to date (10, 40).

Central nervous system reactions

Symptoms referable to the nervous system have been reported in 0.9 to 4.4% of patients receiving the newer fluoroquinolones. Mild headache or dizziness has predominated, followed by sleep disturbance or mood alteration (agitation, anxiety, or depression). With some earlier quinolone analogs dizziness was a major side effect, occurring in as many as half of patients treated with oxolinic acid (88) or rosoxacin (41). Hallucinations and confusion have been uncommon with the newer agents. Seizures have been reported in small numbers of patients receiving nalidixic acid (50, 78), norfloxacin (104), ciprofloxacin (6, 7, 26, 73), and enoxacin (18; K. J. Simpson, Letter, *Lancet* **ii:**161, 1985). In most patients given norfloxacin (104) and in some patients given ciprofloxacin (26), underlying conditions have been sufficient to account for seizures, although these therapies may have lowered the seizure threshold. Concurrent therapy with theophylline and ciprofloxacin or enoxacin and the resultant theophylline accumulation may have accounted for some (6, 18, 73), but not all (Simpson, Letter, *Lancet* **ii:**161, 1985), cases with seizures (see below). Recent reports from Japan also suggest that concurrent therapy with enoxacin and certain nonsteroidal anti-inflammatory agents may predispose to the development of seizures (K. Morikawa, O. Nagata, S. Kubo, H. Kato, and K. Yamamoto, *Program Abstr. 27th Intersci. Conf. Antimicrob. Agents Chemother.*, abstr. no. 255, 1987) (see below).

Also suggesting the potential for central nervous system toxicity in some members of the quinolone class of compounds is amfenolic acid, a compound which was originally investigated as a central nervous system stimulant (1, 52, 59). This compound is a derivative of nalidixic acid containing a methylbenzene substituent at position 7 of the 1,8-naphthyridine ring.

The mechanism by which quinolones affect the nervous system is uncertain, but these effects have been associated with the ability of this class of compounds to inhibit the specific binding of the inhibitory neurotransmitter γ-aminobutyric acid to synaptic membranes prepared from rat brain (102). The ability of penicillins to produce seizures when present in high concentrations has also been suggested to result from their ability to inhibit γ-aminobutyric acid binding to its receptors (5).

Skin and hypersensitivity reactions

Dermatologic reactions to fluoroquinolones have been infrequent, occurring in 0.4 to 2.2% of patients. Unspecified rashes have been most frequent, followed by pruritis. Urticaria has been rare. Photosensitivity has been reported in patients given nalidixic acid (13, 15), pefloxacin (40), ofloxacin (40, 51), and ciprofloxacin (10, 51), but has not yet been reported with norfloxacin (104). Skin reactions in patients receiving nalidixic acid have been reported after exposure to light of both UV (320- to 400-nm) and visible (>400-nm) wavelengths (15), but no information on the wavelength specificity of photosensitive eruptions associated with the newer agents is available. Many of the rashes occurring with nalidixic acid have tended to have subepidermal bulla formation and have been thought to be idiosyncratic hypersensitivity phenomena (13, 15). Skin reactions have been reversible on cessation of drug administration.

Drug fever has been reported rarely (12).

The administration of an intravenous formulation of ciprofloxacin has been associated with local irritative reactions and phlebitis at the site of infusion in 18% of patients in one study (86). Information on these reactions with the intravenous preparations of enoxacin and pefloxacin under development is not yet available. Rapid intravenous infusions of fluoroquinolones have been associated with acute changes in blood pressure (92).

Arthropathy

Rheumatologic adverse experiences during quinolone treatment occur uncommonly. Arthralgias and myalgias have been reported in small numbers of patients receiving nalidixic acid (9), norfloxacin (8, 104), ciprofloxacin (2, 10), ofloxacin (12, 61), and pefloxacin (40), and reversible joint swelling and tendonitis have recurred in two patients rechallenged with norfloxacin (8, 19, 53). In animals quinolones have been shown to produce cartilage erosions in weight-bearing joints of juvenile animals and inhibition of cartilaginous embryonic limb bud growth (16, 17, 49, 61, 83, 99). In a few animals nalidixic acid has also produced some disorganization of the epiphyseal plates (49). Arthropathy has developed with lesser frequency in adult dogs given pefloxacin for prolonged periods and can develop to a lesser degree in immobilized joints (16). The mechanism of these effects in animals is unknown.

The extent to which arthropathy will pose a limitation to fluoroquinolone therapy in children remains unclear. A retrospective case control study of a small number of pediatric patients suggested no difference in the occurrence of arthralgia between those patients receiving nalidixic acid (from 9 to 600 days) and those not receiving the drug (81). No abnormalities in skeletal growth were seen in the nalidixic acid group. Of 30 pediatric patients with cystic fibrosis, 1 is reported to have developed arthropathy while receiving a high dose of ciprofloxacin (1,500 mg/day for a 35-kg patient); all joint symptoms subsided within 2 weeks after cessation of therapy (71). The same report mentioned that 10% of older patients with cystic fibrosis may develop arthropathy, a percentage much higher than reported from adult studies. Others, however, have seen little joint toxicity in small numbers of cystic fibrosis patients younger than 17 years (85). More experience is necessary to assess the likelihood of joint toxicity in children in general and to assess whether toxicity is dose related.

In sum, direct cartilage damage has not yet been reported in humans receiving fluoroquinolone treatment (10), but concerns of the potential for this toxicity based on animal studies have led to recommendations that the fluoroquinolones not be used in patients whose skeletal growth is incomplete or in pregnant women, particularly in cases in which alternative therapies exist.

Laboratory Test Abnormalities

Abnormalities of laboratory tests occurring during fluoroquinolone therapy have been noted in 4.5 to 11.6% of patients (Table 4). As with the reporting of clinical adverse experiences, these values are likely to be upper-limit estimates because of the inclusion of abnormalities possibly, probably, and definitely drug related.

Hematologic reactions

Hematologic abnormalities were found in 0.4 to 5.3% of patients and have included leukopenia, eosinophilia, anemia, and thrombocytosis. Leukopenia was frequently mild and generally did not require cessation of therapy. Even in the rare cases in which neutropenia was severe, it was reversible after therapy was stopped (L. Patoia, R. Guerciolini, F. Menichetti, G. Bucaneve, and A. Del Favero, Letter, *Ann. Intern. Med.* **107:**788–789, 1987). Mild anemia has been seen in 0.4 to 0.6% of patients given ofloxacin or pefloxacin (40, 61), and thrombocytosis has been seen in 0.06 to 0.6% of

Table 4. Laboratory test abnormalities during fluoroquinolone therapy

Category	No. of patients (%)				
	Norfloxacin (104)[a]	Ciprofloxacin (40)	Ofloxacin (40)	Pefloxacin (40)	Enoxacin (25)
Total studied	2,346	4,287	15,641	1,181	2,407
Total with AE[b]	273 (11.6)	261 (6.1)	1,189 (7.6)	53 (4.5)	
Types of AE					
Hematologic	98 (5.3)[c]	17 (0.4)	78 (0.5)	30 (2.5)	
Leukopenia	60 (3.3)	7 (0.2)[d]	78 (0.5)	3 (0.2)	
Eosinophilia	38 (2.1)	10 (0.2)	92 (0.6)	3 (0.2)	
Hepatic	49 (2.7)[c]	193 (4.5)	313 (2.0)	21 (1.8)	
Elevated serum transaminases	49 (2.7)	107 (2.5)[e]	375 (2.4)[e]	21 (1.8)[e]	
Renal	25 (1.4)[c]	17 (0.4)	125 (0.8)	2 (0.2)	
Elevated serum creatinine	15 (0.8)	8 (0.2)	205 (1.3)	2 (0.2)	
Proteinuria	10 (0.5)				

[a]Reference number.
[b]AE, Adverse effects.
[c]Denominator is 1,841 patients.
[d]From reference 7; denominator is 2,829 patients.
[e]Elevations in transaminases and alkaline phosphatase were combined.

patients given ciprofloxacin, pefloxacin, or ofloxacin (10, 40). The mechanism of these abnormalities is not known, but eosinophilia and leukopenia might represent idiosyncratic hypersensitivity reactions.

Hepatic reactions

Abnormalities of liver function have been seen in 1.8 to 2.7% of patients given fluoroquinolones. Elevations of serum transaminases have predominated and have been mild, seldom requiring cessation of treatment. Elevations of serum alkaline phosphatase have also been found, but the frequency of this abnormality is not well defined in the English literature. Hepatic failure secondary to fluoroquinolone therapy has not been reported.

Renal reactions

Mild worsening of renal function during fluoroquinolone therapy has been reported in 0.2 to 1.3% of patients. Crystalluria has been seen rarely in patients given norfloxacin (82, 97, 104) and ciprofloxacin (7, 15a, 87), but the relationship of crystalluria to elevations in serum creatinine is uncertain. Crystalluria appears most likely to occur when these agents are given in high doses and when the urine pH is near neutrality, the point of lowest drug solubility (97, 100). Analysis of the urine crystals induced in animals has revealed a complex of ciprofloxacin (or its metabolites), magnesium, and protein (83). Hematuria (ciprofloxacin [10, 35]), interstitial nephritis (ciprofloxacin [6]), or reversible acute renal failure (ciprofloxacin [42], norfloxacin [J. Boelaert, P. P. de Jaegere, R. Daneels, M. Schurgers, B. Gordts, and H. W. van Landuyt, Letter, *Clin. Nephrol.* **25**:272, 1986]) has been rarely seen during therapy with fluoroquinolones (92).

Rarely hyperuricemia has been encountered during fluoroquinolone therapy (24).

Other Potential Adverse Effects Based on Microbiological or Animal Studies

Ocular reactions

Toxicities of quinolones affecting the eye have been encountered in some animals and rarely in humans. Subcapsular cataracts have developed in rats treated with pefloxacin or dogs treated with rosoxacin for prolonged periods (16, 83). Cataracts have not been seen in other animals given ciprofloxacin or reported in humans receiving fluoroquinolones, however (10, 83). Alterations in retinal electrophysiology and histology have been seen in cats given nalidixic acid, but were not seen when the animals were given norfloxacin (17) or ciprofloxacin (83). Flumequine has produced bulla formation in the maculas of three patients with renal failure (46), but retinal toxicity has not been reported with the use of other fluoroquinolones in humans.

Teratogenicity and safety in pregnancy

Norfloxacin failed to produce malformations in the offspring of pregnant mice, rats, or rabbits, but in high doses producing levels threefold higher in serum than achievable in humans, it produced fetal loss in monkeys (17). No teratogenic effect was seen in monkeys given even higher doses of norfloxacin, and fetal loss was thought to result from the suppression of placental progesterone. Ciprofloxacin also produced no evidence of teratogenicity in rats, mice, and rabbits, but weight loss from gastrointestinal toxicity has been associated with an increased incidence of abortion in rabbits (data on file with Miles, Inc., Pharmaceutical Div., West Haven, Conn.; 15a). Dose-related skeletal abnormalities have been reported in rat fetuses exposed to ofloxacin in utero, but no teratogenicity was seen in rabbits (61, 98). Fetal loss was also seen in rabbits given ofloxacin, as occurred with ciprofloxacin (98). The use of fluoroquinolones in pregnant humans has been infrequent and, because of these and other (see arthropathy above) toxicities, is not currently recommended.

Mutagenicity

In bacteria quinolones interact with DNA gyrase and DNA in a manner that results in the induction of the RecA SOS DNA repair system (66, 68; see chapter 2); this repair is error prone and may thereby increase the frequency of bacterial mutations (65). The Rec assay of mutagenicity in *Bacillus subtilis* has been slightly positive with norfloxacin (17) and ofloxacin (61); other bacterial tests of mutagenicity (*Salmonella typhimurium* reverse mutation test [Ames assay]—norfloxacin [17] and ciprofloxacin [15a]; *Escherichia coli* DNA repair test—ciprofloxacin [15a]) have been negative, however.

In eucaryotic cells the homolog of DNA gyrase, topoisomerase II, is several orders of magnitude more resistant to quinolone action than is the bacterial enzyme (47), suggesting that the ability to produce mutagenic effects by interaction with topoisomerase II and DNA is substantially lower in eucaryotic cells. In most other tests of mutagenicity, norfloxacin, ciprofloxacin, and ofloxacin have produced negative results. Included have been tests of cytotoxicity, sister chromatid exchange, chromosomal aberration, and V-79 mammalian cell mutagenicity (norfloxacin [17]); the mouse micronucleus and dominant lethal tests have also been negative (ciprofloxacin [15a], ofloxacin [89, 90]). One eucaryotic test of mutagenicity, the in vitro rat hepatocyte primary culture-DNA repair test, has given positive results with norfloxacin, ciprofloxacin, ofloxacin, and pefloxacin, but not nalidixic acid (60). The same test conducted in vivo, however, was negative with ciprofloxacin (60); results for the other fluoroquinolones in this test were not given. The reasons for the differences in the results of tests conducted in vitro and in vivo are not certain. The evidence is thus conflicting as to the mutagenicity of the newer fluoroquinolones, although the preponderance of data suggests at this point that the potential for significant mutagenicity in humans is low.

Carcinogenicity

Long-term studies of the carcinogenicity of norfloxacin in rats and mice are negative (16, 17). Information on the carcinogenic potential of the other newer fluoroquinolones has not been reported.

Effects on leukocyte functions

A number of studies have assessed the effects of quinolones on leukocyte functions in vitro or in intact animals (20, 21, 29–32, 36, 37, 47, 56, 64, 69, 76, 79). The newer fluoroquinolones appear to penetrate well into leukocytes (22, 23; M. M. Zweerink and A. M. Edison, Letter, *J. Antimicrob.*

Chemother. **21**:266–267, 1988) and thus might affect their function. The data, however, are conflicting.

Effects on polymorphonuclear leukocytes were reported in three studies (21, 30, 56). Granulocyte chemotaxis in agar was unaffected by norfloxacin, ofloxacin, or ciprofloxacin in concentrations of up to 25 to 50 μg/ml; 10 to 20% decreases were seen at high concentrations of pefloxacin (25 μg/ml) and ciprofloxacin (200 μg/ml) (30, 56). No effect of quinolones on chemiluminescence induced by the treatment of granulocytes with zymosan or the tripeptide formylmethionyl-leucyl-phenylalanine was detected in one study in which norfloxacin, ciprofloxacin, and ofloxacin (0.2 to 25 μg/ml) were used (30), but other workers have noted a 70% increase in the chemiluminescence of granulocytes incubated with norfloxacin or enoxacin (10 to 20 μg/ml) (21). Ciprofloxacin had no effect in the latter study.

Studied most extensively have been the effects of quinolones when incubated with lymphocytes or peripheral blood mononuclear cells (a mixture of lymphocytes and monocytes) in vitro (20, 29, 31, 32, 36, 37, 47, 69, 76, 79). Studies of the incorporation of [^{3}H]thymidine into mitogen-stimulated (using phytohemagglutinin or concanavalin A in most studies) cells as a measure of cell proliferation and DNA synthesis have yielded variable results. Reversible inhibition (10 to 70%) of thymidine uptake into stimulated human peripheral blood mononuclear cells by incubation with difloxacin or A-55620 (5 to 25 μg/ml) (37) or into stimulated human lymphocytes by incubation with nalidixic acid (10 and 100 μg/ml), norfloxacin (3 and 30 μg/ml), or ofloxacin (3 and 30 μg/ml) (20) has been reported. Decreases were also seen with human peripheral blood mononuclear cells incubated with concentrations of pefloxacin, ciprofloxacin, and ofloxacin exceeding, but not below, 25 μg/ml in another study (76). Other workers, however, have reported no effect on human peripheral blood mononuclear cells incubated with ciprofloxacin (5 to 125 μg/ml) (36), amifloxacin, norfloxacin, or rosoxacin (0.1 to 10 μg/ml) (57) or on mouse splenic lymphocytes incubated with ciprofloxacin (0.01 to 10 μg/ml) (79) or ofloxacin (0.1 to 10 μg/ml) (69). Thymidine incorporation of stimulated spleen cells obtained from mice given ciprofloxacin (79) or ofloxacin (1 to 20 mg/kg of body weight) (69) for 7 days was similarly unaffected relative to results from control mice.

In contrast, one group has reported an increase of 45% in thymidine incorporation into stimulated human lymphocytes treated for a prolonged period (5 days) with any of several different quinolones (including norfloxacin, ciprofloxacin, ofloxacin, pefloxacin, difloxacin, and others) at concentrations of between 1.6 and 6.25 μg/ml (32). Subsequent studies by this group suggested that the increase in radiolabeled thymidine uptake seen with ciprofloxacin treatment resulted from an alteration in cellular thymidine pools because of the inhibition of de novo pyrimidine synthesis (31), and thus the actual effect of the drug on DNA synthesis remained unclear.

Proliferation of transformed lymphocyte cell lines, mitogen-stimulated lymphocytes, or lymphoblasts, as measured by cell titer, was inhibited only by high concentrations of ciprofloxacin or ofloxacin ($\geq$20 μg/ml) in vitro (29, 31, 32, 47). The progression of lymphocytes through the cell cycle also appeared to be inhibited by ciprofloxacin at a concentration of 20 μg/ml. Others have studied cellular responses after administration of the drug to live animals. Spleen cells obtained from mice sacrificed after 5 days of treatment with ciprofloxacin (40 mg/kg of body weight twice daily) had proliferative responses that did not differ from those of control mice given saline (36).

Finally, several groups have studied the effects of quinolones on other lymphocyte or monocyte-macrophage functions with similarly varied results. Production of interleukin-1 by murine macrophages (64) was increased by a low concentration of ciprofloxacin (0.5 and 2.5 μg/ml), and that by human monocytes was inhibited by a high concentration of the drug (100 μg/ml) (76). Gamma interferon production by mitogen-stimulated lymphocytes was inhibited by high concentrations of nalidixic acid (100 μg/ml) and norfloxacin (30 μg/ml), but not by ofloxacin (30 μg/ml) (20). Interleukin-2 receptor expression on lymphocytes was not detectably affected by difloxacin (5 to 25 μg/ml) (37).

In vitro, immunoglobulin secretion by lymphocytes was reduced progressively by ciprofloxacin in the concentration range of 5 to 20 μg/ml in one study (32), but no effect on specific antibody production was seen in another study in which ciprofloxacin up to a concentration of 10 μg/ml was used (79). Neither ciprofloxacin nor ofloxacin (1 to 20 mg/kg of body weight) given for 7 days to mice immunized by intraperitoneal injection of sheep erythrocytes inhibited the specific antibody response of spleen cells to this antigen (69, 79), and ciprofloxacin was, in fact, reported to augment antibody production by up to twofold (79).

The currently available array of information summarized above remains confusing, and the mechanism behind each of the alterations of host defense functions reported is uncertain. The concentrations of quinolones that in vitro inhibit by half the eucaryotic homolog of the bacterial target of these drugs, topoisomerase II, exceed 150 μg/ml, and DNA synthesis catalyzed by the purified complex of eucaryotic DNA polymerase α and DNA primase is similarly insensitive to these agents (47). Thus, other drug actions or compartmentalization of the drug within the cell must be postulated.

The clinical significance of the foregoing results is similarly unclear. Many of the effects observed occurred at drug concentrations not achievable in humans. Other antimicrobial agents in clinical use, such as β-lactams, tetracyclines, and gentamicin, may also affect the function of granulocytes in vitro (27, 33, 48), but they have not been associated with clinically important alterations in host defenses when used to treat human infections. It seems unlikely in light of substantial experience with clinical use of the newer fluoroquinolones that efficacy will be compromised by harmful effects on host defenses.

Effects on reproductive functions

Long-term toxicity studies in which high doses (100 mg/kg) of norfloxacin and pefloxacin given for 3 or more months were used have identified testicular atrophy in rats and dogs (16). The mechanisms are unknown, and these effects have not been described for humans.

Drug-Drug Interactions

Xanthines

Theophylline, when administered concurrently with certain fluoroquinolones, may accumulate, sometimes to toxic levels. This interaction has been greatest with enoxacin (62; F. P. V. Maesen, J. P. Teengs, C. Baur, and B. I. Davies, Letter, *Lancet* **ii:**530, 1984; W. J. A. Wijnands, C. L. A. van Herwaarden, and T. B. Vree, Letter, *Lancet* **ii:**109, 1984) and occurred to a lesser extent with ciprofloxacin (63, 74, 84) and pefloxacin (62); ofloxacin (34, 39) and norfloxacin (14, 62) had little effect. Enoxacin and ciprofloxacin appear to prolong the half-life of theophylline elimination, possibly by interfering with hepatic drug metabolism (62). Some adverse experiences (nausea, vomiting, agitation, and dizziness) during quinolone therapy may have resulted from excessive levels of theophylline in serum. Thus, in patients given theophylline with enoxacin, ciprofloxacin, pefloxacin, or other quinolones except norfloxacin and ofloxacin, theophylline levels should be carefully monitored.

Caffeine elimination was similarly prolonged by concurrent administration of enoxacin or ciprofloxacin, but not ofloxacin (93–96), and thus it is possible that in some patients anxiety or nervousness occurring during treatment with fluoroquinolones might have resulted from caffeine accumulation. Patients given enoxacin, ciprofloxacin, or pefloxacin should be cautioned about the possibility of this effect.

Agents affecting gastric acidity or motility

Antacids containing magnesium or aluminum salts, but not those containing calcium salts, have impaired the absorption of orally administered fluoroquinolones (38, 43, 80; L. W. Fleming, T. A. Moreland, W. K. Stewart, and A. C. Scott, Letter, *Lancet* **ii:**294, 1986; F. P. V. Maesen, B. I. Davies, W. H.

Geraedts, and C. A. Sumajow, Letter, *J. Antimicrob. Chemother.* **19:**848–850, 1987; L. C. Preheim, T. A. Cuevas, J. S. Roccaforte, M. A. Mellencamp, and M. J. Bittner, Letter, *Lancet* **ii:**48, 1986). Decreases of 6- to 10-fold in peak concentrations of ciprofloxacin in serum have been seen when a 500-mg dose is given concurrently with antacids. This effect is thought to result from the formation of drug-cation complexes, as occurs with the tetracyclines. It is not clear whether concurrent ingestion of dairy products will produce decreased absorption of fluoroquinolones, but this possibility seems unlikely because calcium is the major divalent cation present in these food products. As noted above, patients should be instructed that when they are taking quinolones they should avoid taking antacids if possible. Histamine receptor antagonists such as cimetidine or ranitidine might be substituted when indicated. If antacids are required, they should be taken no sooner than 2 h after a quinolone dose or only calcium-containing preparations should be used.

The histamine receptor antagonists cimetidine and ranitidine have not altered the absorption or elimination of ciprofloxacin (80) or ofloxacin (61), but have increased the half-life of elimination of pefloxacin, possibly by their inhibition of its hepatic metabolism by the cytochrome P-450 system (80). In contrast to ciprofloxacin and ofloxacin, pefloxacin is eliminated to a large degree by hepatic metabolism (see chapter 4).

Agents that delay gastric emptying (anticholinergics such as pirenzipine and *N*-butylscopolaminium bromide) have been reported to delay the absorption of ciprofloxacin, whereas agents accelerating gastric emptying (metoclopramide) accelerated its absorption (80). Similar effects have been reported with ofloxacin (61). It is doubtful that these effects will be of major clinical significance because quinolone bioavailability (as measured by the area under the curve of concentration in serum plotted versus time) was unaltered.

Nonsteroidal anti-inflammatory agents

The nonsteroidal anti-inflammatory agent fenbufen, when coadministered with enoxacin, has been associated with the development of seizures in patients in Japan (16). This finding in humans correlates with studies with mice in which combinations of fenbufen with quinolones produced seizures at concentrations of over 10-fold lower than those producing seizures with either drug alone (16; Morikawa et al., *27th ICAAC*). The ability of quinolones to displace γ-aminobutyric acid from its receptors on mouse synaptic membranes was also potentiated by some nonsteroidal anti-inflammatory agents (S. Hori, J. Shimada, A. Saito, T. Miyahara, S. Kurioka, and M. Matsuda, *27th ICAAC*, abstr. no. 30, 1987). Patients given fluoroquinolones other than enoxacin concurrently with nonsteroidal agents other than fenbufen have not been reported to develop seizures but should be carefully monitored.

Probenecid

The renal clearances of norfloxacin, enoxacin, ciprofloxacin, and ofloxacin exceed normal glomerular filtration rates by factors of 2 or more (45; see chapter 4), suggesting additional renal tubular secretion as occurs with other organic acids. Probenecid blocks renal tubular secretion and has been found to reduce urinary recovery and renal clearance of norfloxacin and ciprofloxacin by twofold (91, 105). Drug accumulation, however, did not result, presumably because the other pathways of drug elimination (glomerular filtration and hepatic metabolism) were sufficient to compensate.

Financial Toxicities

In the United States only norfloxacin and ciprofloxacin have been released for clinical use. The cost to a pharmacist of a 1-day treatment is $4.26 for norfloxacin (400 mg twice daily) (3) and $3.36 to $6.96 for ciprofloxacin (250 to 750 mg twice daily) (4). The cost to a patient will be higher and may vary. When compared with other generic antimicrobial agents given orally for the treatment of urinary tract infections (e.g., ampicillin, $0.54; trimethoprim-sulfamethoxazole, $0.45; nitrofurantoin, $0.14), these two quinolones are substantially more expensive. When the most suitable alternative agents are parenteral, however, oral quinolone therapy is more cost-effective. Many parenteral agents are more expensive than quinolones (e.g., amikacin, $46.64; ceftazidime, $78.21) (4), and the use of oral agents also avoids the costs of intravenous administration sets and nursing time. Additional savings may be gained if oral therapy allows avoidance of hospitalization, earlier discharge from the hospital, or avoidance of home intravenous therapy services (11, 70).

Conclusions

Contraindications to the use of the fluoroquinolones

The newer fluoroquinolones are contraindicated in patients with a history of allergic reactions to the older (nalidixic acid, cinoxacin, pipemidic acid, and oxolinic acid) analogs or to any of the newer agents. The fluoroquinolones should also be avoided in children with incomplete skeletal growth, pregnant patients, and nursing mothers. Although information is not available for humans, quinolone excretion into breast milk occurs in animals (data on file, Merck Sharp & Dohme, Rahway, N.J., and Miles, Inc., Pharmaceutical Div.). Because the cartilage toxicities of the quinolones have only been well documented for animals and because children have appeared to tolerate nalidixic acid (16) and ciprofloxacin reasonably well (71, 85), in patients for whom other alternative antimicrobial agents lack efficacy or have major contraindications, it may be possible to administer fluoroquinolones safely with careful monitoring. Patients with cystic fibrosis constitute a particular pediatric group in which fluoroquinolone therapy would be considered reasonable. Additional experience is necessary, however, before firm statements can be made.

Patients in whom special caution should be exercised

Particular attention to monitoring for central nervous system toxicities should be paid in patients with a previous history of a seizure disorder or structural lesions of the central nervous system that might predispose to seizures or in patients concurrently receiving nonsteroidal anti-inflammatory agents. In addition, patients receiving concurrent theophylline preparations should have monitoring of theophylline levels in serum when receiving enoxacin, ciprofloxacin, or pefloxacin (monitoring is not needed for norfloxacin or ofloxacin).

Patients receiving higher-than-recommended doses of fluoroquinolones and whose urine has a persistently neutral or alkaline pH because of infection with urea-splitting organisms or because of underlying renal tubular defects should be followed for the development of crystalluria or rises in serum creatinine.

Overall tolerability

Although this chapter has emphasized and detailed the adverse effects of the newer fluoroquinolone agents, the frequencies of toxicities encountered with these agents have been overall quite acceptable when compared with other antimicrobial agents in clinical use, as is emphasized by the data in Table 1. As more extensive use occurs, additional information about rare or late-developing toxicities may become available. The use of fluoroquinolones in doses higher than those used in most studies reviewed in the foregoing chapters on clinical uses is likely to be associated with more adverse effects. With careful dosing, however, studies to date indicate that quinolone intolerance is unlikely to present major limitations to the realization of the clinical potential of these agents as measured by their substantial efficacy in a broad range of infections.

Literature Cited

1. **Aceto, M. D., L. S. Harris, G. Y. Lesher, J. Pearl, and T. G. Brown, Jr.** 1967. Pharmacologic studies with 7-benzyl-1-ethyl-1,4-dihydro-4-oxo-1,8-naphthyridine-3-carboxylic acid. *J. Pharmacol. Exp. Ther.* **158:**286–293.
2. **Alfaham, M., M. E. Holt, and M. C. Goodchild.** 1987. Arthropathy in a patient with cystic fibrosis taking ciprofloxacin. *Br. Med. J.* **295:**699.
3. **Anonymous.** 1987. Norfloxacin (Noroxin). *Med. Lett. Drugs. Ther.* **29:**25–27.
4. **Anonymous.** 1988. Ciprofloxacin. *Med. Lett. Drugs. Ther.* **30:**11–13.
5. **Antoniadis, A., W. E. Muller, and U. Wollert.** 1980. Inhibition of GABA and benzodiazepine receptor binding by penicillins. *Neurosci. Lett.* **18:**309–312.
6. **Arcieri, G., R. August, N. Becker, C. Doyle, E. Griffith, G. Gruenwaldt, A. Heyd, and B. O'Brien.** 1986. Clinical experience with ciprofloxacin in the USA. *Eur. J. Clin. Microbiol.* **5:**220–225.
7. **Arcieri, G., E. Griffith, G. Gruenwaldt, A. Heyd, B. O'Brien, N. Becker, and R. August.** 1987. Ciprofloxacin: an update on clinical experience. *Am. J. Med.* **82**(Suppl. 4A): 381–394.
8. **Bailey, R. R., J. A. Kirk, and B. A. Peddie.** 1983. Norfloxacin-induced rheumatic disease. *N.Z. Med. J.* **96:**590.
9. **Bailey, R. R., R. Natale, and A. L. Linton.** 1972. Nalidixic acid arthralgia. *Can. Med. Assoc. J.* **107:**604–605.
10. **Ball, P.** 1986. Ciprofloxacin: an overview of adverse experiences. *J. Antimicrob. Chemother.* **18**(Suppl. D):187–193.
11. **Barriere, S. L.** 1987. Economic impact of oral ciprofloxacin. A pharmacist's perspective. *Am. J. Med.* **82**(Suppl. 4A):387–390.
12. **Blomer, R., K. Bruch, H. Krauss, and W. Wacheck.** 1986. Safety of ofloxacin—adverse drug reactions reported during phase-II studies in Europe and in Japan. *Infection* **14**(Suppl. 4):S332–S334.
13. **Boisvert, A., and G. Barbeau.** 1981. Nalidixic acid-induced photodermatitis after minimal sun exposure. *Drug Intell. Clin. Pharm.* **15:**126–127.
14. **Bowles, S. K., Z. Popovski, M. J. Rybak, H. B. Beckman, and D. J. Edwards.** 1988. Effect of norfloxacin on theophylline pharmacokinetics at steady state. *Antimicrob. Agents Chemother.* **32:**510–512.
15. **Brauner, G. J.** 1975. Bullous photoreaction to nalidixic acid. *Am. J. Med.* **58:**576–580.

15a. **Campoli-Richards, D. M., J. P. Monk, A. Price, P. Beufield, P. A. Todd, and A. Ward.** 1988. Ciprofloxacin. A review of its antibacterial activity, pharmacokinetic properties and therapeutic use. *Drugs* **35:**373–447.

16. **Christ, W., T. Lehnert, and B. Ulbrich.** 1988. Specific toxicologic aspects of the quinolones. *Rev. Infect. Dis.* **10**(Suppl. 1):S141–S146.
17. **Corrado, M. L., W. E. Struble, C. Peter, V. Hoagland, and J. Sabbaj.** 1987. Norfloxacin: review of safety studies. *Am. J. Med.* **82**(Suppl. 6B):22–26.
18. **Davies, B. I., F. P. V. Maesen, and J. P. Teengs.** 1984. Serum and sputum concentrations of enoxacin after single oral dosing in a clinical and bacteriological study. *J. Antimicrob. Chemother.* **14**(Suppl. C):83–89.
19. **Deaney, N. B., R. Vogel, M. J. Vandenburg, and W. J. C. Currie.** 1984. Norfloxacin in acute urinary tract infections. *Practitioner* **228:**111–117.
20. **De Simone, C., L. Baldinelli, M. Ferrazzi, S. De Santis, L. Pugnaloni, and F. Sorice.** 1986. Influence of ofloxacin, norfloxacin, nalidixic acid, pyromidic acid and pipemidic acid on human gamma-interferon production and blastogenesis. *J. Antimicrob. Chemother.* **17:**811–814.

21. **Duncker, D., and U. Ullmann.** 1986. Influence of various antimicrobial agents on the chemiluminescence of phagocytosing human granulocytes. *Chemotherapy* (Basel) **32:** 18–24.
22. **Easmon, C. S. F., and J. P. Crane.** 1985. Uptake of ciprofloxacin by human neutrophils. *J. Antimicrob. Chemother.* **16:**67–73.
23. **Easmon, C. S. F., and J. P. Crane.** 1985. Uptake of ciprofloxacin by macrophages. *J. Clin. Pathol.* **38:**442–444.
24. **Eron, L. J., L. Harvey, D. L. Hixon, and D. M. Poretz.** 1985. Ciprofloxacin therapy of infections caused by *Pseudomonas aeruginosa* and other resistant bacteria. *Antimicrob. Agents Chemother.* **27:**308–310.
25. **Fass, R. J.** 1987. Adverse reactions associated with quinolones. *Quinolones Bull.* **3:**5–6.
26. **Fass, R. J.** 1987. Efficacy and safety of oral ciprofloxacin in the treatment of serious respiratory infections. *Am. J. Med.* **82**(Suppl. 4A):202–207.
27. **Fietta, A., F. Sacchi, C. Bersani, F. Grassi, P. Mangiarotti, and G. G. Grassi.** 1983. Effect of β-lactam antibiotics on migration and bactericidal activity of human phagocytes. *Antimicrob. Agents Chemother.* **23:**930–931.
28. **Fong, I. W., W. Linton, M. Simbul, R. Thorup, B. McLaughlin, V. Rahm, and P. A. Quinn.** 1987. Treatment of nongonococcal urethritis with ciprofloxacin. *Am. J. Med.* **82**(Suppl. 4A):311–316.
29. **Forsgren, A., A.-K. Bergh, M. Brandt, and G. Hansson.** 1986. Quinolones affect thymidine incorporation into the DNA of human lymphocytes. *Antimicrob. Agents Chemother.* **29:**506–508.
30. **Forsgren, A., and P.-I. Bergkvist.** 1985. Effect of ciprofloxacin on phagocytosis. *Eur. J. Clin. Microbiol.* **4:**575–578.
31. **Forsgren, A., A. Bredberg, A. V. Pardee, S. F. Schlossman, and T. F. Tedder.** 1987. Effects of ciprofloxacin on eucaryotic pyrimidine nucleotide biosynthesis and cell growth. *Antimicrob. Agents Chemother.* **31:**774–779.
32. **Forsgren, A., S. F. Schlossman, and T. F. Tedder.** 1987. 4-Quinolone drugs affect cell cycle progression and function of human lymphocytes in vitro. *Antimicrob. Agents Chemother.* **31:**768–773.
33. **Forsgren, A., D. Schmeling, and P. G. Quie.** 1974. Effect of tetracycline on the phagocytic function of human leucocytes. *J. Infect. Dis.* **130:**412–415.
34. **Fourtillan, J. B., J. Granier, B. Saint-Salvi, J. Salmon, A. Surjus, D. Tremblay, M. Vincent du Laurier, and S. Beck.** 1986. Pharmacokinetics of ofloxacin and theophylline alone and in combination. *Infection* **14**(Suppl. 1):S67–S69.
35. **Garlando, F., M. G. Täuber, B. Joos, O. Oelz, and R. Lüthy.** 1985. Ciprofloxacin-induced hematuria. *Infection* **13:**177–178.
36. **Gollapudi, S. V. S., R. H. Prabhala, and H. Thadepalli.** 1986. Effect of ciprofloxacin on mitogen-stimulated lymphocyte proliferation. *Antimicrob. Agents Chemother.* **29:** 337–338.
37. **Gollapudi, S. V. S., B. Vayuvegula, S. Gupta, M. Fok, and H. Thadepalli.** 1986. Arylfluoroquinolone derivatives A-56619 (difloxacin) and A-56620 inhibit mitogen-induced human mononuclear cell proliferation. *Antimicrob. Agents Chemother.* **30:**390–394.
38. **Golper, T. A., A. I. Hartstein, V. H. Morthland, and J. M. Christensen.** 1987. Effects of antacids and dialysate dwell times on multiple-dose pharmacokinetics of oral ciprofloxacin in patients on continuous ambulatory peritoneal dialysis. *Antimicrob. Agents Chemother.* **31:**1787–1790.
39. **Gregoire, S. L., T. H. Grasela, Jr., J. P. Freer, K. J. Tack, and J. J. Schentag.** 1987. Inhibition of theophylline clearance by coadministered ofloxacin without alteration of theophylline effects. *Antimicrob. Agents Chemother.* **31:**375–378.

40. **Halkin, H.** 1988. Adverse effects of the fluoroquinolones. *Rev. Infect. Dis.* **10**(Suppl. 1):S258–S261.
41. **Handsfield, H. H., F. N. Judson, and K. K. Holmes.** 1981. Treatment of uncomplicated gonorrhea with rosoxacin. *Antimicrob. Agents Chemother.* **20:**625–629.
42. **Hessen, M. T., M. J. Ingerman, D. H. Kaufman, P. Weiner, J. Santoro, O. M. Korzeniowski, J. Boscia, M. Topiel, L. M. Bush, D. Kaye, and M. E. Levison.** 1987. Clinical efficacy of ciprofloxacin therapy for gram-negative bacillary osteomyelitis. *Am. J. Med.* **82**(Suppl. 4A):262–265.
43. **Höffken, G., K. Borner, P. D. Glatzel, P. Koepe, and H. Lode.** 1985. Reduced enteral absorption of ciprofloxacin in the presence of antacids. *Eur. J. Clin. Microbiol.* **4:**345.
44. **Holmes, B., R. N. Brogden, and D. M. Richards.** 1985. Norfloxacin. A review of its antibacterial activity, pharmacokinetic properties and therapeutic use. *Drugs* **30:** 482–513.
45. **Hooper, D. C., and J. S. Wolfson.** 1985. The fluoroquinolones: pharmacology, clinical uses, and toxicities in humans. *Antimicrob. Agents Chemother.* **28:**716–721.
46. **Hurault de Ligny, B., D. Sirbat, M. Kessler, P. Trechot, and J. Chanliau.** 1984. Effects secondaries ocularies de la fluméquine. Trois observations d'atteintes maculaires. *Therapie* **39:**595–600.
47. **Hussy, P., G. Maass, B. Tummler, F. Grosse, and V. Schomburg.** 1986. Effect of 4-quinolones and novobiocin on calf thymus DNA polymerase α primase complex, topoisomerases I and II, and growth of mammalian lymphoblasts. *Antimicrob. Agents Chemother.* **29:**1073–1078.
48. **Iannello, D., D. Delfino, M. Carbone, M. Fera, and T. F. Curo.** 1983. *In vitro* effects of cephalosporin and ampicillin on some human leukocyte functions. *Drugs Exp. Clin. Res.* **9:**67–71.
49. **Ingham, B., D. W. Brentnall, E. A. Dale, and J. A. McFadzean.** 1977. Arthropathy induced by antibacterial fused N-alkyl-4-pyridone-3-carboxylic acids. *Toxicol. Lett.* **1:** 21–26.
50. **Islam, M. A., and T. Sreedharan.** 1965. Convulsions, hyperglycaemia and glycosuria from overdose of nalidixic acid. *J. Am. Med. Assoc.* **192:**1000–1001.
51. **Jensen, T., S. S. Pedersen, C. H. Nielsen, N. Hoiby, and C. Koch.** 1987. The efficacy and safety of ciprofloxacin and ofloxacin in chronic *Pseudomonas aeruginosa* infection in cystic fibrosis. *J. Antimicrob. Chemother.* **20:**585–594.
52. **Juorio, A. V.** 1982. The effects of amfonelic acid and some other central stimulants on mouse striatal tyramine, dopamine and homovanillic acid. *Br. J. Pharmacol.* **77:** 511–515.
53. **Kirby, C. P.** 1984. Treatment of simple urinary tract infections in general practice with a 3-day course of norfloxacin. *J. Antimicrob. Chemother.* **13**(Suppl. B):107–112.
54. **Koverech, A., M. Picari, F. Granata, R. Fostini, D. Toniolo, and G. Recchia.** 1986. Safety profile of ofloxacin: the Italian data base. *Infection* **14**(Suppl. 4):S335–S337.
55. **Kromann-Andersen, B., P. Sommer, C. Pers, V. Larsen, and F. Rasmussen.** 1986. Clinical evaluation of ofloxacin versus ciprofloxacin in complicated urinary tract infections. *Infection* **14**(Suppl. 4):S305–S306.
56. **Lombard, J.-Y., J. Descotes, and J.-C. Evreux.** 1987. Polymorphonuclear leucocyte chemotaxis little affected by three quinolones *in vitro*. *J. Antimicrob. Chemother.* **20:** 614–615.
57. **Manzella, J. P., and J. K. Clark.** 1988. Effects of quinolones on mitogen-stimulated human mononuclear leucocytes. *J. Antimicrob. Chemother.* **21:**183–186.
58. **Mayrer, A. R., and V. T. Andriole.** 1982. Urinary tract antiseptics. *Med. Clin. North Am.* **66:**199–208.

59. **McMillen, B. A., and P. A. Short.** 1978. Amfonelic acid, a non-amphetamine stimulant, has marked effects on brain dopamine metabolism but not noradrenaline metabolism: association with differences in neuronal storage systems. *J. Pharm. Pharmacol.* **30:**464–466.
60. **McQueen, C. A., and G. M. Williams.** 1987. Effects of quinolone antibiotics in tests of genotoxicity. *Am. J. Med.* **82**(Suppl. 4A):94–96.
61. **Monk, J. P., and D. M. Campoli-Richards.** 1987. Ofloxacin. A review of its antibacterial activity, pharmacokinetic properties and therapeutic use. *Drugs* **33:**346–391.
62. **Niki, Y., R. Soejima, H. Kawane, M. Sumi, and S. Umeki.** 1987. New synthetic quinolone antibacterial agents and serum concentration of theophylline. *Chest* **92:** 663–669.
63. **Nix, D. E., J. M. DeVito, M. A. Whitbread, and J. J. Schentag.** 1987. Effect of multiple dose oral ciprofloxacin on the pharmacokinetics of theophylline and indocyanine green. *J. Antimicrob. Chemother.* **19:**263–269.
64. **Petit, J.-C., G.-L. Daguet, G. Richard, and B. Burghoffer.** 1987. Influence of ciprofloxacin and piperacillin on interleukin-1 production by murine macrophages. *J. Antimicrob. Chemother.* **20:**615–617.
65. **Phillips, I.** 1987. Bacterial mutagenicity and the 4-quinolones. *J. Antimicrob. Chemother.* **20:**771–782.
66. **Phillips, I., E. Culebras, F. Moreno, and F. Baquero.** 1987. Induction of the SOS response by new 4-quinolones. *J. Antimicrob. Chemother.* **20:**631–638.
67. **Pichler, H., G. Diridl, and D. Wolf.** 1986. Ciprofloxacin in the treatment of acute bacterial diarrhea: a double blind study. *Eur. J. Clin. Microbiol.* **5:**241–243.
68. **Piddock, L. J. V., and R. Wise.** 1987. Induction of the SOS response in *Escherichia coli* by 4-quinolone antimicrobial agents. *FEMS Microbiol. Lett.* **41:**289–294.
69. **Pulverer, G., W. Roszkowski, H. L. Ko, K. Roszkowski, and J. Jeljaszewicz.** 1986. Tierexperimentelle Untersuchungen zur Beeinflussung des Immunsystems durch Ofloxacin. *Infection* **14**(Suppl. 1):S40–S44.
70. **Quintiliani, R., B. W. Cooper, L. L. Briceland, and C. H. Nightingale.** 1987. Economic impact of streamlining antibiotic administration. *Am. J. Med.* **82**(Suppl. 4A):391–394.
71. **Raeburn, J. A., J. R. W. Govan, W. M. McCrae, A. P. Greening, P. S. Collier, M. E. Hodson, and M. C. Goodchild.** 1987. Ciprofloxacin therapy in cystic fibrosis. *J. Antimicrob. Chemother.* **20:**295–296.
72. **Ramirez-Ronda, C. H., S. Saavedra, and C. R. Rivera-Vázquez.** 1987. Comparative, double-blind study of oral ciprofloxacin and intravenous cefotaxime in skin and skin structure infections. *Am. J. Med.* **82**(Suppl. 4A):220–223.
73. **Raoof, S., C. Wollschlager, and F. Khan.** 1986. Treatment of respiratory tract infections with ciprofloxacin. *J. Antimicrob. Chemother.* **18**(Suppl. D):139–145.
74. **Raoof, S., C. Wollschlager, and F. A. Khan.** 1987. Ciprofloxacin increases serum levels of theophylline. *Am. J. Med.* **82**(Suppl. 4A):115–118.
75. **Renkonen, O.-V., A. Sivonen, and R. Visakorpi.** 1987. Effect of ciprofloxacin on carrier rate of *Neisseria meningitidis* in army recruits in Finland. *Antimicrob. Agents Chemother.* **31:**962–963.
76. **Roche, Y., M.-A. Gougerot-Pocidalo, M. Fay, D. Etienne, N. Forest, and J.-J. Pocidalo.** 1987. Comparative effects of quinolones on human mononuclear leucocyte functions. *J. Antimicrob. Chemother.* **19:**781–790.
77. **Roddy, R. E., H. H. Handsfield, and E. W. Hook III.** 1986. Comparative trial of single-dose ciprofloxacin and ampicillin plus probenecid for treatment of gonococcal urethritis in men. *Antimicrob. Agents Chemother.* **30:**267–269.

78. **Ronald, A. R., M. Turck, and R. G. Petersdorf.** 1966. A critical evaluation of nalidixic acid in urinary tract infections. *N. Engl. J. Med.* **275:**1081–1089.
79. **Roszkowski, W., H. L. Ko, K. Roszkowski, P. Ciborowski, J. Jeljaszewicz, and G. Pulverer.** 1986. Effects of ciprofloxacin on the humoral and cellular immune responses in Balb/c-mice. *Zentralbl. Bakteriol. Mikrobiol. Hyg. Ser. A* **262:**396–402.
80. **Rubinstein, E., and S. Segev.** 1987. Drug interactions of ciprofloxacin with other non-antibiotic agents. *Am. J. Med.* **82**(Suppl. 4A)**:**119–123.
81. **Schaad, U. B., and J. Wedgwood-Krucko.** 1987. Nalidixic acid in children: retrospective matched controlled study for cartilage toxicity. *Infection* **15:**165–168.
82. **Schaeffer, A. J.** 1987. Multiclinic study of norfloxacin for treatment of complicated or uncomplicated urinary tract infections. *Am. J. Med.* **82**(Suppl. 6B)**:**53–64.
83. **Schlüter, G.** 1987. Ciprofloxacin: review of potential toxicologic effects. *Am. J. Med.* **82**(Suppl. 4A)**:**91–93.
84. **Schwartz, J., L. Jauregui, J. Lettieri, and K. Bachmann.** 1988. Impact of ciprofloxacin on theophylline clearance and steady-state concentrations in serum. *Antimicrob. Agents Chemother.* **32:**75–77.
85. **Scully, B. E., M. Nakatomi, C. Ores, S. Davidson, and H. C. Neu.** 1987. Ciprofloxacin therapy in cystic fibrosis. *Am. J. Med.* **82**(Suppl. 4A)**:**196–201.
86. **Scully, B. E., and H. C. Neu.** 1987. Treatment of serious infections with intravenous ciprofloxacin. *Am. J. Med.* **82**(Suppl. 4A)**:**369–375.
87. **Self, P. L., B. A. Zeluff, D. Sollo, and L. O. Gentry.** 1987. Use of ciprofloxacin in the treatment of serious skin and skin structure infections. *Am. J. Med.* **82**(Suppl. 4A)**:**239–241.
88. **Shapera, R. M., and J. M. Matsen.** 1977. Oxolinic acid therapy for urinary tract infections in children. *Am. J. Dis. Child.* **131:**34–37.
89. **Shimada, H., Y. Ebine, Y. Kurosawa, and T. Arauchi.** 1984. Mutagenicity studies of DL-8280, a new antibacterial drug. *Chemotherapy* (Toyko) **32**(Suppl. 1)**:**1162–1170.
90. **Shimada, H., Y. Ebine, T. Sato, Y. Kurosawa, and T. Arauchi.** 1985. Dominant lethal study in male mice treated with ofloxacin, a new antimicrobial drug. *Mutat. Res.* **144:** 51–55.
91. **Shimada, J., T. Yamaji, Y. Ueda, H. Uchida, H. Kusajima, and T. Irikura.** 1983. Mechanism of renal excretion of AM-715, a new quinolonecarboxylic acid derivative, in rabbits, dogs, and humans. *Antimicrob. Agents Chemother.* **23:**1–7.
92. **Smith, C. R.** 1987. The adverse effects of fluoroquinolones. *J. Antimicrob. Chemother.* **19:**709–712.
93. **Staib, A. H., S. Harder, S. Mieke, C. Beer, W. Stille, and P. Shah.** 1987. Gyrase-inhibitors impair caffeine elimination in man. *Methods Find. Exp. Clin. Pharmacol.* **9:**193–198.
94. **Staib, A. H., S. Harder, S. Mieke, C. Beer, W. Stille, P. M. Shah, and K. Frech.** 1986. Gyrase-Hemmer können die Coffein-Elimination verzögern. *Dtsch. Med. Wochenschr.* **111:**1500.
95. **Staib, A. H., S. Harder, A. Papenburg, and W. Stille.** 1987. Gyrasehemmer beeinflussen Metabolisierungsprozesse in der Leber. *Dtsch. Med. Wochenschr.* **112:**1720–1721.
96. **Stille, W., S. Harder, S. Mieke, C. Beer, P. M. Shah, K. Frech, and A. H. Staib.** 1987. Decrease of caffeine elimination in man during co-administration of 4-quinolones. *J. Antimicrob. Chemother.* **20:**729–734.
97. **Swanson, B. N., V. K. Boppana, P. H. Vlasses, H. H. Rotmensch, and R. K. Ferguson.** 1983. Norfloxacin disposition after sequentially increasing oral doses. *Antimicrob. Agents Chemother.* **23:**284–288.
98. **Takayama, S., T. Watanabe, Y. Akiyama, K. Ohura, S. Harada, K. Matsuhashi, K. Mochida, and N. Yamashita.** 1986. Reproductive toxicity of ofloxacin. *Arzneim. Forsch.* **36**(II)**:**1244–1248.

99. **Tatsumi, H. H. Senda, S. Yatera, Y. Takemoto, M. Yamayoshi, and K. Ohnishi.** 1978. Toxicological studies on pipemidic acid. V. Effect on diarthrodial joints of experimental animals. *J. Toxicol. Sci.* **3**:357–367.

100. **Thorsteinsson, S. B., T. Bergan, S. Oddsdottir, R. Rohwedder, and R. Holm.** 1986. Crystalluria and ciprofloxacin, influence of urinary pH and hydration. *Chemotherapy* (Basel) **32**:408–417.

101. **Tsuei, S. E., A. S. Darragh, and I. Brick.** 1984. Pharmacokinetics and tolerance of enoxacin in healthy volunteers administered at a dosage of 400 mg twice daily for 14 days. *J. Antimicrob. Chemother.* **14**(Suppl. C):71–74.

102. **Tsuji, A., H. Sato, Y. Kume, I. Tamai, E. Okezaki, O. Nagata, and H. Kato.** 1988. Inhibitory effects of quinolone antibacterial agents on γ-aminobutyric acid binding to receptor sites in rat brain membranes. *Antimicrob. Agents Chemother.* **32**:190–194.

103. **Urinary Tract Infection Study Group.** 1987. Coordinated multicenter study of norfloxacin versus trimethoprim-sulfamethoxazole treatment of symptomatic urinary tract infections. *J. Infect. Dis.* **155**:170–177.

104. **Wang, C., J. Sabbaj, M. Corrado, and V. Hoagland.** 1986. World-wide clinical experience with norfloxacin: efficacy and safety. *Scand. J. Infect. Dis.* **48**(Suppl.):81–89.

105. **Wingender, W., D. Beermann, D. Foerster, K. H. Graefe, P. Schacht, and V. Scharbrodt.** 1985. Mechanism of renal excretion of ciprofloxacin, a new quinolone carboxylic acid derivative, in humans. *Chemioterapia* **4**(Suppl. 2):403–404.

106. **Wolfson, J. S., and D. C. Hooper.** 1988. Norfloxacin: a new targeted fluoroquinolone antimicrobial agent. *Ann. Intern. Med.* **108**:238–251.

107. **Wollschlager, C. M., S. Raoof, F. A. Khan, J. J. Guarneri, V. LaBombardi, and Q. Afzal.** 1987. Controlled, comparative study of ciprofloxacin versus ampicillin in treatment of bacterial respiratory tract infections. *Am. J. Med.* **82**(Suppl. 4A):164–168.

Chapter 15

Quinolone Antimicrobial Agents: Overview and Conclusions

Robert C. Moellering, Jr.

Department of Medicine, New England Deaconess Hospital, and Harvard Medical School, Boston, Massachusetts 02115

The preceding chapters have clearly defined the properties and promise of the "new quinolones." The rebirth of these compounds is of interest because antimicrobial agents of this family have been known for more than a quarter of a century (5). Nonetheless, the newer agents represent major advances when compared with the older compounds such as nalidixic acid, oxolinic acid, cinoxacin, pipemidic acid, and others. Although these older quinolones did enjoy a measure of success, these agents were never considered first-line antimicrobial agents. Among their major drawbacks is the fact that they have very narrow spectra of activity, encompassing a variety of members of the family *Enterobacteriaceae* and certain other gram-negative bacilli, but not including *Pseudomonas aeruginosa, Serratia marcescens,* any gram-positive organisms, or anaerobes. An even greater problem is the fact that single-step mutations leading to high-level resistance in bacteria occur with frequencies in the range of 10^{-6} to 10^{-8}, a process which may result in the emergence of resistant organisms during therapy. Finally, side effects (particularly those relating to the central nervous system and gastrointestinal tract) are relatively frequent with these compounds. Despite the aforementioned liabilities, the older quinolones share a number of positive features which are characteristic of all members of this class of antimicrobial agents. For instance, they are well absorbed when given by the oral route, they are widely distributed among bodily fluids and tissues, and they are able to penetrate into intracellular locations. In addition, they have relatively long serum half-lives, are bactericidal in therapeutic concentrations, and are devoid of major organ toxicity. With the exception of one, as of yet unconfirmed report (6), there has been no description of transferable or plasmid-mediated resistance to this class of compounds. Finally, the quinolones represent rather simple chemical compounds which are easy to synthesize and potentially inexpensive to manufacture.

The older quinolones did not achieve widespread utility because of the drawbacks noted above, particularly their narrow spectra of activity, lack of intrinsic antimicrobial potency, and propensity to select resistant mutants. However, two chemical modifications of the basic quinolone molecule have resulted in a host of new compounds with markedly improved biologic and toxicologic properties. These changes include the addition of a fluorine atom at position 6 of the molecule (first incorporated in the compound flumequine) and the addition of a piperazine or similar ring structure at position 7 (first seen in pipemidic acid) (1). All of the presently available new quinolones contain these two structural features. As a result, the new quinolones have markedly enhanced spectra of activity which include a wide variety of gram-positive organisms, members of the *Enterobacteriaceae*, and *P. aeruginosa* (but not anaerobes). With the possible exception of *P. aeruginosa*, bacteria tend to be 100 to 1,000 times less likely to develop high-level resistance to these compounds than to the older quinolones. Thus, the emergence of highly resistant mutants during therapy is unlikely to occur. These new compounds are not subject to plasmid-mediated resistance, and under certain circumstances they actually cause the loss of plasmids from bacteria both in vitro and in vivo. In addition, the new quinolones share the favorable pharmacokinetic profiles of their earlier progenitors, and a number of them have exceedingly long serum half-lives. Several of the newer compounds can be administered both intravenously and orally, thus broadening the therapeutic options. The new quinolones are generally exceedingly well tolerated, and in comparative trials to date, the overall rate of side effects seen with these agents has been equivalent to or lower than that seen with older drugs such as trimethoprim-sulfamethoxazole. Most of the data on toxicity obtained thus far, however, have been derived from clinical trials of the drugs in which relatively low doses have been used. It is entirely possible that with more widespread use and an inevitable dose escalation which may be tried by some practitioners, there may be an increased incidence of dose-related toxic effects, such as gastrointestinal intolerance and central nervous system irritability or seizures. Arthropathy of weight-bearing joints has been seen in certain experimental animals given these compounds and, for that reason, it is unlikely that quinolones will be widely used in children or adolescents in the near future. Despite this, there is a considerable body of evidence suggesting that the use of nalidixic acid in children does not produce significant arthropathy. On the basis of this experience, ciprofloxacin has been cautiously tried in a group of children with cystic fibrosis. So far this use has not resulted in irreversible arthropathy. The amount of data available on the use of the newer quinolones in children and adolescents nevertheless is limited. Thus, no firm conclusions can be drawn on their safety in this setting. The new quinolones are weakly mutagenic in some, but not all, bacterial tests (7). This is not surprising, given their ability to interfere with DNA synthesis. Nonetheless, despite the fact that there are conflicting data on mutagenicity in bacteria, the available evidence from studies with eucaryotic cells, animal studies, and clinical trials suggests that the potential for mutagenicity in humans is low (7). Thus far there has likewise been no evidence of teratogenicity, but there are insufficient data available to permit the use of these drugs in pregnant women.

Although the new quinolones have been available for clinical trials for less than a decade, there is already a large and growing body of clinical experience with these compounds. These clinical data are discussed in detail in chapters 5 to 13 of this book, and from the information in these chapters it is now possible to begin to place these compounds in reasonable perspective. It is clear, for example, that the quinolones are most useful agents for the treatment of urinary tract infections of a wide variety of types, as well as for the management of bacterial infections of the gastrointestinal tract. In addition, they are effective for the treatment of certain types of sexually transmitted diseases and for selected infections of the respiratory tract, bones and joints, and skin and soft tissues. Their role in the management of immunocompromised patients remains to be fully defined. In addition, they may have certain limited application in the treatment of infective endocarditis, bacterial meningitis, and various eye infections, but there are simply insufficient clinical data to allow firm conclusions to be drawn concerning the possible utility of these agents for the latter indications.

As noted in chapter 5, the new quinolones have already proved to be effective for the treatment of uncomplicated urinary tract infections. There are, however, also a variety of other antibiotics available for treating these conditions, and for the present the new quinolones should not be considered drugs of choice for the therapy of uncomplicated urinary tract infections. Preliminary data are available on the use of quinolones in the single-dose treatment of uncomplicated urinary tract infections (cystitis). These data suggest that the new quinolones are likely to be effective in this setting. The drugs have not, however, been used in doses equivalent to the high doses of ampicillin and other antimicrobial agents often used for "standard" single-dose therapy of urinary tract infections. Moreover, there are not sufficient data available to determine the optimal duration of therapy of cystitis with these agents (i.e., single-dose, 3-day, or 7- to 10-day course of therapy). A great deal of clinical data has already been accumulated concerning the treatment of complicated urinary tract infections with the newer quinolones. These drugs have proved to be highly effective in this setting especially, because their spectra include *P. aeruginosa* and a broad variety of multiresistant gram-negative bacilli. In addition, the drugs can be given orally and thus have the potential to reduce the length of hospitalization in patients with certain complicated gram-negative urinary tract infections. These compounds seem effective for acute pyelonephritis, but more data are needed before their ultimate place in the therapy of this condition can be fully evaluated.

Because of the ability of the quinolones to penetrate into prostatic tissue and prostatic fluid and because of their spectra of activity, these agents would seem to be ideal agents for treatment of bacterial prostatitis. It is disappointing, therefore, to see the relative lack of data which have been accumulated thus far on the effectiveness of the new quinolones for both acute and chronic prostatitis. It seems highly likely that quinolones will be much more effective for treating acute, rather than chronic, prostatitis, given the nature of the disease processes. Unfortunately, we are in need of much more clinical information before the role of these compounds in the management of prostatitis can be optimally assessed.

Except for furazolidone, the new quinolones are the only antimicrobial agents which are effective against all known causes of bacterial diarrhea (although their activity against *Clostridium difficile* is marginal). Studies to date (summarized in chapter 8) clearly demonstrate the effectiveness of the newer quinolones in shigellosis and traveler's diarrhea and suggest that they will likely be effective in treating salmonella gastroenteritis as well as campylobacteriosis, two conditions for which the value of antimicrobial therapy may be more difficult to demonstrate. Not only will these agents be useful for treating "diarrhea of unknown origin," but they should also prove useful for gastrointestinal infections (especially shigellosis) due to the multiresistant organisms which are currently being seen in areas such as Central Africa, the Indian subcontinent, and Central America. It is not surprising that studies to date suggest that the new quinolones are more effective in treating patients who have fecal leukocytes, which are a marker for invasive disease. Although further data are needed before a final assessment can be made, the preliminary studies on the use of the new quinolones for the treatment of typhoid fever are also very encouraging. These agents may become particularly important as infections due to multiresistant strains of *Salmonella typhi* become more prevalent. A major advantage of the new quinolones in the treatment of both typhoid fever and salmonella gastroenteritis is that they do not lead to a prolongation of the gastrointestinal carriage of salmonellae. In fact, these agents have been shown to be effective in eradicating the carrier state in some patients infected with *S. typhi*. Studies thus far have employed relatively long courses of treatment (28 days), but it seems likely that shorter courses may be equally effective in eradicating the carrier state. Further studies are necessary to assess the role of the new quinolones in the treatment of biliary tract infections, but available data suggest that the drugs may have some utility in this setting as well. However, the lack of superior activity against enterococci may ultimately prove to be a drawback in the therapy of biliary tract sepsis.

As documented in chapter 6, the new quinolones (especially norfloxacin, ciprofloxacin, ofloxacin, and to a lesser extent, enoxacin and rosoxacin) have been shown to be highly effective in the single-dose treatment of uncomplicated gonococcal urethritis and cervicitis. It seems likely that these drugs will also be effective for rectal and pharyngeal gonorrhea, although additional studies are needed. In addition, the new quinolones are very potent against *Haemophilus ducreyi* in vitro and, on the basis of limited trials, appear quite efficacious in the treatment of chancroid. A major drawback pertaining to the use of these compounds for treating sexually transmitted diseases, however, is that patients with sexually transmitted diseases are often infected with more than one organism at the same time. A number of these organisms are relatively or absolutely resistant to the quinolones. For example, it has been clearly demonstrated that single-dose therapy with the new quinolones is not effective in preventing postgonococcal urethritis. Moreover, none of the new quinolones is effective in the single-dose therapy of genital infections with *Chlamydia trachomatis*. Even in multiple doses, norfloxacin and ciprofloxacin are not as effective as doxycycline for treating nongonococcal urethritis. Of the new quinolones, ofloxacin has the greatest potential for the therapy of chlamydial infections and nongonococcal urethritis, but even this quinolone must be used in multidose regimens, and its ultimate utility in this setting remains to be proved. Another potential disadvantage of the quinolones in the treatment of sexually transmitted diseases is their lack of activity against *Treponema pallidum*. There are only limited data on the utility of these compounds in treating syphilis, but such data as exist suggest that they are not effective. Thus, if the new quinolones are used in the treatment of gonorrhea or chancroid, it will be necessary to monitor patients for the subsequent development of syphilis and postgonococcal urethritis. Although the new quinolones may have some utility in the treatment of pelvic inflammatory disease, they are unlikely to be effective as monotherapy in this setting because of their lack of activity against anaerobes and borderline activity against chlamydiae.

The relative lack of efficacy of the new quinolones against streptococci in general and pneumococci in particular means that these compounds should be considered as having only limited utility in the therapy of respiratory tract infections. Even if they are initially successful (as some of the present clinical studies suggest), the widespread use of these compounds for the treatment of respiratory tract infections is likely to lead to some increase in the resistance of streptococci to the new quinolones, and this development, in turn, would be likely to result in increasing numbers of therapeutic failures. For this reason, it seems that none of these compounds should be routinely used for upper respiratory tract infections (unless they can be unequivocally documented to be caused by susceptible organisms, such as *Haemophilus influenzae*). There is a broad and growing literature on the use of the new quinolones for the treatment of lower respiratory tract infections, but the majority of the patients treated thus far have had bronchitis, rather than pneumonia (chapter 7). In this setting, the new quinolones appear to be reasonably effective. These drugs also have been effective in the treatment of pneumonia due to gram-negative bacilli. It is important to note that there are no data on the use of these compounds to treat bacteremic pneumococcal pneumonia. Moreover, there has been a disturbingly high frequency of persistence of *Streptococcus pneumoniae* and *P. aeruginosa* in the sputum of patients treated with these compounds for lower respiratory tract infections (2). The persistence of *P. aeruginosa* is not surprising, but the persistence of *S. pneumoniae* is clearly of concern and stands in striking contrast to observations with penicillin therapy of pneumococcal pneumonia. Several recent studies document the effectiveness of single-dose ciprofloxacin therapy for the eradication of the meningococcal carrier state. The fact that ciprofloxacin (and probably other new quinolones as well) is effective when given as a single dose in this setting makes it particularly attractive and warrants further investigation of the utility of the new quinolones for the eradication of the meningococcal carrier state. They likewise may be given consideration for eradicating *H. influenzae* nasopharyngeal carriage, but the relative contraindications to their administration to children will clearly limit their utility in this setting. Data on their effectiveness in eradicating nasal carriage of staphylococci (especially methicillin-resistant staphylococci) are limited to date, but on the basis of the available data, it does not appear that the presently available compounds are likely to be as effective as rifampin.

Because of their activity against *P. aeruginosa*, the quinolones are being evaluated in the treatment of cystic fibrosis. Studies have progressed slowly thus far because of the concern about potential arthropathy in children. Nonetheless, data (predominantly with ciprofloxacin) suggest that for initial courses of therapy, these compounds may be as effective as standard antipseudomonal penicillin-aminoglycoside therapy. As expected, there has been a fairly high incidence of the emergence of resistant *P. aeruginosa* in the sputum of treated patients, and for unexplained reasons, retreatment does not seem to be as effective as first-course therapy (even though the resistant organisms have been documented to have disappeared from the sputum before retreatment). The fact that these are oral agents clearly makes them attractive for further study in patients with cystic fibrosis, but such investigations must be carried out with considerable caution if done in children or adolescents because of the potential for arthropathy as noted above.

Given their spectrum of activity and ease of administration, the new quinolones are particularly attractive for the treatment of gram-negative bacillary osteomyelitis. Another advantage of the new quinolones for the treatment of osteomyelitis is the fact that they penetrate well into bone and provide prolonged therapeutic levels in this tissue. As outlined in chapter 9, clinical data available so far suggest that the new quinolones (especially oral ciprofloxacin) result in cure rates for gram-negative osteomyelitis of >50 to 70%, and in general, it appears that the new quinolones are at least as effective as standard therapy for gram-negative osteomyelitis. Early data suggest that the new quinolones may be surprisingly useful (particularly in combination with rifampin) for treating staphylococcal osteomyelitis, but at present there are not enough clinical data to allow firm conclusions to be drawn. Because they can be given orally, another potential use for the new quinolones is in the suppression of chronic infections surrounding joint protheses which, for mechanical or other reasons, cannot be removed. In this setting the new quinolones have proved to be effective in controlling symptoms for up to several years. Because data on such use are primarily limited to anecdotal reports thus far, further data will be necessary before firm recommendations can be made.

As documented in chapter 12, there have been four double-blind comparisons of oral ciprofloxacin with intravenous cefotaxime for treating skin and soft tissue infections. These studies revealed essentially no difference in efficacy of ciprofloxacin and cefotaxime, but because most common skin infections are caused by staphylococci and streptococci, the presently available quinolones should not be considered drugs of choice in this setting. This limitation is borne out by reports of failures of the new quinolones to cure some skin and soft tissue infections due to staphylococci (8). The antistaphylococcal penicillins, cephalosporins, and a broad variety of other agents are preferable to the quinolones for the treatment of uncomplicated cellulitis, furunculosis, impetigo, and related infections. There are, however, several settings in which the new quinolones may prove to be particularly useful in treating skin and soft tissue infections. Because of their broad spectrum of activity and coverage of gram-negative organisms, quinolones may be useful for the treatment of infected decubitus ulcers and diabetic foot infections. In both of these settings, it may be optimal to add an agent with activity against anaerobes, such as clindamycin or metronidazole, but the need for combination therapy is not altogether certain. Another setting in which preliminary data suggest that the new quinolones will be particularly useful is for the treatment of malignant otitis externa, a condition usually caused by *P. aeruginosa* in diabetic patients. Several studies have shown that the new quinolones (especially ciprofloxacin) may be highly effective in this setting.

The quinolones have certain characteristics which make them particularly interesting as agents for the prophylaxis of bacterial infections in severely neutropenic and immunocompromised patients (chapter 10). They have striking activity against members of the *Enterobacteriaceae* and *P. aeruginosa* and reasonable activity against *Staphylococcus aureus*, while lacking activity against anaerobes. Given the latter characteristic, the drugs have the potential to maintain "colonization resistance" in the gastrointestinal tract, although the latter phenomenon has not been clearly demonstrated to be important in humans, as it has in a number of animal models. The fact that quinolones yield high drug levels in feces and do not cause plasmid-mediated resistance represents further incentive for use in immunocompromised patients. Clinical trials to date suggest that the new quinolones are as or more effective than control regimens in preventing infections in neutropenic patients, particularly in preventing gram-negative bacteremia. The use of quinolones has not, however, resulted in a significant decrease in gram-positive infections (particularly those due to coagulase-negative staphylococci and viridans group streptococci). The drugs have been well tolerated, and in the limited studies reported thus far, the emergence of resistance has not been a major problem. With more widespread use of these drugs, however, especially for prophylaxis in neutropenic patients, the emergence of stepwise resistant mutants seems a worrisome possibility and may ultimately impair the utility of these compounds in this setting. Only limited data are available on the use of the new quinolones for the treatment of febrile episodes and documented infections in neutropenic patients. These data thus far allow a few generalizations. Treatment with quinolones alone or quinolones plus aminoglycosides is effective but is associated with a relatively high prevalence of gram-positive (especially streptococcal) failures and superinfections. The use of quinolones with or without aminoglycosides for treating *P. aeruginosa* infections has also been associated with the emergence of resistant organisms. Data available thus far suggest that if the new quinolones are to be used for the treatment of febrile episodes in neutropenic patients, it is probably best to combine the drugs with another agent, such as an antipseudomonal penicillin or other beta-lactam antibiotic.

Chapters 11 and 13 provide the available data on the use or potential use of the new quinolones for the treatment of endocarditis, meningitis, and ophthalmologic infections. In all three settings we are hampered by a lack of clinical data in humans. It is clear, however, that the presently available compounds will not be drugs of choice for the routine treatment of endocarditis or meningitis because of their lack of activity against streptococci. On the basis of animal studies and limited human data, it is possible that the new quinolones may have some utility in highly selected cases of gram-negative endocarditis and meningitis. It is further possible that, in combination with rifampin, the new compounds might be useful in oral regimens for treating selected cases of right-sided endocarditis due to *S. aureus*. On the basis of data available thus far, however, it appears that the potential of these compounds for both endocarditis and meningitis is limited.

There are several other settings in which the new quinolones may be potentially useful, but so far there are not sufficient data to allow any definitive statements. For instance, given their activity in uncomplicated gonorrhea and the levels in serum which are obtained after standard doses of these drugs, it seems likely that quinolones will be effective in the treatment of disseminated gonococcemia. Both in vitro and animal data suggest that the new quinolones may be effective in the treatment of *Legionella* pneumonia. Although quinolones have activity against mycobacteria, including some potency against the *Mycobacterium avium-Mycobacterium intracellulare* complex, early studies on the use of ciprofloxacin and related quinolones for disseminated infection with the *M. avium-M. intracellulare* complex in patients with acquired immunodeficiency syndrome have not been encouraging.

It is clear from the foregoing that the quinolones presently available represent a significant therapeutic advance, and it is likely that the drugs will find a major place in the therapeutic armamentarium. Given the progress which has been made to date in developing improved members of this class of antimicrobial agents, it seems reasonable to ask about the potential for future advances in this field. As the knowledge of structure-activity relationships has expanded, it has become clear that it is possible to develop compounds with even broader spectra of activity, including improved activity against gram-positive organisms, anaerobes, chlamydiae, and legionellae (9; D. T. W. Chu and P. B. Fernandes, *Antimicrob. Agents Chemother.*, in press). Substitution of a pyrrolidine ring for the piperazine moiety at position 7, for instance, results in striking enhancement of activity against gram-positive bacteria (9). The addition of a fluorophenyl (or difluorophenyl) substituent to moieties at position 1 results in improved potency against anaerobes, chlamydiae, and legionellae, as seen with difloxacin (3; Chu and Fernandes, in press). The addition of further halogen atoms and various substitutions at position 3 also alters the biologic properties of the new quinolones. Monocyclic quinolones with antibacterial activity have recently been synthesized. Another novel approach is the production of fused cephalosporin-quinolone prodrugs (N. H. Georgopapadakow, H. A. Albrecht, A. Bertasso, L. M. Cummings, D. D. Kieth, and D. A. Russo, *Program Abstr. 28th Intersci. Conf. Antimicrob. Agents Chemother.*, abstr. no. 442, 1988). Such drugs potentially have a very broad spectrum of activity which is not unleashed until the compounds are actually taken up by bacterial cells. The use of computer programs has further aided the ability to design new and improved quinolones and provides considerable promise for further development in this area (4). Thus, it appears likely that we will see newer and improved versions of the presently available quinolones. It must be realized, however, that currently the toxic potential for most of the newest compounds has not been delineated, and it is entirely possible that many, if not all, of them will fail to realize their therapeutic potential because of currently unrecognized defects other than spectrum of activity.

The quinolone antimicrobial agents represent an area of intense basic and clinical investigation. Large numbers of compounds are presently available, and even larger numbers will soon be released for clinical use in virtually all parts of the world. Because of the relatively low toxicity, broad spectrum of activity, and ease of administration of these drugs, the quinolones share immense potential for abuse. Of greatest immediate concern is that widespread overuse of these compounds for conditions such as respiratory tract infections will eventually lead to a significant emergence of resistance, even if plasmid-mediated resistance never becomes a problem. On the basis of currently available data, it is most likely that gram-positive organisms, especially the streptococci and possibly staphylococci, as well as *P. aeruginosa* will become relatively resistant first. Unfortunately, the stepwise mutations which are likely to cause resistance result in cross resistance to all of the known quinolone agents and, in some instances, produce pleiotropic resistance to unrelated drugs such as chloramphenicol, tetracycline, trimethoprim, and beta-lactams.

In summary, the new quinolones represent a major advance in the chemotherapy of infectious diseases. Unfortunately, these compounds also have a great potential for overuse and misuse, which, in turn, may result in the emergence of resistance among a number of the organisms currently susceptible to the quinolones. Therefore, the future viability of the quinolones will in large part depend on the ability of the medical community to use them wisely.

Literature Cited

1. **Crumplin, G. C.** 1988. Aspects of chemistry in the development of the 4-quinolone antibacterial agents. *Rev. Infect. Dis.* **10**(Suppl. 1):S2–S9.
2. **Davies, B. I., F. P. V. Maesen, and C. Baur.** 1986. Ciprofloxacin in the treatment of acute exacerbations of chronic bronchitis. *Eur. J. Clin. Microbiol.* **5**:226–231.
3. **Fernandes, P. B., D. T. W. Chu, R. N. Swanson, N. R. Ramer, C. W. Hanson, R. W. Bower, J. M. Stamm, and D. J. Hardy.** 1988. A-61827 (A-60969), a new fluoronaphthyridine with ac-

tivity against both aerobic and anaerobic bacteria. *Antimicrob. Agents Chemother.* **32:**27–32.
4. **Klopman, G., O. T. Macina, M. E. Levinson, and H. S. Rosenkranz.** 1987. Computer automated structure evaluation of quinolone antibacterial agents. *Antimicrob. Agents Chemother.* **31:**1831–1840.
5. **Lesher, G. Y., E. J. Froehlich, M. D. Gruett, J. H. Bailey, and R. P. Brundage.** 1962. 1,8-Naphthyridine derivatives: a new class of chemotherapeutic agents. *J. Med. Pharm. Chem.* **5:**1063–1065.
6. **Munshi, M. H., K. Haider, M. Rahaman, D. A. Sack, Z. U. Ahmed, and M. G. Morshed.** 1987. Plasmid-mediated resistance to nalidixic acid in *Shigella dysenteriae* type 1. *Lancet* **ii:**419–421.
7. **Phillips, I.** 1987. Bacterial mutagenicity and the 4-quinolones. *J. Antimicrob. Chemother.* **20:**771–773.
8. **Righter, J.** 1987. Ciprofloxacin treatment of *Staphylococcus aureus* infections. *J. Antimicrob. Chemother.* **20:**595–597.
9. **Wise, R., J. P. Ashby, and J. M. Andrews.** 1988. In vitro activity of PD 127,391, an enhanced-spectrum quinolone. *Antimicrob. Agents Chemother.* **32:**1251–1256.

Index